Surveying

Surveying for Construction

Fifth Edition

William Irvine and Finlay Maclennan

The **McGraw·Hill** *Companies*

London • Burr Ridge IL • New York • St Louis • San Francisco • Auckland
Bogotá • Caracas • Lisbon • Madrid • Mexico • Milan
Montreal • New Delhi • Panama • Paris • San Juan • São Paulo
Singapore • Sydney • Tokyo • Toronto

Surveying for Construction
William Irvine and Finlay Maclennan
ISBN-13 978–0–07–711114–4
ISBN-10 0–07–711114–1

Published by McGraw-Hill Education
Shoppenhangers Road
Maidenhead
Berkshire
SL6 2QL
Telephone: 44 (0) 1628 502 500
Fax: 44 (0) 1628 770 224
Website: *www.mcgraw-hill.co.uk*

British Library Cataloguing in Publication Data
A catalogue record for this book is available from the British Library

Library of Congress Cataloguing in Publication Data
The Library of Congress data for this book has been applied for from the Library of Congress

New Editions Editor: Kirsty Reade
Editorial Assistant: Laura Dent
Marketing Manager: Alice Duijser
Production Editor: James Bishop

Text Design by RefineCatch Ltd, Bungay, Suffolk
Cover design by Paul Fielding Design Ltd
Typeset by RefineCatch Ltd, Bungay, Suffolk
Printed and bound in Singapore by Markono Print Media Pte Ltd

Fourth Edition published in 1995 by McGraw-Hill Education

ISBN-10: 0–07–711114–1
ISBN-13: 978–0–07–711114–4

The McGraw-Hill Companies

Brief Table of Contents

Detailed Table of Contents

Preface

This is the fifth edition of *Surveying for Construction*. It has been necessitated principally by the unprecedented technological advances in survey instrumentation and computing techniques.

Nowadays, qualified surveyors are obliged to keep abreast with modern developments due to the introduction of continuous professional development, by reading modern literature, in the form of professional journals, technical papers and books.

This book is not written for them. It is written as a course for the aspiring potential surveyor who is at the beginning of his or her career and who requires a solid grounding in the fundamental principles of land surveying.

This book therefore concentrates on teaching the basic principles of surveying and follows the format of previous successful editions by (a) introducing theories in a clear, hopefully unambiguous manner, (b) exemplifying the theories in a series of well-structured examples and (c) providing self-assessment exercises, with answers, at frequent intervals throughout every subject area. In short, it follows the principles of a well-designed lecture.

This edition has been completely revised and rewritten wherever necessary to reflect the huge changes in surveying practice which have taken place throughout the past decade. In doing so, however, the authors have attempted to keep the sections on new technology non-specific and have concentrated on the wider spectrum rather than on particular models of surveying instruments and software packages. There are literally hundreds of new models of surveying equipment which are being constantly up-graded and we have endeavoured to include references that are as up to date as possible.

This edition has been reviewed independently by eleven university or college lecturers who have made several suggestions, which we have endeavoured to incorporate. Some were critical of the continued inclusion of linear surveying, taped measurement in traversing and the use of non-electronic theodolites, while others expressed a wish for their retention and even further amplification. We think these forms of surveying are still valid because they continue to be a part of the syllabi of many Educational Qualification Boards and the instruments are still in widespread use with small construction companies and educational establishments. They have therefore of necessity been included in addition to other forms of surveying, thus emphasizing the flexible nature of the book.

Some reviewers pointed out that the inclusion of a chapter on computing and a chapter on a student project were valuable but, due to time constraints on college and university courses, were seldom used. The authors have therefore deemed it wise to remove them and give only some introductory guidance to the use of spreadsheets in the text. The project will be made available on the book's Online Learning Centre, www.mcgraw-hill.co.uk/textbooks/irvine

This edition is aimed at undergraduate university courses in civil engineering and building and environmental studies, as well as college surveying modules in construction, town planning, engineering and topographic studies. The responsibility for the drawing and accuracy of all diagrams and the compilation and solutions to all questions is entirely our own.

Guided Tour

Learning Objectives

Each chapter opens with a set of learning objectives, summarising what readers should learn from each chapter.

Figures and Tables

Each chapter provides a number of detailed and practical figures and tables to help you to visualise the various methods and to illustrate and summarise important concepts.

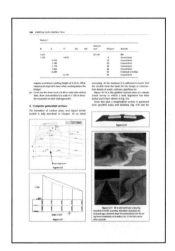

Examples

Throughout the book these short examples give you opportunities to test your understanding of the section material.

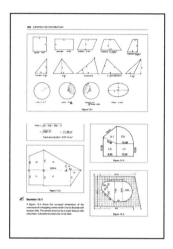

Exercises

The exercises interspersed throughout the chapter material provide a more in-depth chance to practice your understanding of surveying methodology and computation, and to understand real-world surveying practice.

Answers

The answers to the exercises are provided at the end of each chapter.

Chapter summary

This briefly reviews and reinforces the main topics you will have covered in each chapter to ensure you have acquired a solid understanding of the key topics.

Technology to enhance learning and teaching

Visit www.mcgraw-hill.co.uk/textbooks/irvine today

Online Learning Centre (OLC)

After completing each chapter, log on to the supporting Online Learning Centre website. Take advantage of the study tools offered to reinforce the material you have read in the text, and to develop your knowledge in a fun and effective way.

Resources for students include:

- *A surveying project allowing you to practise the techniques you have learnt.*
- *Coverage of alternative methodologies.*
- *Useful weblinks for land surveying.*

Lecturers: Customise Content for your Courses using the McGraw-Hill Primis Content Centre

Now it's incredibly easy to create a flexible, customised solution for your course, using content from both US and European McGraw-Hill Education textbooks, content from our

Professional list including Harvard Business Press titles, as well as a selection of over 9,000 cases from Harvard, Insead and Darden. In addition, we can incorporate your own material and course notes.

For more information, please contact your local rep who will discuss the right delivery options for your custom publication – including printed readers, e-Books and CDROMs. To see what McGraw-Hill content you can choose from, visit **www.primisonline.com**.

Study Skills

Open University Press publishes guides to study, research and exam skills to help undergraduate and postgraduate students through their university studies.

Visit www.openup.co.uk/ss to see the full selection of study skills titles, and get a **£2 discount** by entering the promotional code **study** when buying online!

Computing Skills
If you'd like to brush up on your Computing skills, we have a range of titles covering MS Office applications such as Word, Excel, PowerPoint, Access and more.

Get a £2 discount off these titles by entering the promotional code **app** when ordering online at www.mcgraw-hill.co.uk/app

Acknowledgements

Our thanks go to the following reviewers for their comments at various stages in the text's development:

Morteza Alani	University of Portsmouth
John Arthur	University College London
John Ashton-Yamnikar	University of Central Lancashire
Paul Barber	University of Glamorgan
Mark Davison	Nottingham Trent University
William Evans	Kingston University
Mike Hoxley	Anglia Ruskin University
Iwan Morris	Swansea Institute
John Rafferty	Cork Institute of Technology
Mike Young	Coventry University

Authors' Acknowledgements

Our sincere thanks are due to the following organizations and individuals:

Rachel Letts, Marketing Manager and Sokkia Coy. Ltd; Topcorn Europe B.V.; Leica Geosystems; Pentax Ltd; Hilti (Gt Britain); Craig Muir, Peter Houghton and Trimble Ltd; Autodesk; Survey Solutions (Scotland); LazerCAD; and Riegl Ltd. – for permission to use material from their various advertising literature and information from their websites.

The City and Guilds of London Institute; The Institute of Building; The Royal Institution of Chartered Surveyors; and The Scottish Qualification Authority – for permission to use questions from their various examination papers.

CHAPTER 1 Surveying fundamentals

> **In this chapter you will learn about:**
>
> - the accuracy of surveyed quantities
> - the classification of errors and the difference between accuracy and precision
> - the units of measurement used in surveying
> - the use of scales and the methods of showing scales on maps and plans
> - drawing to scale
> - the fundamental trigonometrical formulae used in surveying and some applications of these formulae

Every day of the year, in many walks of life, maps and plans are in common use. These maps and plans include road maps, charts of lakes and rivers, construction site plans and architectural plans.

All of these plans are drawn to scale by cartographers, engineers, architectural draughtsmen or surveyors, using conventional drawing tools and materials or computer aided drawing packages, from measurements of distance, heights and angles. The measurements are made by surveyors or engineers, using instruments such as tapes, levels, theodolites, electromagnetic distance measuring instruments (EDM) and global positioning system instruments (GPS). They are employed in the field of land surveying, which is the science and art of measuring the size and shape of natural and man-made features on the surface of the earth.

1. Accuracy of surveyed quantities

In carrying out their work, surveyors' primary objective is to achieve accuracy in their measurements. No matter how speedily or economically they conduct the survey or how neatly or pleasingly they present the results, the survey is of little value if it is not accurate.

In surveying, the most common task is to find the three-dimensional positions of a series of points, the x, y and z coordinates. Physical measurements are therefore required in the form of linear, angular, and height dimensions. There will be errors in these quantities which must be eliminated, discounted or distributed.

(a) Classification of errors

Errors in surveying are classified under the following headings:

Gross. Gross errors are simply mistakes. They arise mainly due to the inexperience, ignorance or carelessness of the surveyor. Simple examples are (i) reading the tape wrongly, (ii) recording a wrong dimension when booking, (iii) turning the wrong screw on an instrument. These errors cannot be accommodated and observations have to repeated.

Systematic. These are errors which arise unavoidably in surveying and follow some fixed law. Their sources are well known. A simple example is illustrated by the temperature error in tape measurements. A tape is only correct at a certain standard temperature; therefore if the ambient temperature on a certain day is higher than standard, the tape will expand and cause an error which will be the same no matter how often the line is measured. Conversely, if the temperature is lower than the standard, the tape will contract and similarly cause an error but of the opposite sign.

Constant. These are errors which do not vary at any time, in other words they have the same sign, either positive or negative. As an illustration, consider the nature of the dimensions required for plotting on maps and plans. These must be horizontal and if no attention is paid to the slope of the ground when making a measurement, the dimension so obtained will be too long. No matter how often the measurement is made or how many other slopes are measured, the results will always be too long.

Likewise, if a tape has stretched through continuous use or abuse, any resultant measurements will be always be too short.

These and all other sources of constant error are well known and appropriate corrections are applied to obtain the correct result.

Accidental or random. These are the small errors which inevitably remain after the others have been eliminated. There are three main causes, (i) imperfections of human sight and touch (indeed they are often called human errors), (ii) imperfections of the instruments being used at the time and (iii) changing atmospheric conditions.

A good example arises in the measurement of an angle using an instrument called a theodolite (Chapter 7). Such an instrument can measure angles to one second (= 1/3600th part of a degree). If ten measurements are made of an angle, the results will differ slightly, due to (i) the inability of the surveyor to sight a point exactly in the same way each time, (ii) the instrument not being in perfect adjustment and (iii) the change in temperature and wind pressure throughout the measurement procedure, which will affect the stability of the instrument.

These are random errors and are reduced, though never quite eliminated, by repeating the measurement of whatever quantity is being measured.

Summary: Gross and systematic errors are largely, but not necessarily completely, eliminated by sound observational techniques, frequent checking and application of corrections, so that the only errors remaining are random errors. These errors follow the laws of probability. They can be treated statistically and it can be shown that the most probable value (mpv) of a set of observations is the arithmetic mean. The statistical treatment of errors is beyond the scope or remit of this textbook and will not be further pursued.

(b) Accuracy and precision

Accuracy and precision are not synonymous. The difference is illustrated in the following example.

A line of a survey is measured six times and all six measurements lie within an error band of ±2 millimetres. However, the tape, when checked, is found to have stretched by 10 millimetres through continuous use.

The results of the six measurements are precise in that they have little scatter, but they are not accurate because each is in error by 10 mm.

Precision is therefore relative in that each measurement is close to any other, whereas accuracy is closeness to the truth.

In surveying, accuracy is defined by specifying the limits between which the error of a measured quantity may lie, so, for example, the accuracy of the measured height of a building might be (20.54 ± 0.01) metres.

2. Units of measurement

In most countries of the world (there are some notable exceptions), the metric system of measurement is used for the linear measurement of distance and height, while the sexagesimal system is used for angular measurement.

(a) Linear measurement

The following units are the most commonly used units in surveying.

Quantity	Symbol	Unit
Length	metre	m
Area	square metre	m²
Volume	cubic metre	m³
Mass	kilogramme	kg
Capacity	litre	l

Taking the metre as a basic unit, Table 1.1 shows how multiples and sub-multiples of the unit are derived.

Table 1.1

Prefix	Multiplication factor	Derived unit	SI recommended unit
kilo	1000	kilometre	kilometre (km)
hecto	100	hectometre	
deca	10	decametre	
		metre	metre (m)
deci	0.10	decimetre	
centi	0.01	centimetre	
milli	0.001	millimetre	millimetre (mm)

In Table 1.1, it should be noted that only three units are recommended for general use. This holds good for other units and Table 1.2 shows the small selection of units included in the Système Internationale (SI) which are in common use.

Table 1.2

Quantity	Recommended SI unit	Other units that may be used
Length	kilometre (km) metre (m) millimetre (mm)	centimetre (cm)
Area	square metre (m²) square millimetre (mm²)	square centimetre (cm²) hectare (100 m × 100 m) (ha)
Volume	cubic metre (m³) cubic millimetre (mm³)	cubic decimetre (dm³) cubic centimetre (cm³)
Mass	kilogramme (kg) gramme (g) milligramme (mg)	
Capacity	cubic metre (m³) cubic millimetre (mm³)	litre (l) millilitre (ml)

Finally, Table 1.3 shows the basic relationship between volume, mass and capacity (of water), from which others may be deduced.

Table 1.3

Volume	Mass	Capacity
1 cubic metre	1000 kilogrammes	1000 litres
1 cubic decimetre	1 kilogramme	1 litre
1 cubic centimetre	1 gramme	1 millilitre

(b) Angular measurement

Angular measurements are made using surveying instruments which measure both horizontally and vertically in degrees. Degrees are sexagesimal units which are subdivided into minutes and seconds in exactly the same manner as time. There are sixty minutes in one degree and sixty seconds in one minute.

It is worth noting that in many European countries (e.g. Switzerland, Austria) angles are measured in grades. There are four hundred grades in one complete revolution, whereas there are three hundred and sixty degrees in one revolution.

EXAMPLES

Linear measurements should be written to three decimal places to avoid confusion, unless required otherwise.

1 Find the sum of the following measurements:
(a) 1 metre and 560 millimetres

> **Answer:**   1.000
> + 0.560
> = 1.560 m

(b) 15 metres and 31 centimetres

> **Answer:**   15.000
> + 0.310
> = 15.310 m

(c) 25 m, 9 cm and 8 mm.

> **Answer:**   25.000
> + 0.090
> + 0.008
> = 25.098 m

2 In surveying, angles have to added or subtracted frequently. Their values are written in sexagesimal form as follows:

10 degs, 23 mins, 18secs = 10° 23′ 18″

Find the value of:

(a) 23° 24′ 30″ − 10° 18′ 15″

> **Answer:**   23° 24′ 30″
> − 10° 18′ 15″
> = 13° 06′ 15″

(b) 56° 35′ 20″ − 15° 19′ 45″

> **Answer:**   36° 35′ 20″
> − 15° 19′ 45″

Since 1 min = 60 s, the calculation becomes

> 56° 34′ 80″
> − 15° 19′ 45″
> = 41° 15′ 35″

Adding the figure 60 to either minutes or seconds is, of course, done mentally.

The calculations in example 2 can be done easily, on a pocket calculator using the DMS key. Unfortunately, not all calculators use the same logic, so users should consult their calculator manual for full instructions.

Exercise 1.1

1 Write the following measurements in metres, to three decimal places:
(a) 4 metres and 500 millimetres
(b) 3 m and 17 mm
(c) 3 m, 40 cm and 67 mm
(d) 100 m and 9 mm
(e) 10 cm

2 Calculate the following, giving answers to three decimal places:
(a) 3 m + 4 cm + 250 mm + 35 cm
(b) 10 m − 82 cm + 140 mm + 120 cm
(c) 435 mm + 965 mm + 8 mm
(d) 10.326 m + 9 mm − 126 mm + 17.826 m
(e) 10 m + 10 cm + 100 mm − 15 cm − 50 mm

3 Calculate the area (in hectares) of the following rectangular plots of land:
(a) 100 m long by 150 m wide
(b) 90 m long by 53 m wide
(c) 90.326 m long by 265.112 m wide

4
(a) Find the sum of the three angles of a triangle, measured using a theodolite (a theodolite is a surveying instrument used for the accurate measurement of angles).
Angle ABC: 58° 17′ 40″, angle BCA: 67° 23′ 20″ and angle CAB: 54° 18′ 20″.
(b) Find the error of the survey.

3. Understanding scale

The end product of a survey is usually the production of a scaled drawing and throughout the various chapters of this book scaled drawings will have to be made. A scale is a ratio between the drawing of an object and the actual object itself.

EXAMPLE

3 Figure 1.1 is the drawing of a two pence piece. The diameter of the coin on the drawing is 13 mm. An actual two pence piece has a diameter of 26 mm; hence the scale of the drawing is

$$\frac{\text{Plan size}}{\text{Actual size}} = \frac{13\ \text{mm}}{26\ \text{mm}} = \frac{1}{2}$$

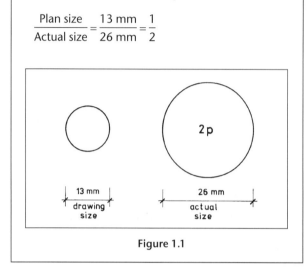

Figure 1.1

(a) Methods of showing scale

The scale of a map or plan can be shown in three ways:

1. *It may simply be expressed in words* For example, 1 centimetre represents 1 metre. By definition of scale, this simply means that one centimetre on the plan represents 1 metre on the ground.

2. *By a drawn scale* A line is drawn on the plan and is divided into convenient intervals such that distances on the map can be easily obtained from it. If the scale of 1 centimetre represents 1 metre is used, the scale drawn in Fig. 1.2 would be obtained.

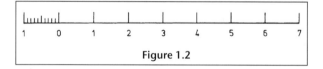

Figure 1.2

Figure 1.2 is an example of an open divided scale in which the primary divisions (1.0 metre) are shown on the right of the zero. The zero is positioned one unit from the left of the scale and this unit is subdivided into secondary divisions. An alternative method of showing a drawn scale is to fill in the divisions, thus making a filled line scale, an example of which is shown in Fig. 1.3.

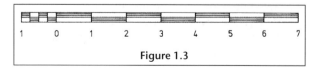

Figure 1.3

3. *By a representative fraction* With this method of showing scale, a fraction is used in which the numerator represents the number of units on the map (always 1) and the denominator represents the number of the same units on the ground. With a scale of 1 centimetre representing 1 metre, the representative fraction will be 1/100, shown as 1:100, since there are 100 centimetres in 1 metre.

A representative fraction (RF) is the international way of showing scale. Any person looking at the RF on a map thinks of the scale in the units to which he is accustomed.

EXAMPLES

4 Calculate the scale of a plan where 1 mm represents 0.5 m.

$$\text{Scale} = \frac{\text{plan size}}{\text{actual size}} = \frac{1\ \text{mm}}{0.5\ \text{m}} = \frac{1\ \text{mm}}{500\ \text{mm}} = \frac{1}{500}$$

5 Figure 1.4 is the scale drawing of a badminton court drawn to a scale of 1:250. Using the open divided scale provided, measure:
(a) the overall length and breadth of the court,
(b) the distance between the tram lines,
(c) the size of the service courts.

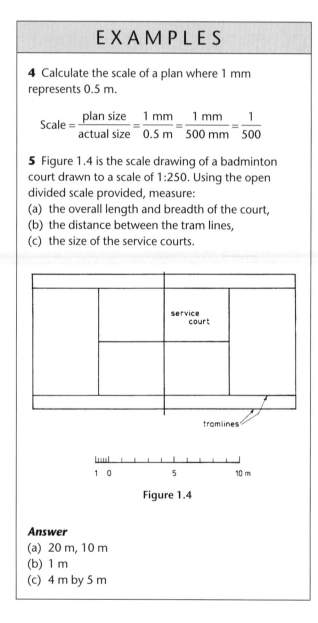

Figure 1.4

Answer
(a) 20 m, 10 m
(b) 1 m
(c) 4 m by 5 m

(b) Conversion of area by representative fractions

Frequently, in survey work, the area of a parcel of land is measured on a scaled plan, using an instrument called a planimeter (Chapter 13). A planimeter measures areas in square centimetres (cm²) and the actual ground area has to be calculated. If the RF on a plan is very large, say 1:4, one unit on the plan will represent four units on the ground. A square of 1 unit on the plan will therefore represent a ground area of 4 units × 4 units. From these facts emerges a simple formula:

$$\text{Plan scale} = 1{:}4$$
$$\text{Plan area} = 1 \times 1 \text{ sq. units}$$
$$\text{Therefore ground area} = (1 \times 4) \times (1 \times 4) \text{ sq. units}$$
$$= 1 \times (4 \times 4) \text{ sq. units}$$
$$= 1 \times 4^2 \text{ sq. units}$$
$$\text{i.e. ground area} = \text{plan area} \times 4^2$$
$$= \text{plan area} \times (\text{scale factor})^2$$

$$\text{Therefore plan area} = \frac{\text{ground area}}{4}$$
$$= \text{ground area} \times (\text{RF})^2$$

EXAMPLE

6 An area of 250 cm^2 was measured on a plan, using a planimeter. Given that the plan scale is 1:500, calculate the ground area in m^2.

Answer

$$\text{Plan area} = 250 \text{ cm}^2$$
$$\text{RF (scale)} = 1{:}500$$
$$\text{Ground area} = \text{Plan area} \times (\text{scale factor})^2$$
$$= (250 \times 500^2) \text{ cm}^2$$
$$= 250 \times \frac{500^2}{100^2} \text{ m}^2$$
$$= (250 \times 25) \text{ m}^2$$
$$= 6250 \text{ m}^2$$

Exercise 1.2

1 Calculate the scale of a plan where 1 cm represents 20 m.

2 Figure 1.5 is the plan view of a house drawn to scale 1:100. Beginning at point A and moving in a clockwise direction, measure the lengths of the various walls, using a scale rule.

3 A parcel of ground was measured on a 1:250 scale map, using a planimeter, and found to be 51.25 cm^2. Calculate the ground area in hectares.

4 The dimensions of a room on a 1:50 scale plan are 60 mm × 85 mm. Calculate the area of the room in m^2.

4. Drawing to scale

A surveyor's main objective is to achieve accuracy in field operations. However, unless the results can be depicted accurately, legibly and pleasingly on paper, field proficiency loses much of its value. Nowadays, the majority of plans produced by surveyors are drawn automatically using some survey drawing package and are frequently enhanced by Autocad. *However, on College and University courses of study, particularly in early projects, much drawing has to be done by hand.* The fol-

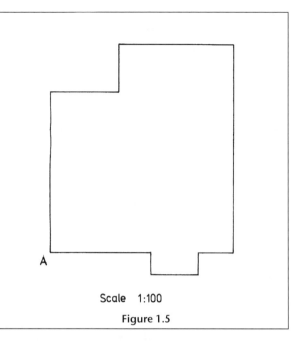

Scale   1:100

Figure 1.5

lowing short section is intended to be an introduction to the art of plan production, using only basic equipment and plotted from a few survey field notes. The student should follow the steps in Example 7 to produce his or her own plan.

(a) Equipment required for plotting

1. *Paper*   The survey when plotted may have to be referred to frequently, over a number of years, and it is essential therefore that the material on which it is plotted should be stable. Modern drawing materials are excellent in that respect and most show little or no shrinkage over a long period.

2. *Scale rule*   Most scale rules are made of plastic. They are double sided and are usually manufactured with eight scales:

    1:1, 1:5, 1:20, 1:50, 1:100, 1:200, 1:1250 and 1:2500

Other common plotting scales, 1:500 and 1:1000, can be derived simply by multiplying the scale units of the 1:50 and 1:100 scales by the factor 10.

3. *Other equipment*   This includes two set squares, 45° and 60°, varying grades of pencils, 4H to H, sharpened to a fine point, paper weights, curves, inking pens, pricker pencil, erasers, steel straight edge and spring-bow compasses.

(b) Procedure in plotting

EXAMPLE

7 Figure 1.6 is an example of a surveyor's field measurements of a small ornamental garden. These field measurements are to be plotted to scale 1:100 on an A4 size drawing sheet, to produce a finished drawing.

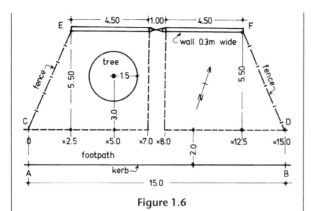

Figure 1.6

Solution

1. *Orientate the survey*   The North direction on any plan should point towards the top of the paper and the subject of the plan should lie in the centre of the sheet.

2. *Centre the drawing on the paper*   The overall size of the garden, as determined from the field measurements, is 15 m long and 8 m wide. At scale 1:100, the overall size of an A4 sheet excluding the border is 28 m by 19 m. The starting point A should therefore lie 6.5 m from the left edge and 5.5 m from the bottom of the paper (Fig. 1.7).

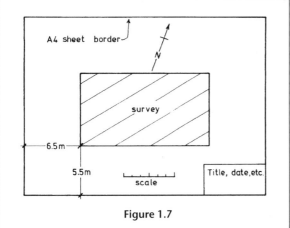

Figure 1.7

3. *Plot a base line* (Fig. 1.8)   Line AB, being the longest line, is chosen as the base line. Beginning at point A, the line is drawn parallel to the bottom of the sheet and scaled accurately at 15.0 m to represent the toe of the kerb.

4. *Complete the survey*   Using two set squares, line CD is drawn parallel to AB, at a distance of 2 m to the north of AB. Between points C and D, the various running dimensions from 2.5 to 12.5 m are accurately scaled and, using a set square, the remaining dimensions to E, F and the tree are scaled at right angles to line CD.

5. *Make a finished drawing*   The surveyed details are drawn in ink, using a drawing pen, set square and

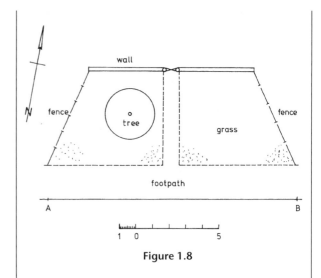

Figure 1.8

compasses, to produce a finished drawing. Suitable symbols are shown in Fig. 2.2 in Chapter 2. A scale is drawn parallel to the bottom edge of the sheet and a title, date, etc., are added in a box in the bottom right-hand corner of the sheet.

Exercise 1.3

1 Plot the field notes of Fig. 1.6 to scale 1:100 and make a finished drawing of the result. Compare the drawing with Fig. 1.8.

(c) Computer aided design/draughting (CAD)

The manual skills of draughtsmanship are much rarer today, and it is likely that the final survey drawing will be produced using CAD software.

AutoCAD from Autodesk and Microstation from Bentley are the most widely used packages. These are expensive and complicated, having many functions which are not required for drawing a survey plot. Simpler, less expensive software such as Design CAD 3DMAX would suffice, but because of the popularity of AutoCAD and Microstation, many survey drawings are provided to the client in one of these formats.

Instruction on the use of CAD software is beyond the scope of this book, but as a rough guide, the following procedure could be used to plot Fig. 1.6 using AutoCAD.

1. Set NEW LAYERS for 'kerb', 'fence', 'grass', 'wall', and 'tree', and LOAD LINETYPES to suit.
2. Make the 'kerb' layer CURRENT.
3. With the ORTHO on, draw LINE AB 15m long.
4. Make layer 'grass' CURRENT.
5. Draw LINE segments for CD.
6. Draw LINE segments of 5.5 m at right angles to CD, using AUTOSNAP to start the line at the 7 m and 8 m points.
7. Make layer 'wall' CURRENT.

8. Draw LINE segments of 4.5 m, snapping to the ends of the perpendiculars drawn in step 6.
9. OFFSET these lines by 0.3 m.
10. Make layer 'fence' CURRENT.
11. With the ORTHO off, draw line segments EC and FD, snapping to the end points.
12. Make layer 'tree' current.
13. Draw a LINE 3 m long perpendicular from the 5 m point along CD.
14. Draw a CIRCLE centred on the end of this perpendicular with a radius of 1.5 m.
15. ERASE the perpendicular line.
16. Use TEXT to annotate the drawing.

5. Basic principles of surveying

The surface of the earth over any construction site is really only a collection of points lying at different heights. The objective of a survey is to take sufficient measurements, linear and/or angular, to relate any one unknown point to any two other known points. Since all points lie at different elevations, distances measured between them will not be horizontal. These distances are called slope lengths. They can be converted to their horizontal equivalents very easily. *For the moment, the distances in the following sections will be considered to be horizontal.*

In Figs 1.9(a) and 1.9(b), the three points X, Y and Z are taken to lie on a horizontal plane on account of the assumption above. Their relative positions can be fixed in a number of ways, depending largely upon the area covered by the points. In Fig. 1.9(a), the area is small and can be surveyed by using linear measurements only. In Fig. 1.9(b), the area is much larger in extent and angle measurements would be required.

(a) Using linear methods only

1. *Trilateration* The word means measurement of three sides. When the principle is applied to Fig. 1.9(a) the lengths XY, YZ, and XZ are all measured in the field. Length XY is then drawn to scale on paper, and arcs representing the lengths YZ and XZ are drawn using compasses to intersect in the point Z.

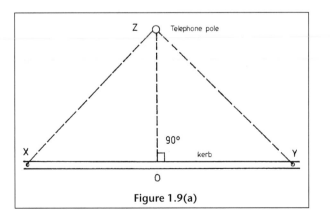

Figure 1.9(a)

2. *Lines at right angles* In the field, lengths XO and OY are measured along line XY, and line OZ is measured exactly at right angles to XY (Fig. 1.9(a)). Using a set square and scale rule, point Z can again be plotted in its correct relationship with X and Y. This method of surveying is known as offsetting.

(b) Using linear and angular methods

1. *Triangulation* In the field (Fig. 1.9(b)), the line XY, known as the baseline, is measured by tape or by electromagnetic means (EDM). Angles ZXY and XYZ are measured using a theodolite. The survey could then be drawn to scale with scale rule and protractor but would almost certainly be calculated and plotted using rectangular coordinates (Chapter 9).

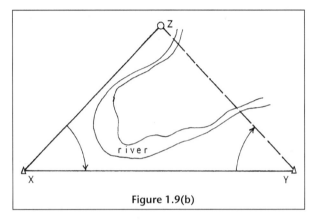

Figure 1.9(b)

2. *Polar coordinates* In the field (Fig. 1.9(b)), lines XY and XZ are measured by tape or by EDM, while angle ZXY is measured by theodolite. Again the survey could be drawn to scale with scale rule and protractor but would almost certainly be plotted by means of rectangular coordinates. Where two lines and the included angle are measured, the survey method is known as traversing.

All of the above principles are used in building and engineering surveys in some form or other and, indeed, a survey of a relatively small building site will almost certainly combine some of them.

(c) Application of principles

The principles outlined above are used in some form in making a site survey. Figure 1.10 shows a house and garden covering about one hectare of ground, which would be considered to be a medium sized site. The survey could be carried out by any of the methods described above. This survey (and all others) is carried out in two parts, (1) a framework and (2) a detail, survey.

1. Framework survey

The framework is first established over the whole site to form a sound geometrical figure and is carried out using one of the following methods:

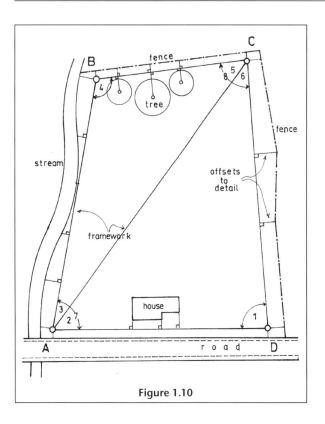

Figure 1.10

methods use different survey techniques. The rectangular coordinates (x, y, z) of every point are computed automatically by dedicated surveying computer programs and plotted using survey drawing packages supplemented with AutoCAD.

A cautionary note must be introduced at this point. Modern surveying instruments can perform many of the tasks which are traditionally done by hand and surveyors are in danger of being unable to comprehend the values produced by such processes. Basic surveying rules are easily overridden or neglected and consequent errors may pass undetected.

As a result, the remainder of this textbook deals with the fundamental methods of surveying and checking and emphasizes the importance of being able to manually compute results and construct drawings from these results.

6. Surveying mathematics

Surveying is the practical application, on the ground, of the principles of geometry. The surveyor should therefore be a practical mathematician. The branch of mathematics in which he or she must be proficient is trigonometry, which deals with the solution of triangles, using a number of mathematical formulae.

The following formulae are given because of their general usefulness in everyday situations. They deal with (a) right angled triangles and (b) angles of any magnitude right-angled triangles. No attempt is made to prove them. Should any proof be required, a good mathematical textbook should be consulted.

(a) Fundamental trigonometrical ratios

In Fig. 1.11 angle H is a right angle. Sides o, a and h lie opposite the angles O, A and H.

1. $\sin O = o/h$ 4. $\operatorname{cosec} O = h/o = 1/\sin O$
2. $\cos O = a/h$ 5. $\sec O = h/a = 1/\cos O$
3. $\tan O = o/a$ 6. $\cot O = a/o = 1/\tan O$

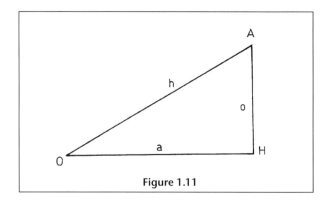

Figure 1.11

1. *Trilateration* (linear survey) The sides of the triangles ABC and ACD are accurately measured using a steel tape. These dimensions are then plotted to scale using appropriate drawing instruments.

2. *Triangulation* The angles 1, 2, 3, 4, 5 and 6 are measured using a theodolite. One side of the survey, say AD, is measured by tape to serve as a baseline. All other sides are calculated and the whole survey is plotted by rectangular coordinates.

3. *Traversing* The angles 1, 7, 4 and 8 are measured using a theodolite, together with the lengths of the sides DA, AB, BC and CD. The figure would again be calculated and plotted by coordinates (Chapter 9).

2. Detail survey
The fences, stream, buildings and path are the details which are added to the framework survey by means of offsets, which are short lines at right angles to the main framework lines.

Alternatively, detail can be added to a framework survey by means of EDM tacheometry. This more advanced technique is fully considered in Chapter 9 (Detail surveying).

All surveys are conducted in this manner, i.e. a two-stage process in which the whole area is covered by a framework within which the detail is added. This ensures that the survey rule of 'working from the whole to the part' is obeyed.

Survey methods using electromagnetic distance measurement (EDM) or the global positioning system (GPS) have become the 'norm' in surveying. These

EXAMPLES

8 In Fig. 1.12. A and B lie at different elevations. The slope length along line AB is 50.00 m and the angle of slope is 10°. Calculate the horizontal length AC and the difference in height BC, between the points.

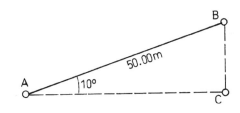

Figure 1.12

Answer

AC/AB = cos 10°	BC/AB = sin 10°
AC = 50 × cos 10°	BC = 50 × sin 10°
AC = <u>49.24 m</u>	BC = <u>8.68 m</u>

9 In Fig. 1.13. P and Q are two survey points. The direction from North (the bearing) of line PQ is 49° and the horizontal length of line PQ is 45.50 m. These two quantities are the polar coordinates of point P. Calculate the values of the rectangular coordinates x and y of point Q.

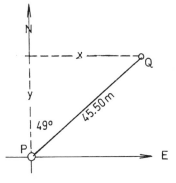

Figure 1.13

Answer

Easting (x) = PQ × sin NPQ
x = 45.50 × sin 49°
x = <u>34.34 m</u>

Northing (y) = PQ × cos NPQ
y = PQ × cos 49°
y = <u>29.85 m</u>

Given that the coordinates of point P are known from a previous survey, or are simply assumed to be 150.00 m E and 200.00 m N, calculate the final coordinates of point Q.

Answer

Point P	= 150.00 East		200.00 North
ΔE_{PQ}	= <u>34.34</u>	ΔN_{PQ} =	<u>29.85</u>
Point Q	= <u>184.34 E</u>		= <u>229.85 N</u>

Exercise 1.4

1 Calculate the x, y, and z coordinates of a survey point E, relative to survey point D, given the following survey data.

Slope length DE	= 210.50 m
Angle of slope of line DE	= 5° 30′ 00″
Bearing of line DE	= 75° 00′ 00″

In everyday surveying situations, the reverse calculation of that in Example 9 is frequently required. In this situation, the coordinates of two points are known and the bearing and distance between them are required, i.e. the rectangular coordinates are known and the polar coordinates are to be calculated.

EXAMPLE

10 In Fig. 1.13, the coordinates of points P and Q are as follows:

Point	East (m)	North (m)
P	150.00	200.00
Q	184.34	229.85

Calculate the bearing and distance of line PQ.

Answers

The bearing of the line is from P to Q. Therefore the coordinates of point P are subtracted from those of point Q.

	East	North
Q	184.34	229.85
P	150.00	200.00
ΔE =	+34.34	ΔN = +29.85

Bearing PQ $= \tan^{-1}(\Delta E / \Delta N)$
$= \tan^{-1}(+34.34 / +29.85)$
$= \tan^{-1} 1.150\,418\,8$
$= \underline{49°\ 00'}$

Distance PQ $= (34.34^2 + 29.85^2)^{1/2}$
$= (2070.2581)^{1/2}$
$= \underline{45.50\ m}$

(b) Trigonometrical ratios of any magnitude

In many surveying situations, e.g. triangulation and traversing, angles greater than 90° are common.

Figure 1.14 shows the 360° circle divided into four quadrants for surveying purposes. The layout differs from the recognized mathematical configuration in that North has a value of 0° and the angular value increases clockwise through East, South and West to return to North at 360°.

Figure 1.15 shows that in quadrant 1 (angles between 0° and 90°), the values of *sin, cos* and *tan* are all *positive*.

In the second quadrant (90° to 180°), the trigonometrical value of any angle $y°$ assumes the value of its

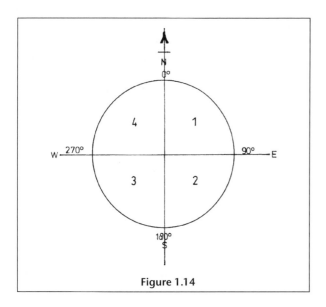

Figure 1.14

corresponding angle $x°$, which equals $(180 - y)°$. In Fig. 1.15 the trigonometrical ratios for *sin, cos* and *tan* are clearly shown. Only the *sin* is positive in quadrant 2.

In quadrant 3, (180° to 270°), the value of angle $y°$ is the value of its corresponding angle $x°$ which equals $(y - 180)°$. Figure 1.15 shows that only the *tan* is *positive* in this quadrant.

Quadrant 4 (270° to 360°) shows that the value of any angle $y°$ is the value of its corresponding angle $x°$ which equals $(360 - y)°$. Only the *cos* is *positive* in quadrant 4.

(c) Sine rule and cosine rule

In later chapters of this book, certain surveying methods (triangulation, intersection, resection, etc.)

will be described or explained. The associated calculations will involve obtuse angles. The solution of triangles involving such angles make use of the following formulae:

1. *Sine rule* Figure 1.16 shows a triangle where the values of the three angles A,B,C and one side *a* are known. The values of the other two sides *b* and *c* are required. The sine rule is used, where the general formula is:

$$\frac{a}{\sin A} = \frac{b}{\sin B} = \frac{c}{\sin C}$$

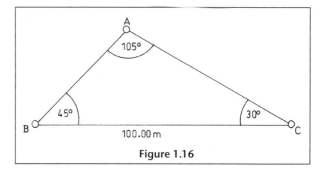

Figure 1.16

In $\triangle ABC$, $a = 100$ m, $A = 105°$, $B = 45°$, $C = 30°$

$$b = \frac{a \sin B}{\sin A} \qquad \text{and} \qquad c = \frac{a \sin C}{\sin A}$$

$$b = \frac{100 \sin 45°}{\sin 105°} \qquad\qquad c = \frac{100 \sin 30°}{\sin 105°}$$

$$b = \frac{100 \times 0.7070}{0.9659} \qquad\qquad c = \frac{100 \times 0.5000}{0.9659}$$

$$b = \underline{73.2 \text{ m}} \qquad\qquad c = \underline{51.8 \text{ m}}$$

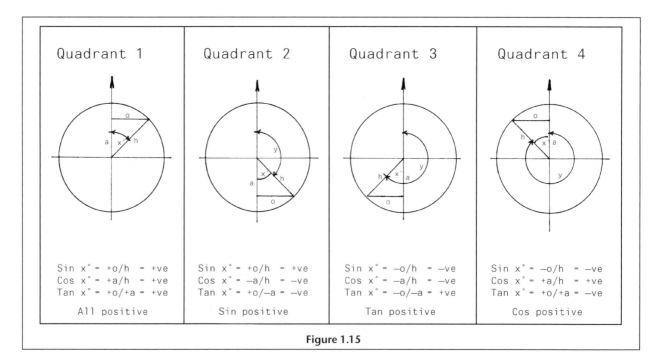

Figure 1.15

2. *Cosine rule* Figure 1.17 shows a triangle in which the values of two sides b and c and one angle A are known. The value of side a is required, as a first step. The cosine rule is used where the general formula is:

$$a^2 = b^2 + c^2 - 2bc \cos A$$

In $\triangle ABC$, $b = 73.20$ m, $c = 51.77$ m A = 105°

$$\begin{aligned} a^2 &= 73.20^2 + 51.77^2 - (2 \times 73.20 \times 51.77 \times \cos 105°) \\ &= 5358.24 + 2680.13 - (-1961.62) \\ &= 9999.99 \\ a &= \underline{100.00 \text{ m}} \end{aligned}$$

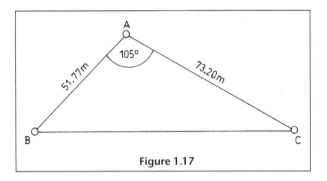

Figure 1.17

In order to calculate angles B and C either the sine rule or cosine rule may be used.

Calculating angle B by transforming the sine rule gives

$$\begin{aligned} \sin B &= \frac{b \sin A}{a} \\ &= \frac{73.20 \times \sin 105°}{100.00} \\ B &= \underline{45°} \end{aligned}$$

Calculating angle C by transforming the cosine rule gives

$$\begin{aligned} \cos C &= \frac{a^2 + b^2 - c^2}{2ab} \\ &= \frac{100.00^2 + 73.20^2 - 51.77^2}{2 \times 100.00 \times 73.20} \\ &= \frac{12\,678.11}{14\,640.00} \\ &= 0.8660 \\ C &= \underline{30°} \end{aligned}$$

Exercise 1.5

1 Figure 1.18 shows a triangle ABC in which the values of all angles and the side AB are as follows:

Angle A = 30° 40′ 16″
angle B = 98° 18′ 33″   side AB = 1217.19 m
angle C = 51° 01′ 11″

Calculate the values of sides BC and AC, using the sine rule.

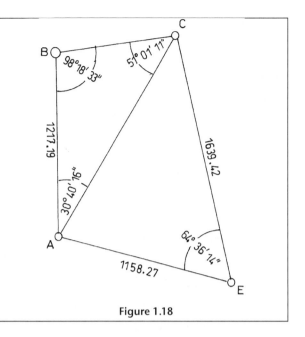

Figure 1.18

2 Figure 1.18 shows a second triangle ACE in which the values of sides AE and CE and the included angle AEC are known as follows:

Angle AEC = 64° 19′ 55″
Side AE = 1158.27 m   and   side CE = 1639.42 m

Calculate the values of angle CAE and side AC using the cosine rule. (Side AC should check with the answer to question 1.)

3 Five points of a survey have the following coordinates:

Point	East (m)	North (m)
A	1000.00	1000.00
B	1420.96	2142.08
C	2202.45	1977.05
D	2881.69	640.69
E	1962.27	355.31

Calculate the bearings and distances of lines AE, ED, DC, CB and BA, following the method detailed in Example 10.

7. Answers
Exercise 1.1

1 (a) 4.500 m
(b) 3.017 m
(c) 3.467 m
(d) 100.009 m
(e) 0.100 m
2 (a) 3.640m
(b) 10.520 m
(c) 1.408 m
(d) 28.035 m
(e) 10.000 m
3 (a) 1.500 ha
(b) 0.477 ha
(c) 2.395 ha

4 (a) 179° 59′ 20″
   (b) 00° 00′ 40″

Easting of point E (x)   = 202.39 m
Northing if point E (y) = 54.23 m

Exercise 1.2

1 1:2000
2 4.20 m, 1.90 m, 1.25 m, 3.15 m, 5.45 m, 1.00 m, 0.60 m, 1.30 m, 0.60 m, 2.75 m
3 0.032 ha
4 12.75 m²

Exercise 1.4

1 Horizontal length DE   = 209.53 m
   Height of point E (z)   = 20.18 m

Exercise 1.5

1 BC = 798.73 m
   AC = 1549.36 m
2 Angle CAE = 72° 54′ 57″
            AC = 1549.36 m

3 Line	Bearing	Distance
AE	123° 49′ 14″	1158.27
ED	72° 45′ 22″	962.68
DC	333° 03′ 25″	1499.07
CB	281° 55′ 27″	798.73
BA	200° 14′ 00″	1217.19

Chapter summary

In this chapter, the following are the most important points:

- Surveying is an exact science in the mathematical sense but, practically, all surveyed quantities have to be accurate and precise. These two terms are not synonymous. Accuracy is relative to the true answer whereas precision is relative to other measures of the same quantity.

- All surveyed quantities are subject to errors which fall into different classes, namely gross, constant, systematic and random. The effects of these errors are significantly different. A variety of examples is given on pages 1 and 2.

- Survey measurements of distance and direction are made in metres and degrees and their various sub-divisions. They are used to determine three dimensional positions, areas and volumes. Tables of metric sub-divisions are shown on pages 2 and 3.

- Surveyed dimensions are mainly used to produce scaled drawings. Scale is the ratio between the actual object and the drawn version of the object and can be shown in three ways, namely, in words, as a drawn scale or by a representative fraction. Plans were traditionally drawn by hand but manual methods have now been superseded by computer aided draughting methods (CAD).

- Surveying is the practical application of geometry and trigonometry on the ground. It is neither random nor haphazard but follows strict mathematical rules. The principles of triangulation, trilateration, polar and rectangular coordinates are used to make land surveys.

- Integrated with those methods are many trigonometrical formulae which must be mastered for use in subsequent chapters. The study of this section is a valuable step in the education of an aspiring surveyor as it promotes a feel for survey work and gives an insight into the intricacies of survey calculations, traversing in particular.

The most useful formulae are detailed on pages 8 to 11 together with a series of worked examples.

CHAPTER 2 Understanding maps and plans

In this chapter you will learn about:

- the difference between a map and a plan and various substantive items which, together, form a construction site plan
- Ordnance Survey (OS) maps and plans: the National Grid and its use in referencing OS plans drawn to various scales
- the national topographic database and digital map formats
- Local Scale Factor (LSF) and Mean Sea Level (MSL) corrections required to connect local surveys to the National Grid

---

Maps and plans are representations, on paper, of physical features on the ground. It is important to note the difference between a map and a plan. A plan accurately defines widths of roads, sizes of buildings etc; in other words every feature is true to scale. Site plans are large-scale productions and are essential to the planning and development of a construction project.

A map is a representation of an object, no matter how accurately it may be shown. As an example, a winding country road about the width of a car measures almost a metre on a 1:50 000 map. This represents 50 metres on the ground, far in excess of the actual width of the road.

1. Site plans

Figure 2.1 (overleaf) is part of a plan of a proposed development, drawn to scale 1:200.

(a) Code of symbols

In order to read a site plan, it is necessary to understand the code of signs and symbols used to depict ground features. Figure 2.2 shows some of the symbols of the British Standard Code of Signs, together with alternatives commonly used by architects and engineers.

The signs are, as far as possible, plan views of the actual objects being portrayed, thus a tree would appear as a circle, the diameter of which represents the spread of the branches. Other features require some

annotation, for example a small circle with the letters LP represents a lamppost and a small square with the letters MH represents a manhole.

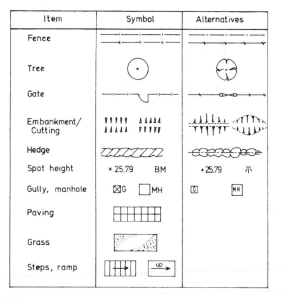

Figure 2.2

(b) Scale

A full explanation of scales and scaling was given in Chapter 1 and should be revised, if required.

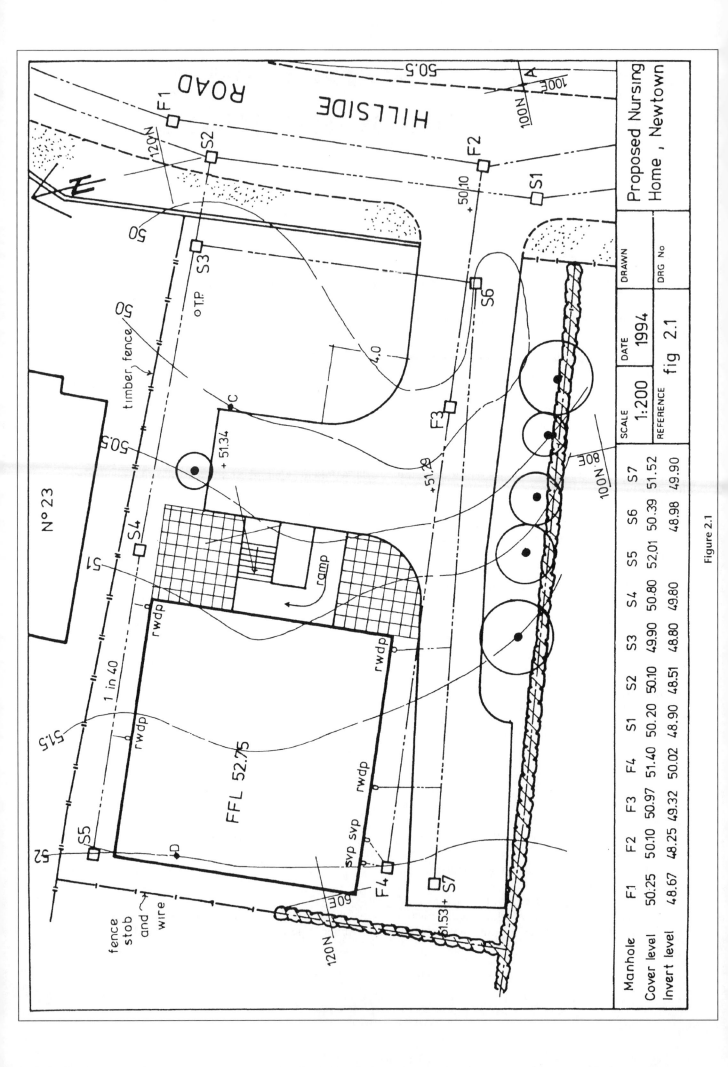

Manhole	F1	F2	F3	F4	S1	S2	S3	S4	S5	S6	S7
Cover level	50.25	50.10	50.97	51.40	50.20	50.10	49.90	50.80	52.01	50.39	51.52
Invert level	48.67	48.25	49.32	50.02	48.90	48.51	48.80	49.80		48.98	49.90

Figure 2.1

Proposed Nursing Home , Newtown

SCALE	DATE	DRAWN
1:200	1994	
REFERENCE fig 2.1		DRG No

E X A M P L E S

1 Using the code of symbols of Fig. 2.2, list the features forming the boundaries of the site in Fig. 2.1.

Answer

South boundary . . . . a hedge
West boundary . . . . partly a hedge,
                           partly a stob and
                           wire fence
North boundary . . . . a timber fence
East boundary . . . . partly a brick wall,
                           partly a stob and
                           wire fence

2 State the number of trees on the site.

Answer

Six

3 State the number and type of manholes on Hillside Road.

Answer

Two foul manholes (sewerage) and two storm water manholes (rainfall)

4 Using the scale stated on the plan, measure the overall size of the building and the various widths of the proposed driveway.

Answer

Building   14 m × 13.5 m
Driveway   5.5 m. 3.5 m, 5.5 m, 5.5 m

(c) Grid lines

On almost every site plan there is, or should be, a north point indicating the direction of north chosen for that particular plan. There is also a family of lines, drawn parallel to, and perpendicular to, the north direction at some convenient interval. These lines form a grid on which it is possible to position features by coordinates referenced to some origin.

On Fig. 2.1 point A is the origin of the grid, but instead of having coordinates of zero metres east and zero metres north, the point has been given coordinates of 100 m east and 100 m north. The grid lines are drawn at 20 m intervals. Any point may be coordinated by scaling the distance to the point from the grid lines. Thus manhole F1 has coordinates of 102 m east, 119 m north, stated to the nearest metre.

E X A M P L E S

5 State the coordinates, to one metre, of the following features:
(a) storm manhole S4,
(b) the south-west corner of the proposed building,
(c) the single tree to the east of the building.

Answer

(a) 81 m, 126 m
(b) 60 m, 118 m
(c) 84 m, 122 m

6 State the features that have the following coordinates:
(a) 84, 108
(b) 71, 107
(c) 90, 99

Answer

(a) Manhole F3
(b) Tree
(c) Junction of fences

(d) Surface relief

Undulations in the ground surface are shown on a plan by contour lines and spot heights. Spot heights are called levels. A full study of contours and levels is made in succeeding chapters of this book. Using contour lines and levels, it is possible to deduce the slope of the ground.

E X A M P L E S

7 From a study of Fig. 2.1. state the value of:
(a) the highest contour line,
(b) the lowest contour line,
(c) the height interval between successive contour lines.

Answer

(a) 52.0
(b) 50.0
(c) 0.5

8 Describe the general slope of the ground from east to west.

Answer

The ground falls from a height of 50.5 m at Hill side Road to a height of 50 m and then rises to 52 m at the western end of the site.

9 State the level of the proposed driveway at:
(a) the short side road,
(b) the western parking area.

Answer

(a) 51.29
(b) 51.53

(e) Gradients

The gradient between any two points on the ground, a road or sewer can be calculated by (a) computing the difference in height between the points and (b) scaling the horizontal distance between them. From these dimensions, the gradient between the points is expressed as the rise or fall over the horizontal length.

EXAMPLE

10 Calculate the average gradient of the ground between points C and D on the plan.

Answer

Gradient = rise (C to D) over horizontal length CD
= (52 – 50)m over 24 m
= 2 m over 24 m
= 1 in 12 or 1/12

The level of the inside of the bottom of a sewer pipe is called the invert level. In order to function properly a sewer pipe must be laid on a gradient, between two invert levels.

EXAMPLE

11 Calculate the gradient of the sewer pipe between foul manholes 3 and 4.

Answer

Invert level manhole 3 = 49.32
Invert level manhole 4 = 50.02
Rise (manhole 4 to 3) = 0.70
Length (manhole 4 to 3) = 24.6 m
Gradient = 0.70 in 24.6
= 1 in 35 or 1/35

It is now commonplace to express gradients as percentages. In order to change any vulgar fraction into a percentage, the fraction is simply multiplied by 100.

EXAMPLES

12 Express the following gradients as percentages:
(a) 1 in 10
(b) 1 in 50
(c) 1 in 16.67

Answer

(a) 1 in 10 = (1/10 × 100)% = 10%
(b) 1 in 50 = (1/50 × 100)% = 2%
(c) 1 in 16.6 = (1/16.67 × 100)% = 6%

13 Express the following percentage gradients as fractions:
(a) 25% (b) 1.667% (c) 6.535%

Answer

(a) 25% = 1 in (100/25) = 1 in 4
(b) 1.667% = 1 in (100/1.667) = 1 in 60
(c) 6.535% = 1 in (100/6.535) = 1 in 15.3

Exercise 2.1

Using site plan of Fig. 2.1:

1 Measure the length of the fence on the western boundary, using a scale rule.

2 Estimate or interpolate, using the contour lines, the existing ground level at the four corners of the proposed building.

3 State the floor level of the proposed building.

4 Calculate the invert level of the storm water manhole 5.

5 Calculate the gradient of the storm water drain between storm water manholes 6 and 3.

6 State the coordinates of:
(a) storm water manhole 3
(b) foul water manhole 4
(c) the telephone pole in the north-east corner of the site

7 State the features that have coordinates of:
(a) 88.6, 113.2.
(b) 78.0, 120.0.
(c) 65.0, 115.8.

8 Count the number of soil ventilation pipes and rain water downpipes that serve the proposed building.

2. Ordnance Survey maps and plans

Positions on the spherical earth are represented as positions on a flat map sheet by using some form of map projection. The Ordnance Survey (OS) uses a projection called the Transverse Mercator Projection (TMP) for the conversion. The basic principles of the projection are explained more fully in section (h), following.

Great Britain lies in the northern hemisphere of the Earth and is situated north of the 49° N line of latitude and almost bisected by the 2° W line of longitude. As is well known, lines of latitude form circles around the Earth, the values of which increase northwards from 0° at the Equator to 90° at the North Pole and similarly southwards to the South Pole. Lines at right angles to the Equator are Great Circles, called meridians of longitude. They all meet at the North and South Poles. Their values are measured east or west of the zero degrees (0°) line of longitude which passes through London and is called the Greenwich Meridian.

(a) National Grid

The 2° west line of longitude is called the Central Meridian since it almost bisects the country (Fig. 2.3). Its intersection with the 49° North line of latitude is the True Origin of a rectangular grid, based on the Transverse Mercator Projection, which covers the whole of Great Britain. The grid, shown in Fig. 2.4. is called the National Grid.

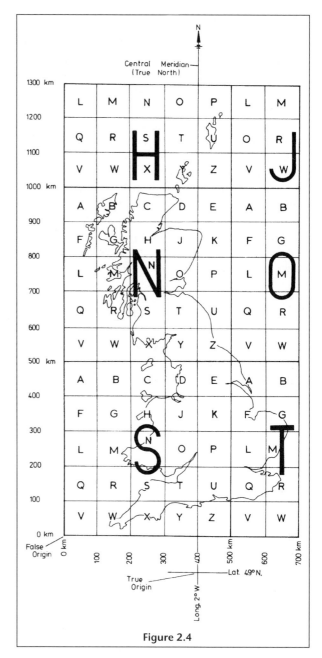

Figure 2.4

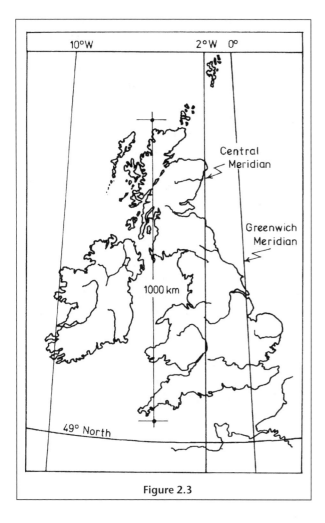

Figure 2.3

The north–south lines of the grid are parallel to the Central Meridian and cannot therefore point to true north. The direction in which they do point is called Grid North.

Since all east–west lines of the National Grid are parallel to the 49° north line of latitude they point truly east–west.

The 49° north line of latitude was chosen as the axis for the National Grid because all points in Great Britain lie to the north of it, while the 2° west line of longitude was chosen because it runs roughly centrally through the country. The use of these two lines as axes means that grid coordinates of points west of the Central Meridian would be negative while points on the mainland of the extreme north of Scotland would have north coordinates in excess of 1000 kilometres.

In order to keep all east–west coordinates positive and all north coordinates less than 1000 kilometres, the origin of the National Grid was moved northwards by 100 km and westwards by 400 km to a point south-west of the Scilly Isles. This point is called the false origin (Fig. 2.4). Any position in Great Britain is therefore known by its easting followed by its northing, which are respectively the distances east and north of the false origin.

(b) Grid references

Commencing at the false origin, the National Grid of 100 km squares covers the country as shown in Fig. 2.4. Each 100 km square is given a separate letter of the alphabet (I being excepted); therefore there are 25 squares of 100 km side in any 500 km block. In order to differentiate between 500 km blocks, each is given a prefix letter H, J, N, O, S or T (anagram ST JOHN or JOHN ST).

Figure 2.5 shows how the National Grid reference is given for any point in the British Isles. The 100 km easting figure is followed by the 100 km northing figure and translated into letters of the alphabet. The letters are followed by the remaining easting figures and northing figures, quoted to any degree of accuracy, ranging from 10 kilometres to 1 metre.

EXAMPLE

14 The National Grid coordinates of Ben Nevis are 216 745E, 771 270N while those of Cardiff Castle are 318 100E, 176 610N. Quote (a) the NG 1 metre and (b) the NG 100 metre reference of each point.

Solution

Ben Nevis
(a) NN16745 71270
(b) NN167 712

Cardiff Castle
(a) ST18100 76610
(b) ST181 766

(c) Types and scales of OS plans

Ordnance Survey maps are topographic maps which show details of the terrain, i.e. the hills and valleys, roads, rivers and man-made features. Some of the maps also show contour lines. Traditionally, they have been published as paper map series at scales from 1:1250 to 1:625 000. The only plans which can be considered to be of relevance in the construction industry are the large-scale 1:1250 plans which cover urban areas; the 1:2500 plans which cover the main populated areas; and the medium-scale 1:10 000 maps covering the whole country. These maps and plans are used in Town and Country planning matters and in applications for planning permission and building warrant. Each map is referenced by a sheet numbering system based on the grid coordinates of its southwest corner.

(d) Reference numbers of maps and plans

Figure 2.4 shows that the major blocks of the National Grid have sides of 100 km. Block NS is bounded on the north and south by the 700 km and 600 km northing lines and on the west and east by the 200 km and 300 km easting lines, respectively.

Figure 2.6(a) shows the block subdivided into 100 squares, each measuring 10 km square.

1.25000 scale maps

Taking a 10 km square block bounded by Grid lines 670 km north, 660 km north, 250 km and 260 km east (Fig. 2.6(b)) and drawing it to scale 1:25 000 produces a square of 400 mm side. This is the typical format for the small-scale 1:25 000 maps. Points of detail are shown by conventional signs and variation in height by contours at 5 m vertical intervals, with the National Grid superimposed at 1 km intervals.

Each map has a unique reference referred to its south-west corner. Since the side of the map is 10 km long, the reference must be given to the 10 km figure. Thus the reference of the map shown in Fig. 2.6(b) is derived as follows:

Coordinates of south-west corner
250 000 m east: **660** 000 m north
= **250** km **east 660** km north
100 km Grid reference = **26** = **NS**

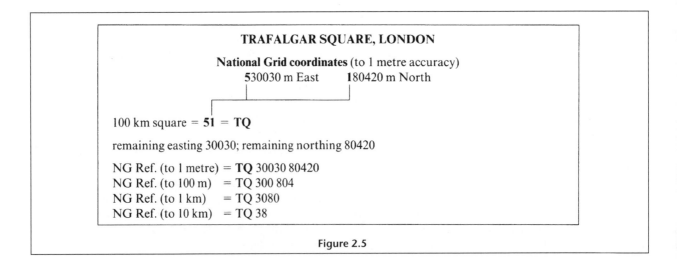

TRAFALGAR SQUARE, LONDON

National Grid coordinates (to 1 metre accuracy)
530030 m East   180420 m North

100 km square = **51** = **TQ**

remaining easting 30030; remaining northing 80420

NG Ref. (to 1 metre) = **TQ** 30030 80420
NG Ref. (to 100 m) = TQ 300 804
NG Ref. (to 1 km)   = TQ 3080
NG Ref. (to 10 km) = TQ 38

Figure 2.5

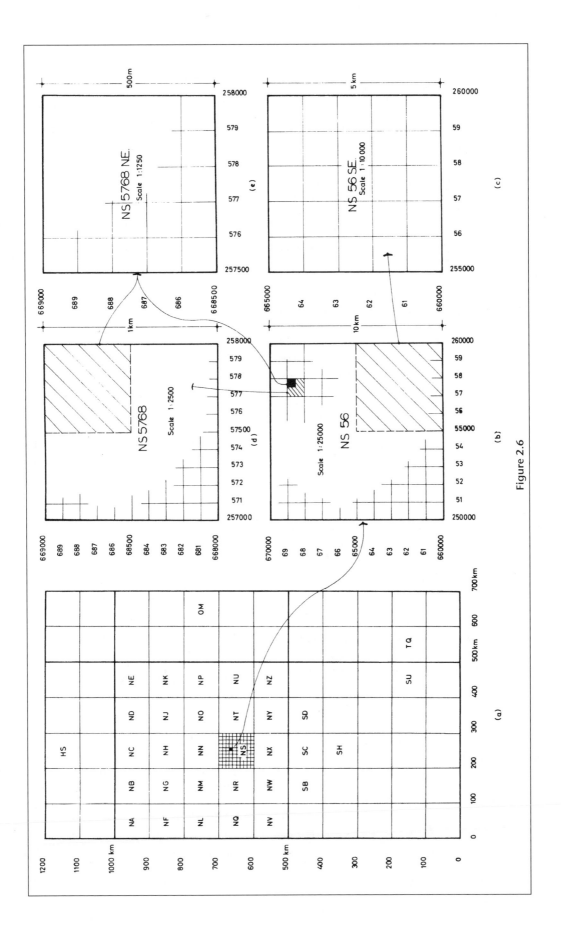

Figure 2.6

Therefore 10 km Grid reference = **NS 56** = NS 56

The 1:25 000 map is the base map on which the various series of larger-scale maps and plans are built.

1:10 000 scale maps

Figure 2.6(c) shows a ground square of 5 km side drawn to a scale of 1:10 000 to produce the format for the 1:10 000 scale series of maps. Certain Town and Country Planning matters and Development proposals are shown on this scale. The map really represents one-quarter of the area of the 1:25 000 scale map shown in Fig. 2.6.

Details on these maps are shown true to scale and only the widths of narrow streets are exaggerated. Surface relief is shown by contour lines at 10 metre vertical intervals in mountainous areas and 5 metre vertical intervals in the rest of the country. The National Grid is superimposed at 1 km intervals.

The map is referenced as a quadrant of the 1:25 000 map of which it is part. Thus the 1:10 000 scale map of Fig. 2.6(c), being the south-east quadrant of the 1:25 000 map **NS 56**, is given the reference **NS 56 SE**.

1:2500 scale plans

Figure 2.6(d) shows a 1 km square taken from the 1:25 000 map NS 56 and enlarged to scale 1:2500 to produce the format for the 1:2500 National Grid series of plans.

The whole of Great Britain, except moorland and mountain area, is covered by this series. All details are true to scale, and areas of parcels of land are given in acres and hectares. Surface relief is shown by means of bench marks and spot heights, and the National Grid is superimposed at 100 metre intervals.

This plan is probably the most commonly used plan in the construction industry. Most site and location plans are shown on this scale. The plan reference is again made to the south-west corner and is given to 1 km. Thus the reference of the map shown in Fig. 2.6(d) is derived as follows:

Coordinates of south-west corner
257 000 m east: **668** 000 m north
= **257** km east: **668** km north
100 km Grid reference **26** = **NS**

Therefore 1 km Grid reference = **NS 5768.**

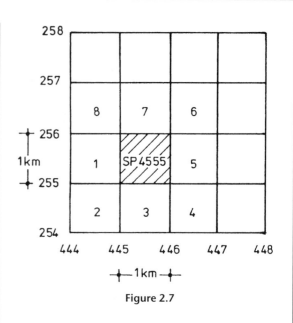

Figure 2.7

Answer

Map	1 km easting	1 km northing	Plan reference
1	444	255	SP 4455
2	444	254	SP 4454
3	445	254	SP 4554
4	446	254	SP 4654
5	446	255	SP 4655
6	446	256	SP 4656
7	445	256	SP 4556
8	444	256	SP 4456

It has been found that it is more convenient and economical to produce the 1:2500 plans in pairs. The large-scale 1:2500 plan covers an area of 2 kilometres east–west by one kilometre north–south. The grid line forming the western edge of the sheet is always an even number. Thus plan **NS 5768** (Fig. 2.6(d)) would be the eastern half of sheet **NS 5668–5768**. This reference may be shortened to read **NS 56/5768**.

1:1250 scale plans

Figure 2.6(e) shows a ground square of 500 m side drawn to a scale of 1:1250 to produce the format for the 1:1250 National Grid series of plans. The plan represents one-quarter of the area of the 1:2500 plan of Fig. 2.6(d).

These maps are the largest scale published by the Ordnance Survey and cover only urban areas. All details are true to scale. Surface relief is shown by bench marks and spot heights and the National Grid is carried at 100 m intervals.

The plan is referenced as a quadrant of the 1:2500 plan of which it is a part. Thus the 1:1250 scale plan of Fig. 2.6(e), being the north-east quadrant of the 1:2500 plan **NS 5768**, is given the reference **NS 5768 NE**.

EXAMPLE

15 Figure 2.7 shows a 1:2500 scale OS plan hatched in black. The south-west corner has a 1 km easting figure of **445** and a 1 km northing figure of **255**. Its 1 km reference is therefore **SP 4555**. List the reference numbers of plans 1 to 8 immediately adjoining it.

Exercise 2.2

1 Figure 2.8(a) shows diagrammatically the 1:2500 scale Ordnance Survey plan **SK 5265**. List the reference numbers of the eight plans immediately adjacent to it.

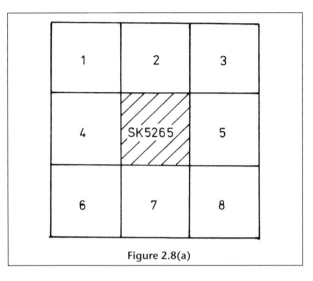

Figure 2.8(a)

2 Figure 2.8(b) shows diagrammatically the 1:1250 scale Ordnance Survey plan **SK 3657 SE**. List the reference numbers of the fifteen plans adjacent to it.

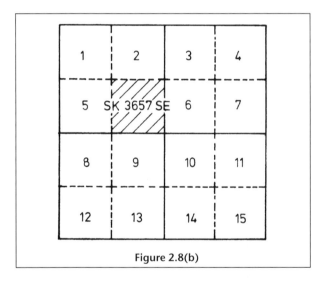

Figure 2.8(b)

(e) OS digital map data. The National topographic database

The origination of all OS maps is now from digital data. Only the 1:25 000 map series is published as a traditional paper map. Plans at a larger scale are available from OS Options Agents. In order to procure a large-scale plan, the OS coordinates, or postcode, of the relevant area are specified, viewed on a computer screen, then printed off as a paper copy. (The cost has increased considerably from the days of paper maps.) The appearance of digital maps is very similar to the paper maps which they replace since they are based on these maps. A facsimile example of a digital map is shown in Fig. 2.13.

1. OS digital formats
OS digital data is supplied in two formats.

1. *Vector data* The data consists of lines and points usually in what is termed 'link and node' structure, where the junction of three or more lines is called a node. This enables polygons to be identified. The data is obtained by tracing from the master copy of the old paper maps.

2. *Raster data* Raster data is stored as pixels and is obtained by scanning the original maps. It is therefore unintelligent, whereas vector data has attributes, i.e. a fence knows it is a fence.

2. OS digital databases
1. *Landline data* The vector data has 37 feature codes in NTF format and 43 layers in DXF format for CAD systems. The data is used in 1:1250 urban areas, 1:2500 rural areas and 1:10 000 mountain and moorland areas.

2. *Landline Plus* Landline Plus has an additional 26 feature codes for vegetation and topography. It is published as tiles equivalent to the old paper map sheets.

When purchasing OS digital maps, the options are (i) Superplan – a site-centred option, specified by coordinates or postcode, (ii) Siteplan – a cheaper raster image of landline data for use in planning applications and land registration, and (iii) Mastermap – a development of Landline data for use with geographical information systems (GIS).

3. Accuracy of data
Since the digital data has been produced from the original paper masters there can be no improvement in accuracy. Tests have shown that points on the maps have a standard positional error of about 0.3 mm on paper. (The actual ground error is 0.3 mm at the relevant scale.) However, the Positional Accuracy Improvement programme (PAI) is gradually being extended across the country to reduce the error in the maps.

(f) Surface relief

On Ordnance Survey large-scale and medium-scale maps there are three methods of showing height.

1. *Contour lines* A contour line is a line joining all points of equal height. The 1:10 000 scale map is the largest-scale map which shows contours.

2. *Spot levels* A spot level is simply a dot on the plan, the level of which is printed alongside. This method is used on the 1:1250 and 1:2500 scale plans.

3. *Bench marks* Bench marks (BM) are permanent marks established throughout the country by the Ordnance Survey, using a network of precise levelling lines.

Conventional Signs — Scales 1:1250 and 1:2500

.... Bracken

.... Coniferous Tree (Surveyed)

.... Coniferous Trees (Not Surveyed)

.... Coppice, Osier

.... Non-coniferous Tree (Surveyed)

.... Non-coniferous Trees (Not Surveyed)

.... Antiquity (site of)

.... Direction of water flow

↑ B M .... Bench Mark (Normal)

.... Electricity Pylon

E T L .... Electricity Transmission Line

.... Marsh, Saltings

.... Orchard Tree

.... Reeds

.... Rough Grassland

.... Scrub

.... Heath

+ .... Surface Level

△ .... Triangulation Station

∫ .... Area Brace (1:2500 scale only)

.... Perimeter of built-up area with single acreage (1:2500 scale only)

Roofed Building

Slopes
Top

Boundaries

England, Wales & Scotland

· · · · · · .... Civil Parish Boundary

Boro (or Burgh) Const   Co Const .... Parly & Ward Boundaries based on civil parish

Boro (or Burgh) Const & Ward Bdy
⎯⎯⎯⎯⎯⎯⎯⎯⎯
Co Const Bdy .... Parly & Ward Boundaries not based on civil parish

Examples of Boundary Mereings

F F
R H .... Symbol marking point where boundary mereing changes

· · · · Und · · · · .... Undefined boundary

· · · Def · · · .... Original boundary feature destroyed or defaced

C B	Centre of Bank	E K	Edge of Kerb
C C	Centre of Canal, etc.	F F	Face of Fence
C D	Centre of Ditch, etc.	F W	Face of Wall
C R	Centre of Road, etc.	S R	Side of River, etc.
C S	Centre of Stream, etc.	T B	Top of Bank
C O C S	Centre of Old Course of Stream	Tk H	Track of Hedge
C C S	Centre of Covered Stream	Tk S	Track of Stream
4ft R H	4 feet from Root of Hedge		

Abbreviations

B H	Beer House	P	Pillar, Pole or Post
B P, B S	Boundary Post, Boundary Stone	P C	Public Convenience
Cn, C	Capstan, Crane	P C B	Police Call Box
Chy	Chimney	P T P	Police Telephone Pillar
D Fn	Drinking Fountain	P O	Post Office
El P	Electricity Pillar or Post	P H	Public House
E T L	Electricity Transmission Line	Pp	Pump
F A P	Fire Alarm Pillar	S B, S Br	Signal Box, Signal Bridge
F S	Flagstaff	S P, S L	Signal Post, Signal Light
F B	Foot Bridge	Spr	Spring
G P	Guide Post	S, S D	Stone, Sundial
H	Hydrant or Hydraulic	Tk	Tank or Track
L B	Letter Box	T C B	Telephone Call Box
L C	Level Crossing	T C P	Telephone Call Post
L Twr	Lighting Tower	Tr	Trough
L G	Loading Gauge	Wr Pt, Wr T	Water Point, Water Tap
Meml	Memorial	W B	Weighbridge
M P U	Mail Pick-up	W	Well
M H	Manhole	Wd Pp	Wind Pump
M P	Mile Post or Mooring Post	M H or L W	Mean High or Low Water (England and Wales)
M S	Mile Stone		
N T	National Trust	M H or L W S	Mean High or Low Water Springs (Scotland)
N T L	Normal Tidal Limit		

Figure 2.10

The datum for the levelling is the tide gauge at Newlyn harbour, Cornwall. The network consists of about 200 fundamental BMs which were guaranteed to be error free on the original levelling. From that network about half a million more BMs were established by less precise levelling techniques. These marks are identified by being cut into buildings as in Fig. 2.9. The value of each BM is shown to the nearest centimetre on the 1:1250 and 1:2500 scale plans.

Care must be exercised in the use of these BMs since the network has not been maintained on a regular basis. They are prone to subsidence, particularly in mining areas where they are known to have sunk by several metres in some cases.

These bench marks are no longer maintained by the Ordnance Survey. Nowadays, precise GPS ellipsoid heighting techniques are used to establish new bench marks.

(g) Conventional signs

A selection of signs, symbols and abbreviations used on the 1:1250 and 1:2500 scale paper plans is given in Fig. 2.10. A further selection of those used on digital plans is shown in Fig. 2.11.

Exercise 2.3

Using Fig. 2.12, which is a *facsimile* OS 1:2500 scale paper plan, and the lists of conventional and symbol signs, answer the following questions (overleaf):

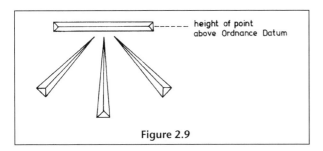

height of point above Ordnance Datum

Figure 2.9

Boundary information

Boundary post or stone	O
Boundary mereing symbol	⦵
Point feature	O

Vegetation

Coniferous trees	🌲
Coniferous trees (scattered)	🌲
Positioned coniferous trees	🌲
Non-coniferous trees	🌳
Non-coniferous trees (scattered)	🌳
Positioned non-coniferous trees	🌳
Coppice or osiers	⟱
Heath	,،וווווו,,
Marsh, salt marsh or reeds	⌐וווⵊ⌐
Orchard	🍃🍃🍃
Rough grassland	,،ווווווⵊ,,
Scrub	℘

Buildings

Roofed building or glasshouse indicator	✕

Other features

Pylon	⊠
Upper level of communication indicator	✕

Water features

Flow arrow	⟵—≪
Water indicator	⌇⌇⌇

Landforms

Boulders	🪨
Boulders (scattered)	🪨
Cliff indicator	⫟
Rock	🪨
Rock (scattered)	🪨
Scree	▲
Slope indicator	◿

The representation of a road, track or path is no evidence of a right of way.

Height information

Bench mark	↑
Spot height	+
Triangulation point	△
Copyright	©

Common abbreviations

Boundary information
UA Bdy	Unitary authority
Dist Bdy	District
Met Dist Bdy	Metropolitan district
C	Community
CP Bdy	Civil parish
Boro Const Bdy	Borough constituency
Co Const Bdy	County contituency
Burgh Const Bdy	Burgh constituency
Euro Const Bdy	European constituency
EER Bdy	European electoral region
LB Bdy	London Borough
Asly Const Bdy	Assembly consituency
Asly ER	Assembly electoral region
P Const Bdy	Scottish parliamentary constituency
PER Bdy	Scottish parliamentary electoral region
GL Asly Const	Greater London authority assembly constituency

Other information
CG	Cattle grid
CHY	Chimney
Coll	College
Ct	Court
El Sub Sta	Electricity sub station
FB	Footbridge
Fl Sk	Flare stack
Fn	Fountain
FS	Flagstaff
GP	Guide post
LC	Level crossing
Liby	Library
Meml	Memorial
MHW(s)	Mean high water (springs)
MLW(s)	Mean low water (springs)
Mon	Monument
MP, MS	Mile post or stone
NTL	Normal tidal limit
P, Ps	Post(s) or pole(s)
PH	Public house
PO	Post Office
Pol Sta	Police station
PW	Place of worship
Sch	School
Spr	Spring
Sta	Station
TK	Tank or track
W	Well

Figure 2.11

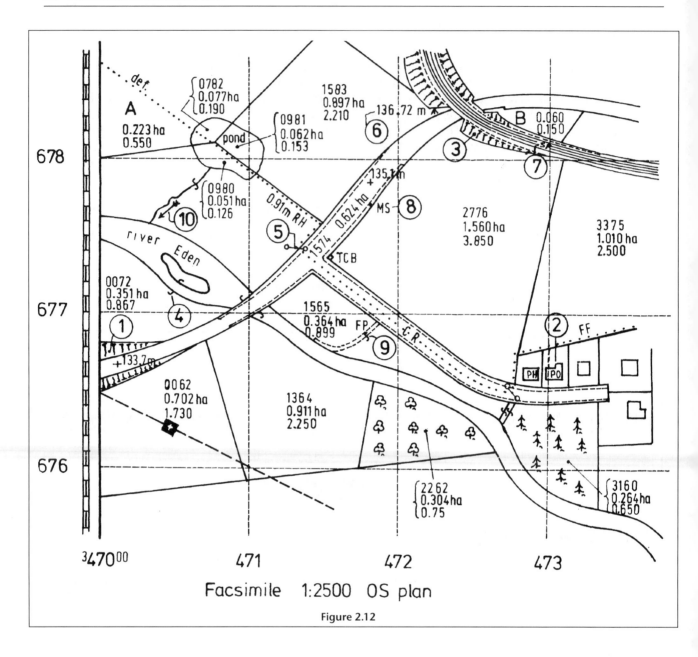

Facsimile 1:2500 OS plan

Figure 2.12

1 State the values of the two spot levels along the main roadway and calculate the gradient of the roadway between these points.

2 State the value of the bench mark on the railway bridge.

3 State the features that have the following coordinates:
(a) 4730, 6767　　(b) 4729, 6781
(c) 4719, 6778　　(d) 4704, 6777

4 State the coordinates (to the nearest 10 metres) of:
(a) the centre of the orchard
(b) the telephone call box
(c) the public house
(d) the electricity pylon

5 State the direction in which the electricity transmission line runs.

6 Make a list of the conventional signs, circled and numbered from 1 to 10.

Exercise 2.4

Using Fig. 2.13, which is a *facsimile* OS 1:2500 scale digital map, and the lists of signs answer the following questions:

1 State the approximate coordinates and heights of the bench mark at Langloan farm.

2 State the direction of flow of the stream and the meaning of the abbreviations cs, ccs and und.

3 Calculate the difference in height between the spot levels on either side of Langloan farm.

4 Identify the features which have the following coordinates:
(a) 207 363, 647 350　　(b) 207 221, 647 496
(c) 207 335, 647 510

5 State the number of buildings at Langloan farm.

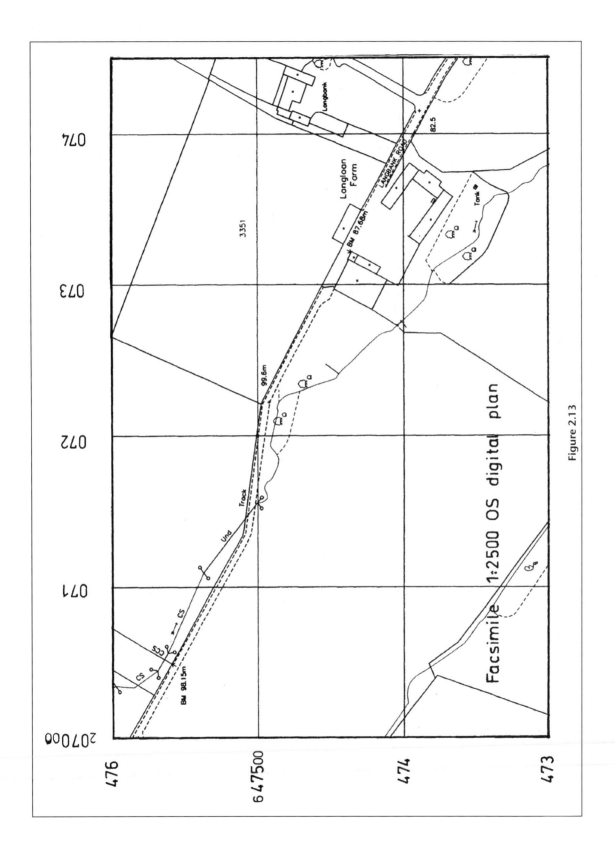

Facsimile 1:2500 OS digital plan

Figure 2.13

(h) Local scale factor (LSF) and Mean Sea Level (MSL) corrections.

Local scale factor (LSF)

The shape of the Earth is almost spherical. Technically it is an oblate spheroid, meaning that the Earth is flattened slightly at the Poles. Since positions on the spherical Earth have to be represented on a flat map sheet, there is a problem which is solved by the use of map projections.

The Ordnance Survey uses a map projection called the Transverse Mercator Projection (TMP) for the conversion. The subject is highly complex and mathematical but the construction surveyor needs only a basic knowledge of the principles of the subject.

Consider firstly the various curves and definitions of these which are used to state the positions of points on the Earth's surface. A Great circle is a circle traced out on the surface of a sphere by a plane passing through the centre of the Earth. It is therefore the circle of greatest diameter which can be drawn on a sphere. All other circles are called small circles. The importance of a Great circle is that it is the shortest path between two points and can be compared to a straight line on a map. Meridians of longitude are all Great circles but the only line of latitude which is a Great circle is the Equator. (Fig. 2.14)

which in the case of Great Britain is the 2 degree West line of longitude. This circle of longitude, as was explained in section (a), is called the Central Meridian as it almost bisects the country from north to south. Being a Great circle it is drawn true to scale as a straight line on a map sheet.

Any other Great circle which crosses the Central Meridian at right angles may also, rightfully, be drawn as a straight line so a group of Great circles will appear as a series of parallel straight lines perpendicular to the Central Meridian. These circles in reality converge at two points on the Equator, namely 88° East and 92° West. (Fig. 2.14)

By this method an attempt has been made to promote a true representation of the surface of the Earth on a flat map sheet but in so doing a distortion has been introduced. The parallel straight lines cannot meet but the circles do meet at the Equator. The distance between any two parallel lines remains constant but the distance represented on the Earth is reducing until at the Equator the distance is zero. This therefore means that the scale of the map is increasing with distance from the Central Meridian.

The maximum error of scale which is produced is approximately 1 part in 1250. On an Ordnance Survey 1/2500 scale plan the length from corner to corner across the plan is 565 mm, so an error of (565/1250) = 0.5 mm approx, would be present. This is definitely plottable so to reduce the error, the cylinder of the TMP, instead of being tangential to the Central Meridian, is made to cut the Earth's sphere at two points on either side of it (Fig. 2.15).

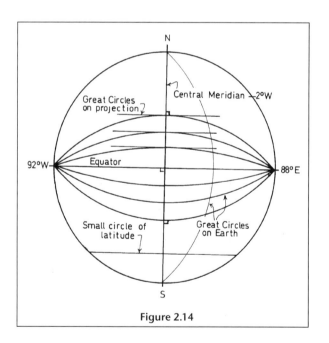

Figure 2.14

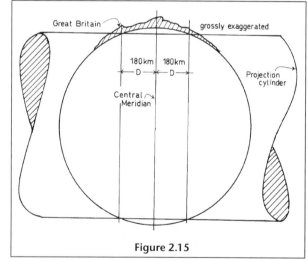

Figure 2.15

The Transverse Mercator Projection is used in countries which have their greatest extent in latitude but are comparatively narrow in longitude. Egypt and Great Britain are good examples. The TMP is a cylindrical projection, in which the axis of the cylinder which encloses the Earth is horizontal. The cylinder therefore contacts the Earth along a meridian of longitude,

This action reduces the scale error by half, i.e. to 1 part in 2500. The scale at the Central Meridian reduces to (2499/2500 = 0.9996) while at the east and west edges of the country it increases to (2501/2500 = 1.0004). In other words, an actual distance of 2500 units measured on the Earth's surface along the Central Meridian would translate to 2499 units on the projection while at the edges the distance on the projection would be 2501 units.

At the two points where the cylinder cuts the Earth's sphere the scale is of course correct. The two points occur at 180 km on either side of the Central Meridian.

The above effect produces a local scale factor (LSF) by which any distance measured on the ground must be multiplied to find the projection length. The factor is 0.9996 at the Central Meridian, 1.0000 at 180 km on either side and 1.0004 at the edges. The LSF varies therefore with distance east or west of the Central Meridian.

The local scale error at the Central Meridian is $(1.0000 - 0.9996) = 0.0004 = 0.04\%$ and is in fact equal to $(k.D^2)$ where D represents the distance in kilometres from the Central Meridian.

Since the scale at 180 km east/west is 1.0000
$$0.0004 = k \times 180^2$$
$$\text{therefore } k = 0.0004/180^2$$
$$= \underline{1.228 \times 10^{-8}}$$

The LSF at any point is therefore the value of the LSF at the Central Meridian plus kD^2

i.e. LSF (any point) $= 0.9996 + [1.228 \times 10^{-8} \times D^2]$

where D is the distance (km) from the Central Meridian.

As an example, the LSF at a point 80 km east of the Central Meridian would be computed as follows:

$$\text{LSF} = 0.999\,6 + [1.228 \times 10^{-8} \times D^2]$$
$$= 0.999\,6 + [1.228 \times 10^{-8} \times 80^2]$$
$$= 0.999\,6 + 0.000\,08$$
$$= \underline{0.999\,68}$$

An actual distance of 500 m in that area would therefore translate to 499.84 m on the National Grid.

Application of LSF

The practical effect of LSF for the construction surveyor occurs only when surveys are to be connected to OS National Grid coordinates. The horizontal distances on any such survey must be multiplied by the appropriate value of LSF. Likewise, when distances are to be set out on site (Chapter 12) the distance obtained from the plan coordinates must be divided by LSF to produce horizontal ground distances.

Different local scale factors are required only when the extent of a survey exceeds 20 km in width from east to west; otherwise it is considered to remain constant for the complete survey.

Mean sea level correction (MSL)

Complementary to the local scale factor correction is a second correction called Mean Sea Level correction (MSL). In simple terms, the projection calculations are done as if the area being computed lay at sea level, i.e. on the theoretical spherical surface of the Earth. The

mean radius of the Earth is 6384.100 km and this figure is usually used as in the following example.

Figure 2.16 illustrates the situation where a survey is being conducted at an average height of 280 m above sea level. The distance D measures 3500 metres. Before applying the Local Scale Factor correction above, the distance D must firstly be reduced to its mean sea level equivalent d metres.

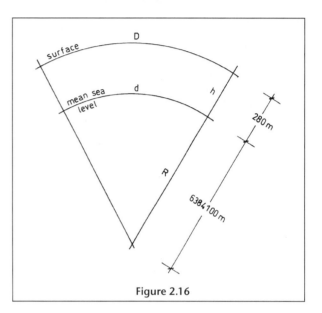

Figure 2.16

In Figure 2.16,

proportionally, $d/D = R/(R + h)$
$$\text{therefore } d = D \times R/(R + h)$$
$$= \frac{3500 \times 638\,410\,0}{638\,438\,0}$$
$$= \underline{3499.85 \text{ m}}$$

The local scale factor is then applied to produce the equivalent Grid distance.

Exercise 2.5

Note: This question is of an advanced nature and is intended only for students of Engineering courses which include computation involving the National Grid.

1 Two stations A and B have the following NG coordinates (m):
  A: 438 455.66 E   634 564.99 N   238.45 AOD
  B: 433 245.43 E   630 443.12 N   212.55 AOD

Calculate:
(a) the NG bearing and length of line AB,
(b) the average height above datum of line AB, hence the length of line AB corrected for mean sea level,
(c) the mid Easting of line AB, hence the length of line AB corrected for LSF,
(d) the actual horizontal and slope distance of line AB.

3. Answers
Exercise 2.1

1 12 m
2 SE 51.2 m, SW 52.2 m, NW 52.0 m, NE 51.1 m
3 FL 52.75 m
4 Invert level S5 is 50.20 m
5 Gradient 1 in 86
6 (a) 96 119   (b) 61 116   (c) 92 120
7 (a) Centre of curve   (b) Stairs   (c) RWDP
8 SVPs 2, RWDPs 4

Exercise 2.2

1 1. SK5166   2. SK5266   3. SK5366   4. SK5165
   5. SK5365   6. SK5164   7. SK5264   8. SK5364
2 1. SK3657NW   2. SK3657NE   3. SK3757NW
   4. SK3757NE   5. SK3657SW   6. SK3757SW
   7. SK3757SE   8. SK3656NW   9. SK3656NE
   10. SK3756NW   11. SK3756NE   12. SK3656SW
   13. SK3656SE   14. SK3756SW   15. SK3756SE

Exercise 2.3

1 133.7 m, 135.1 m, 1 in 153
2 136.72 m
3 (a) Post office   (b) signal post   (c) milestone
   (d) direction of flow of stream
4 (a) 4722, 6762   (b) 4715, 6773   (c) 4729, 6766
   (d) 4705, 6763
5 ESE
6 1. Roadway cutting
   2. Post office
   3. Railway embankment
   4. Area brace (joins areas of land)
   5. Boundary (change of description)
   6. Bench mark value

7. Signal post
8. Mile stone
9. Footpath
10. Direction of flow of stream.

Exercise 2.4

1 Coords 207 321 mE 647 437 m N Height 87.68 m
2 The stream flows approximately south eastwards. It forms a boundary which runs along the centre of the stream (cs): the centre of the stream covered (ccs) then along an undefined line (und).
3 17.10 m
4 (a) Tank (b) Junction of fences (c) Field No. 3351
5 Ten

Exercise 2.5

(a) $\Delta E_{AB} = -5210.23$ $\Delta N_{AB} = -4121.87$
   Length AB = 6643.52 m
   Bearing AB = 231° 39′ 08″

(b) Average height of line AB = 225.50
   $AB = 6643.52 \times (R + h)/R$
   AB = 6643.75 m (Horz)

(c) Mid Easting of line AB = 435 850.55 m
   Distance from Central Meridian
   $= (435\,850.55 - 400\,000.00)$ m = 35.85 km
   $LSF = 0.9996 + (1.228 \times 10^{-8} \times D^2)$
      $= 0.999\,615\,7$
   $AB = 6643.75/0.999\,615\,7$
      = 6646.30 m

(d) $\Delta H_{AB} = 25.90$ m
   Slope $AB = (6646.30^2 + 25.90^2)^{1/2}$
      = 6646.35 m

Chapter summary

In this chapter, the following are the most important points:

● Site plans are essential for showing how the existing landscape and environment is to be transformed by the planning and execution of construction works. It is essential to understand the methods of depicting the existing topography by the signs, symbols and contour lines explained on pages 13 to 16.

● The Ordnance Survey is the principal body charged with mapping Great Britain and Ireland in a global context. Since the maps and plans are the basis of those used by construction engineers, the relevance of the types and scales of the maps, bench marks, mean sea level correction, local scale factor and map references to enable connections with OS services to be made is emphasized.

CHAPTER 3 Linear surveying

In this chapter you will learn about:

- the mathematical principles involved in making a linear survey
- the equipment required to make a linear survey
- the methods used to obtain accurate results
- the errors encountered in measuring and the methods used to eliminate or minimize these errors
- the procedure in planning, conducting and recording a linear survey
- the preparatory work and procedure in plotting the results manually and with the use of a CAD program

---

Introduction (chapter context)

The one piece of surveying equipment universally used throughout the construction industry is a tape. It is carried by every architect, site agent, engineer, planner and professional surveyor and is in use every day. Although everyone will claim to be able to use a tape, few can use it with accuracy because of the many errors which arise during usage.

A tape is required to set out steel columns on large buildings where EDM is not accurate enough. It is also a common method of setting out the centre lines of curved roadways and railways and is used on accurate traverse surveys, all of which are fully explained in later chapters.

Linear surveying: One area where a tape is used exclusively is in linear surveying. It is claimed in many circles that linear surveying is obsolete and in commercial circles that may well be true. However, as a method of imparting the principles of geometry and as a starting point for teaching young aspiring surveyors the rudiments of measuring, it is invaluable.

The overriding reason for its inclusion in this book, however, is the fact that it is part of several modules in courses of the Scottish and some of the numerous English Qualification Boards including the Higher National Diploma in Building.

Traverse surveying: A traverse survey involves the measurement of both angles and distances. Measuring the lengths of lines of a traverse survey with the required standard of accuracy is a very skilled operation and is probably the most difficult task in surveying. The surveyor must be aware of a variety of errors of every class (described in Chapter 1) and be able to physically overcome the terrain difficulties, present in most measurements. The sources of error and subsequent corrections to measurements are much more onerous in traversing than in linear surveying, hence the decision to split the use of tapes between two chapters. This chapter will deal with measurement in linear surveying, while a more in-depth treatment will follow in Chapter 8 (Traverse surveys).

Students who are pursuing courses of study which do not include linear surveying should simply choose to ignore this chapter. Nothing will be lost. By reading it, however, much will be gained.

1. Principles of linear surveying

Consider Fig. 3.1. It shows a small site adjacent to a roadway which is to be developed as a building plot for two houses. An accurate plan of the site is required for the development.

In order to make an accurate scaled drawing of the site (Fig. 3.1) the student needs to understand:

(a) the geometric principles involved in making the survey,
(b) the techniques of measuring distance using a tape and the methods of handling the errors which arise each time a tape is used,
(c) the method of conducting the survey and recording the many measurements, and
(d) the method of plotting these measurements to scale.

In linear surveying, the principles of trilateration and offsetting are used. (These principles are outlined in Chapter 1 Sec. 5 and should be revised at this juncture.)

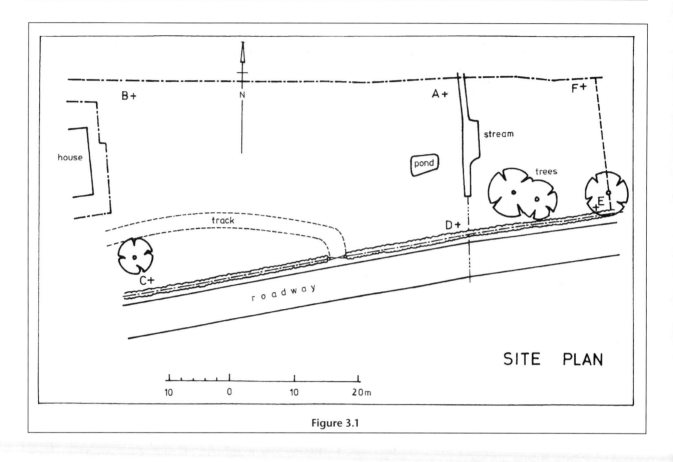

Figure 3.1

(a) Trilateration

In Fig. 3.2 (see page 31) a trilateration framework is clearly shown covering the whole area with as few well-conditioned triangles as possible. The framework consists of four triangles, ABD, BCD, DAF and FDE. The sides of these triangles are measured using a steel tape, the correct usage of which is described in the following Sec. 2.

(b) Offsetting (lines at right angles)

In Fig. 3.2, the details of the site, namely fences, trees, hedges, stream, etc., are surveyed by use of offsets. As already explained in Chapter 1, offsets are short measurements made at right angles to the main framework lines.

2. Linear measuring techniques

(a) Equipment

All linear measuring equipment should conform to the British Standards specification. Such equipment will then have clear legible graduations and unambiguous figuring. The instruments required on a linear survey vary from the folding one metre rule to the long fifty metre steel tape.

Table 3.1 shows the lengths, graduations and method of figuring of the principal linear measuring instruments.

Steel tapes

These tapes are made from hardened tempered steel sheathed in durable white plastic. The tapes are 10 mm wide with black and red figures and black graduations. They are very tough and the graduations almost indestructible. The tapes are housed in open-frame plastic winders fitted with a quick rewind handle. They are subdivided in millimetres throughout, figured at every 10 mm with quick-reading metre figures in red at every 100 mm, and whole metre figures are in red (Fig. 3.3).

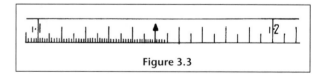

Figure 3.3

Synthetic tapes

These tapes are manufactured from multiple strands of fibreglass and coated with PVC. Fibron is impervious to water and can be wound back into the case without damage, even when wet. The tapes are graduated throughout in metres and decimals, the finest graduation being 10 mm (Fig. 3.4).

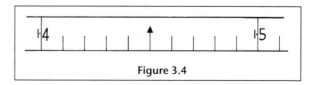

Figure 3.4

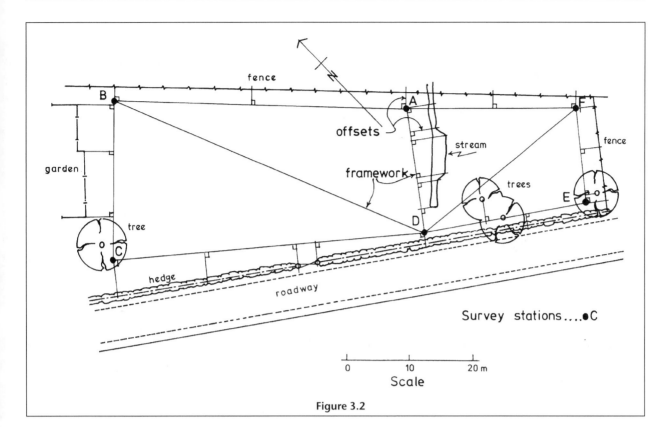

Figure 3.2

Table 3.1

Measuring instrument	Length (m)	Graduations			Method of figuring
		Major	Inter	Fine	
Folding rule	1	10 mm	5 mm	1 mm	10 mm intervals using 3-digit numbers
Folding and multi-folding rods	1, 1.5, 2	10 mm	5 mm	1 mm	10 mm intervals using 3-digit numbers
Steel pocket rule	2	10 mm	5 mm	1 mm	10 mm intervals using 3-digit numbers
Steel tapes	10, 20, 30	100 mm	10 mm	5 mm	100 mm intervals in decimals of a metre
		First and last metres further subdivided into divisions of			
Synthetic tapes	10, 20, 30	100 mm	50 mm	10 mm	50 mm marks are denoted by arrows

(b) Use of the tape

Reading the tape

On every survey, there will inevitably be a variety of long and short, flat and inclined, lines to be measured accurately. Figure 3.5(a) shows a short survey line AB marked on the ground by two pegs. The distance AB is shorter than one length of tape. The measurement of the line AB is obtained by unreeling the tape and straightening it along the line between the pegs. The zero point of the tape (usually the end of the handle) is held against station A by the rear tape-person (called the follower). The forward end of the tape is read against station B by the forward tape person (called the leader) after it has been carefully tightened.

EXAMPLE

1 Figure 3.5(b) shows tape readings A, B, C and D. Read the metres and decimals from the tape, count the centimetres and estimate the millimetres to give the correct reading at each point.

Answer

A. 11.580 m   B. 0.088 m   C. 29.818 m   D. 15.003 m

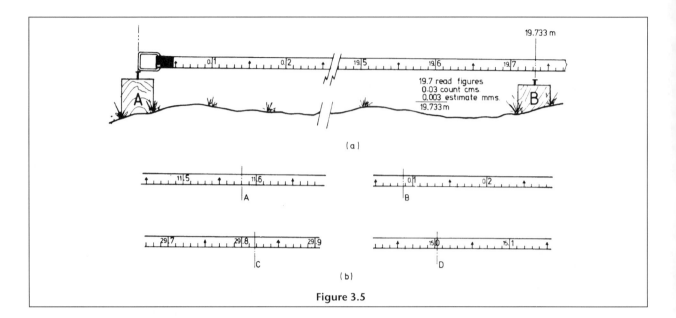

19.733 m

19.7 read figures
0.03 count cms.
0.003 estimate mms.
19.733 m

(a)

(b)

Figure 3.5

On surveys most of the lines will be considerably longer than one tape length and a sound operational technique is required. Two ancillary pieces of equipment are necessary, namely ranging rods and marking arrows.

Ranging rods are 2 metres long, round, wooden poles, graduated into 500 mm divisions and painted alternately red and white. They have a pointed metal shoe for penetration into the earth.

Marking arrows are made from steel wire, 375 mm long, pointed at one end, a 30 mm loop at the other and painted in fluorescent paint. They are made up in sets of ten. Both instruments are shown in Fig. 3.6.

Two surveyors are required to measure a long line. The leader's job is to pull the tape in the required direction and mark each tape length. A known number of arrows and a ranging rod are carried by the leader. The follower's job is to align the tape and count the tape lengths.

In Fig. 3.7, a line AB is to be measured across a gently sloping grassy field. The follower holds the zero end of

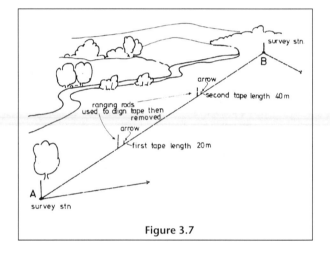

Figure 3.7

the tape against station A and the leader pulls the tape towards station B. When the tape has been laid out the leader holds the ranging rod vertically approximately on the line. The follower signals to the leader to move it until it is exactly on the line AB. The tape is tightened between the newly erected rod and station A and an arrow is pushed into the ground at the 20 mark of the tape.

The follower moves forward to this new point and the whole procedure is repeated for the remainder of the line until station B is reached. The follower gathers the marking arrows and the number of tape lengths measured is the number of arrows carried by the follower. The portion of tape between the last arrow and station B is then measured and added to the number of complete tape lengths to produce the total length of the line.

Inclined measurements

When any measured distance is to be shown on a plan, the horizontal distance is required and any inclined

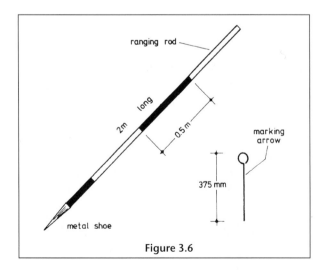

Figure 3.6

distance must be converted to its horizontal equivalent before plotting.

Figure 3.8 shows a survey line measured between two stations A and B. The line is not horizontal. Trigonometrically, the inclined distance is the hypotenuse of a right-angled triangle ABC.

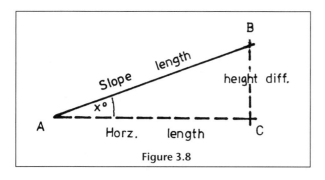

Figure 3.8

In $\triangle$ABC
$$AC/AB = \cos x$$
Therefore $AC = AB \times \cos x$
i.e. horz. length = slope length $\times$ cos (slope angle)

The angle of inclination is measured in the field with less accuracy than in traversing, hence a simple hand-held clinometer is used. Several survey companies manufacture these instruments. Amongst the best is the Suunto PM 5, an instrument produced for use by surveyors, engineers, cartographers and architects, which can measure angles of inclination (slope angles) to 0.5 degrees. Simple operating instructions are provided with the instrument.

A much more accurate method of obtaining a horizontal length is to measure the slope length as normal and find the difference in height between the stations by using a levelling instrument. (The subject of levelling is completely covered in Chapter 4.) The

horizontal length is then easily calculated, using the theorem of Pythagoras. In Fig. 3.8, the slope length AB and the difference in height BC have been measured. The horizontal length AC is found as follows:
In $\triangle$ABC,

$$AC = \sqrt{AB^2 - BC^2}$$

EXAMPLE

2 In Fig. 3.8, the slope length of line AB is 48.25 m and the angle of slope is 4.5°. Calculate the horizontal length AC of the line.

Answer

$AC = AB \times \cos 4.5°$
$\quad = 48.25 \times 0.9969$
$\quad = \underline{48.10 \text{ m}}$

3 In Fig. 3.8, the slope length of line AB is 48.25 m and the difference in level BC is 3.80 m. Calculate the horizontal length AC of the line.

Answer

$$AC = \sqrt{AB^2 - BC^2}$$

$\quad = \underline{48.10 \text{ m}}$

Step taping
An alternative but less accurate method of obtaining horizontal measurements is step taping. This is a field method where the horizontal distances are obtained directly. It is less accurate and more difficult to carry out than other methods and should only be used as a last resort where slopes are short and fairly steep.

Figure 3.9 illustrates the principle. Ranging rods are

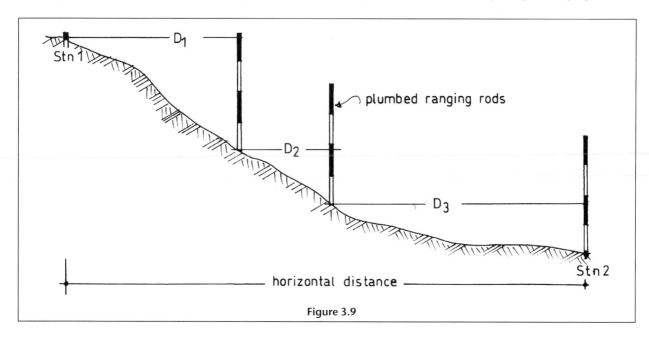

Figure 3.9

aligned between the end stations at intervals dependent upon the gradient and accurately plumbed using a long builder's spirit level. Three persons are required to measure the line, one leader, one follower and one observer. The follower holds the end of the tape against the rods, the leader holds the tape horizontally and the observer estimates the horizontal position of the tape, thus forming a right-angled step. The leader then reads the tape.

Considerable tension is required to straighten the tape and avoid sagging, so short steps are desirable.

When all horizontal lengths have been measured the summation gives the length of the line.

(c) Errors in taped measurements

All taped measurements are subject to error, no matter how carefully any line is measured.

Chapter 1, Sec. 1, lists the classes of error as follows:

(a) Gross
(b) Constant
(c) Systematic
(d) Random (accidental or human).

All of these errors are present in linear surveying but because this survey method is less accurate than others, some of the errors may be discounted.

Gross errors
Gross errors are blunders, arising from inexperience, carelessness or lack of concentration. The main gross errors are:

1. *Misreading the tape* This is probably the most common error in surveying, particularly with inexperienced young surveyors. For example, 6 metres and 40 millimetres is 6.040 m not 6.400 m.

2. *Miscounting the number of tape lengths* The surveyor may lose count of the number of full tape lengths measured along a line, even if there are only two, from personal observational experience.

3. *Booking errors* The booker simply records the wrong measurement.

These errors are usually detected by measuring each line twice.

Constant errors
Constant errors are those which occur no matter how often the line is measured. The error is always of the same sign for any one tape or for any given set of circumstances.

1. *Misalignment* The measurement simply must be made in a straight line between the two end stations, otherwise the distance measured will always be too long and any offsets measured from the tape will be wrong.

2. *Standardization* The tape should always be checked at the start of a survey to determine whether it has distorted in any way. It may have stretched due to continuous rough usage. If a tape has stretched, the resultant measurement will be shorter than the correct measurement. Tapes are expensive but if any tape has stretched by more than 10 mm it should be discarded.

3. *Slope* All distances measured on a survey are slope lengths and must be converted into plan lengths before plotting.

The formula (derived previously) for slope correction is as follows:

Plan length = slope length × cos (slope angle)

Some survey lines may have several changes of gradient along their lengths. Each inclined section is treated separately and its plan length is calculated as in Example 4. The plan lengths are added to produce the final horizontal length of the whole line.

EXAMPLE

4 Figure 3.10 shows a straight line ABCD which has three changes of gradient along its length. Calculate the plan length of the line AD from the following field measurements.

Line	Section	Slope length (m)	Inclination (deg)
AD	AB	84.40	−5.0
	BC	47.21	+2.0
	CD	39.47	+6.5

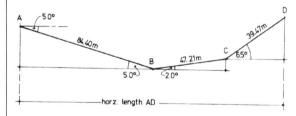

Figure 3.10

Solution

Plan length   AB = 84.40 × cos 5° = 84.08
              BC = 47.21 × cos 2° = 47.18
              CD = 39.47 × cos 6.5° = 39.22
Total length   AD = 170.48 m

Systematic errors
Systematic errors are errors which arise from the application of some fixed law; for example, a tape will change length due to the effects of (i) temperature at the time of measurement, (ii) tension applied when pulling the tape during measurement, (iii) allowing the tape to sag.

Steel tapes are made from material which stretches like many other metals. The tape is only correct at a standard temperature of 20°C when pulled at standard tension 50 N and supported throughout its length.

Since linear survey is the least accurate form of surveying, generally less than 1/5000, it would be unusual to make any allowances for error due to these sources.

Using a 30 m tape on a day when the temperature is unusually high at 30 degrees and the tape is pulled excessively at 75 N would produce an error of 5 mm, which is a relative error of 1/6000.

If the survey schedule recommends that systematic errors be eliminated, then a semirigorous approach would suffice. This suggests that (i) the tape be prevented from sagging, i.e. it should be kept reasonably in contact with the measuring surface, (ii) a constant-tension handle be employed to tension the tape (this instrument is clipped to the tape and when activated always applies the correct tension), (iii) a simple version of temperature correction be applied as follows:

Temp correction = ± [1 mm per 10 m per 10°C] difference from standard temperature.

If, on a freezing winter's day when the temperature was 0°C, a distance of 30.00 m was measured between two points, the correction would be

$$C = -(1 \times 3 \times 2) \text{ mm}$$
$$= -6 \text{ mm}$$
$$\text{Corrected Distance} = 56.21 - 0.006$$
$$= 56.20 \text{ m}$$

The strict mathematical correction is 6.72 mm.

Note: Systematic errors are important in traversing and setting out and must be applied. A detailed example is given in Chapter 8 (Traverse surveys).

Random (or accidental) errors
This class of error arises from defects of human sight and touch, when marking the various tape or chain lengths and when estimating the readings on the tape when they do not quite coincide with a graduation mark. The chances are that two people will mark or estimate slightly differently. It is also reasonable to assume that no one person will overestimate or underestimate every reading, nor mark every tape length too far forward. These small errors tend to be compensatory and have relatively little significance at this level of surveying.

Conclusions
In every measured length there will be errors that fall into one of three classes.

1. *Gross errors* These are mistakes that should not occur in practice provided each line is measured at least twice.

2. *Constant errors* All of these errors are cumulative and have a very great effect on the accuracy of any length. Every tape should be checked to ensure that it is the correct length, and each tape length must be accurately aligned. The gradient must be determined along each measured length. In linear surveying, temperature and tension corrections are *not* necessary. They are applied in theodolite traversing (Chapter 8) where accuracy is essential.

3. *Random errors* These are small residual errors of human sight and touch and tend to be compensating.

Exercise 3.1

1 Calculate the corrected plan length of line AB measured in 3 sections as follows:

Line AB	Section 1	Section 2	Section 3
Measured length	36.50	19.26	52.77
Angle of slope	2°	3.5°	5°

3. Procedure in linear surveying

Where comparatively small areas have to be surveyed a linear survey might be used. As already mentioned, the principle of linear surveying is to divide the area into a number of triangles, all the sides of which are measured. The errors that can arise when measuring have already been discussed. It is obvious therefore that great care must be exercised for every measurement.

(a) Reconnaissance survey

On arrival at the site (Fig. 3.1), the survey team's first task is to make a reconnaissance survey of the area, i.e. the team simply walks over the area with a view to establishing the best sites for survey stations. The sites must be chosen with care and are in fact governed by a considerable number of factors.

Working from the whole to the part
The area to be surveyed is treated as a whole and is then broken down into several triangles rather than the reverse.

Formation of well-conditioned triangles
The triangles into which the area is broken should be well conditioned, i.e. they should have no angle less then 30° nor greater than 120°. These are minimum conditions. The ideal figure is an equilateral triangle and every effort should be made to have triangles whose angles are all around 45° to 75°.

Good measuring conditions
All of the lines of the survey must be accurately measured and it is sound practice to select lines that are going to be physically easy to measure. Roads and paths are usually constructed along even gradients and present good measuring conditions. Lines that change gradient frequently are best avoided.

Permanency of the stations

The survey stations may have to be used at some future date when setting out operations take place. They may, therefore, have to be of a permanent nature. Examples of permanent marks are shown in Fig. 3.11.

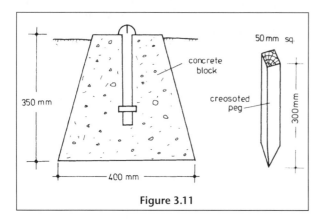

Figure 3.11

The marks must be sited in places that do not inconvenience anyone: for example, concrete blocks can do considerable damage to ploughs, etc., and cannot be placed in the middle of a field.

Referencing the stations

When the stations have to be used again, it is wise to position them such that they can be found easily. Each station should be referenced to nearby permanent objects. Figure 3.12 illustrates a typical situation.

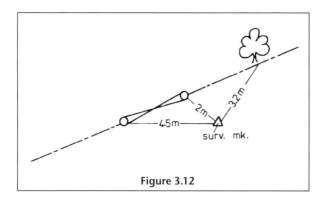

Figure 3.12

Obstructions to measuring

When siting any station with consideration for the factors outlined above it may be found that the survey line will cross a pond, river or railway cutting, which will present a considerable problem to measurement. While such problems can be overcome they should be avoided if at all possible.

Intervisibility of stations

Strictly speaking it is necessary to be able to see only from any one station to the other two stations of any triangle. Check measurement have to be made, however, and wherever possible an attempt should be made to see as many stations as possible from any one station.

Check measurements or tie lines

Before the survey is complete, check measurements must be made. A check line is a dimension that will prove the accuracy or part, or all, of a survey. In Fig. 3.13(a), line CF has been measured to check triangles ABD, BCD and ADF.

On completion of the plotting, the scaled distance of this line must agree with its actual measured length. If it does, the survey is satisfactory, but if not, there is an error in one or more lines and each must therefore be remeasured until the error is found – hence the necessity for referencing each station. This procedure could take a considerable time and a better method is to check each triangle in turn by providing it with its own check line. When plotted triangle by triangle, each must check; if it does not, it is then necessary to measure only the sides of the faulty triangle. Figures 3.13(b) and 3.13(c) show two methods of checking a triangle.

(b) Conducting a survey

Surveying the framework

Consider Fig 3.1 where a plot of land is to be developed as a small private housing estate. Fig 3.13 shows the survey layout, consisting of a trilateration framework and a series of offsets.

Once the trilateration stations have been selected, the various lines are measured, using the methods previously described. It must be borne in mind continuously that the plan length is required and all gradients must be carefully observed and measured, or step taping must be employed.

The three sides and check line of each triangle should be measured before another triangle is attempted. The survey may begin on any station. In Fig. 3.13(a) the order of measuring the sides is AB, BD, DA and check line AG. In measuring line BD a ranging pole should be left at station G. It will then serve as a check point for triangle BCD, producing a check line CG. Only the lines BC and CD remain to be measured to complete the second triangle. There is therefore economy of movement of the survey party in using this technique, resulting in a considerable saving of time.

In Fig. 3.13(a) the survey lines are sited as close as possible to the details that have to be surveyed. These details include the hedges, trees, fences, building and stream.

Offsetting

The principle of offsetting has already been explained in Sec. 1(b). Offsets are short lengths measured to all points of detail from points along the main frame-work lines. These latter points are called chainages. Thus, any point of detail must have at least two measurements to fix its position, namely a chainage and an offset. Figure 3.14 shows surveyors measuring an offset to a tree which lies close to a main framework line.

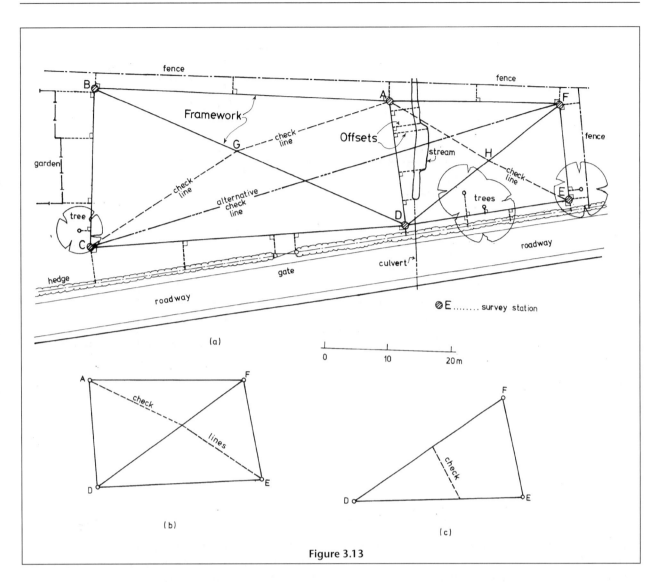

Figure 3.13

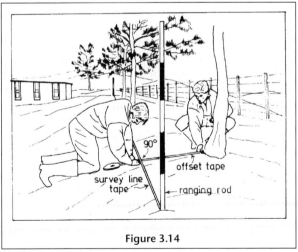

Figure 3.14

Wherever possible, the offsets are measured at right angles to the survey lines, the right angle being judged by eye.

The maximum length of offset is determined by the scale to which the plan is to be plotted.

On average, the naked eye can detect a size of 0.25 mm on paper. When a survey is plotted to a scale of 1/1000, the eye will be able to detect $(0.25 \times 1000 \text{ mm})$ = 0.25 m actual ground size. Thus any error over 0.25 m will be detectable in the plotting.

Tests have shown that the average person is able to judge a right angle to ±3 degrees. In Fig. 3.15 (a), the detail point A will therefore be fixed to within ±x metres. If x exceeds 0.25 m, it will be plottable, thus the maximum offset at which the error will be indiscernible will be

$$(0.25/\tan 3°) \text{ m} = (0.25/0.0541) \text{ m}$$
$$= 5.0 \text{ m approx.}$$

It may be necessary or desirable to fix certain points of detail by more than one offset, whereupon the estimation of right angles ceases. In Fig. 3.15(b) certain details are fixed from two chainage points by 'oblique' offsets. This is a more accurate method and is used to fix important details, like the corners of a house. It is necessary to use this method when the maximum allowable right-angled offset is to be exceeded.

The principle of well-conditioned triangles should again be adhered to, and the distance between chain-

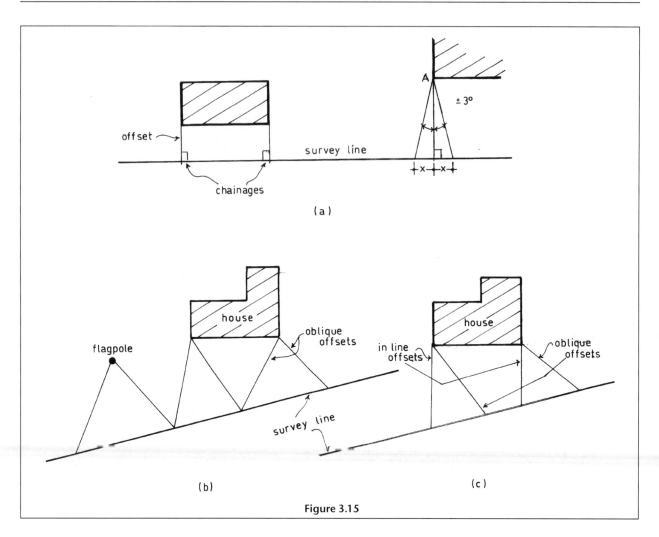

Figure 3.15

age points should be approximately equal to the oblique offsets.

When a building or wall, etc., lies at an angle to a survey line, it may be desirable to use 'in line' offsets. Such offsets are very similar to 'oblique' offsets but have the advantage that they are measured on the line of the detail feature, and when plotting is being done a 'bonus' is thereby obtained (Fig. 3.15(c)).

(c) Recording the survey notes

One of the most important points in a linear survey is recording the information. There is little point in measuring accurately if the survey notes are not clear, legible and intelligible.

Referencing the survey
The first task in booking is to make a reference sketch of the survey as a whole. The sketch is drawn to show the main survey stations in their correct relationship. Figure 3.16 is the reference sketch of the survey of the building site shown in Figs 3.1 and 3.2.

As each line is measured, its length is written alongside the line, together with any gradient values. The measurements are always written in the direction of travel and arrows, indicating gradients, always point downhill. From this reference sketch, the basic framework can be plotted.

Booking the details
In Fig. 3.13, the details to be surveyed from the main survey lines include the road, hedges, fences, building and stream. These are surveyed by taking offsets, at selected chainages, along the main survey lines. Figure 3.17 shows line BC being measured in the direction of B towards C. The field book is opened at the back; station B is positioned on the survey line (double red centre lines) at the bottom of the page and the booking proceeds up the page.

This method of booking is called the 'double line' method and all chainages are noted within the red lines. The various features to which offsets are taken are drawn in their relative positions and the offsets to right or left of the survey line are noted graphically. Virtually no attempt is made to draw to scale. The emphasis is placed rather on relative positioning of the features and on clear legible figuring.

A second method of booking, the 'single line' method, finds favour among a large number of surveyors. A blank field book is used, whereon a single pencil line is drawn to represent the survey line. Chainages

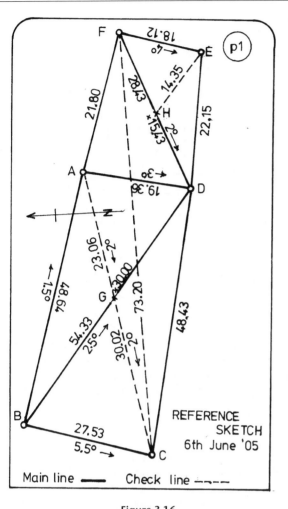

Figure 3.16

are noted alongside the survey line and offsets shown graphically to right and left, as before. Figure 3.18 shows the same information as in Fig. 3.17 but as booked in 'single line' style.

EXAMPLE

5 Figure 3.19 shows line AD of the site survey of Fig 3.1 in its correct position in relation to the stream and wildlife pond. Using a 1:200 scale rule as the equivalent of a tape, make an offset survey of the stream and pond, showing how the dimensions would be recorded in a field book.

Answer: Fig. 3.20

4. Plotting the survey

Undoubtedly the surveyor's main objective is to achieve accuracy in the field operations. However, unless results can be depicted accurately, legibly and pleasingly on paper, proficiency in the field is robbed of much of its value. A systematic approach is therefore required.

(a) Results of the fieldwork

Figures 3.21 and 3.22 show the complete set of survey results of the site shown in Fig. 3.1. They are to be plotted to a suitable scale on paper.

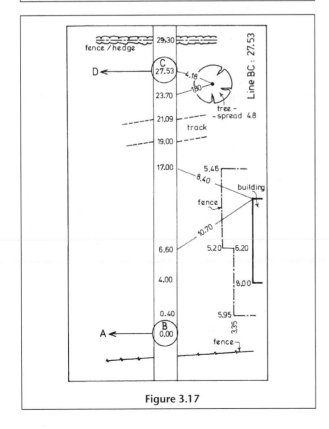

Figure 3.17

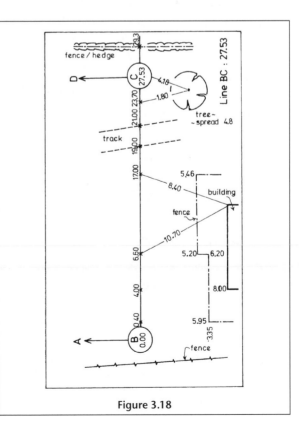

Figure 3.18

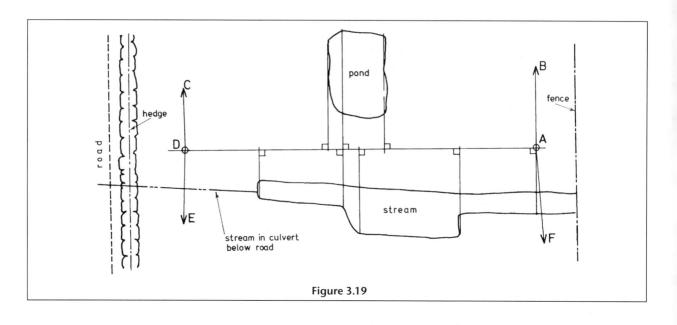

Figure 3.19

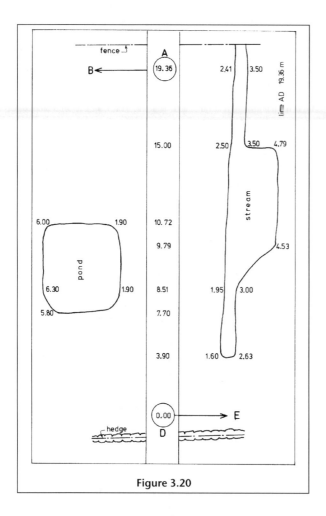

Figure 3.20

(b) Plotting equipment

1. Paper. The survey when plotted may have to be referred to over a number of years. It is therefore essential that the material on which it is plotted should be stable and not prone to stretching.

2. Scale rules. Most modern scale rules are made of plastic. The British Standard scale rule is double sided with two scales along each of its four edges – 1:1, 1:5, 1:20, 1:50, 1:100, 1:200, 1:1250 and 1:2500.

3. Compasses. Since the lines of linear surveys might be long, large-radius compasses are required.

4. Other equipment includes set squares, varying grades of pencils, paper weights or tape, curves, pens, erasers, steel straight edge and small-radius compasses.

(c) Procedure in plotting

Orientation

Most maps and plans are drawn and interpreted looking north towards the top of the paper. It is customary on linear surveys to find north by some means, e.g. by compass or more roughly from an O. S. map. The plotting material should always be oriented to north such that the top and bottom of the paper are respectively north and south.

Rough sketch

If a sketch of the survey is roughly drawn to scale it will greatly facilitate centralizing the survey on the drawing paper and will result in a much more balanced appearance.

Scale

A line scale, filled or open divided, is next drawn along the bottom. This scale drawn on the paper is necessary to detect possible shrinkage or expansion of the drawing material.

Calculation of plan lengths

Before any line of any survey can be plotted, the hori-

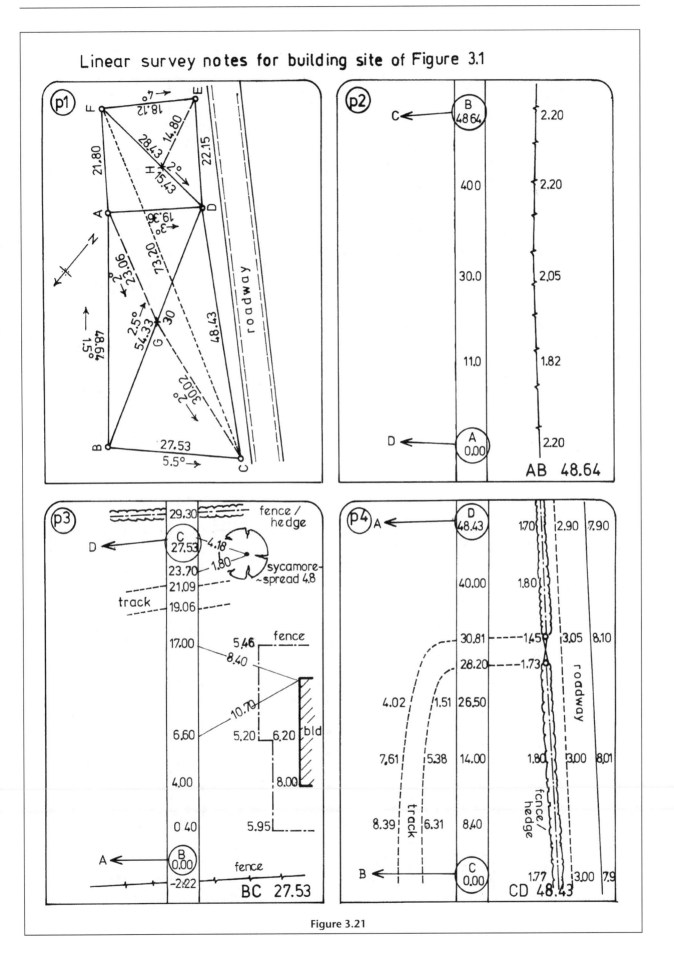

Figure 3.21

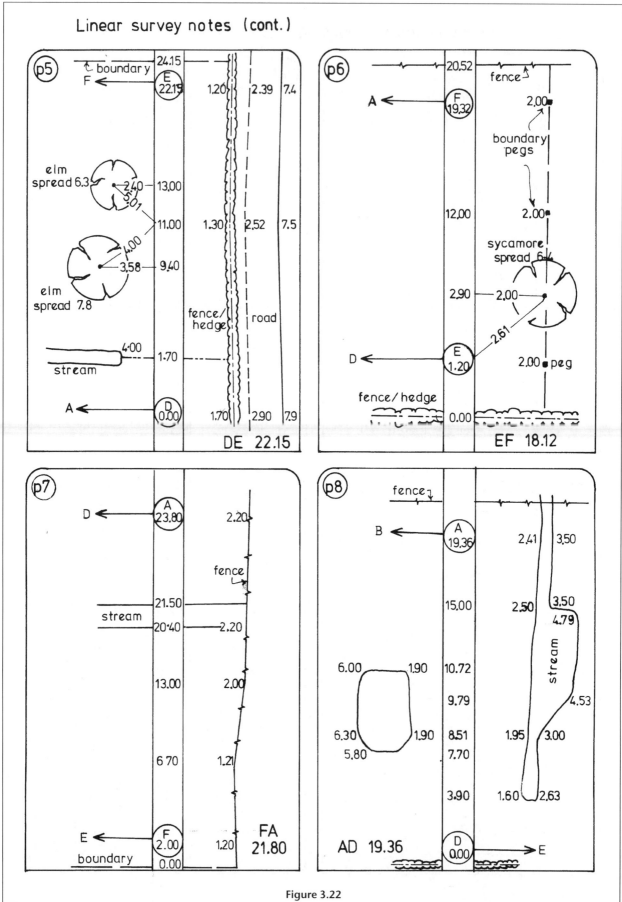

Figure 3.22

zontal (plan) length must be calculated. In Sec. 2(b) it was shown that

Plan length = slope length × cos inclination

In the reference sketch of Fig. 3.21 (page 41) the slope lengths of all lines of the survey are shown together with their angles of slope. The plan lengths of lines AB and BC are calculated in example 6.

EXAMPLE			
6 Line	Slope (S) length (m)	Angle (A) of slope (deg)	Plan (P) length (= S cosA)
AB	48.64	1.5	48.62
BC	27.53	5.5	27.40

Exercise 3.2

1 Calculate the plan lengths of the remaining framework lines of Fig. 3.21. The relevant data is shown in Table 3.2 following.

Table 3.2

Line	Slope length	Angle of inclination (deg)
BD	54.33	2.5
BG	30.00	2.5
CG	30.02	2.0
GA	23.06	2.0
AD	19.36	3.0
FE	18.12	4.0
FH	15.43	2.0

Plotting the framework
The details in Figs 3.21 and 3.22 are to be plotted to a scale of 1:200 on an A3 size sheet of drawing paper. The reference sketch (Fig. 3.21) indicates that the line DA of the survey points northwards. Line DA should, therefore, point towards the top of the sheet. From the survey sketches, the overall size of the survey is approximately 70 m east–west and 38 m north–south. The A3 plotting sheet measures (at scale 1:200) 84 m by 60 m. In order that the plotted survey finishes in the middle of the drawing sheet, point B will have to be placed 11 m from the left edge and 16 m from the top edge of the sheet (Fig. 3.23), to allow for offsets.

Station B is now fixed as the starting point of the plotting. From B, the line BA is drawn parallel to the top edge of the sheet and the horizontal length of 48.62 m is accurately scaled. An arc of a circle of length BD (54.28 m) is drawn using the compasses and a second arc of length DA (19.33 m) is drawn to intersect the first at point D. The check point G, along the line BD, is marked and the distance GA is scaled. The scaled length should agree closely with the actual plan length of GA (23.05 m). Should it fail to do so, it indicates that either the plotting of one or more lines of the triangle ABD is in error or the field measurement of one or more of the lines is wrong. This is precisely the purpose of the tie line GA.

Assuming that the scaled length GA agrees with the field length, the triangle BCD is then plotted from the line BD in the same way as was the triangle ABD, and the length of its check line is compared with the actual plan length of 30.00 m.

The triangles AFD and DFE are plotted in like manner, thereby completing the construction of the framework. The various survey lines are then lightly drawn in ink and each survey station marked by a small circle.

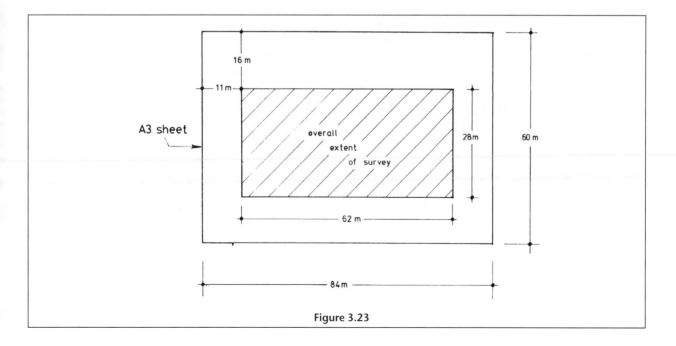

Figure 3.23

Plotting the details
It now remains to plot the details along each chain line. On the line AB, for example, the various chainages of 0, 11, 30 and 40 m are marked off. Short right-angled lines, lying to the right of the survey line, are marked off and the various offsets are scaled along them. At 11 m, an offset of 1.82 m is scaled to the right and at 30 m, a 2.05 m offset is scaled. When these points are themselves joined by a straight line, they form the line of the fence, lying to the right of the survey line AB.

Along line BC, there are two sets of oblique offsets, which fix the positions of a tree and a corner of a house. These offsets are arced using a pair of compasses, the intersection of the 10.70 m arc from chainage 6.60 m and the 8.40 m arc from chainage 17.00 m forming a corner of the house.

Along each chain line the plotting of the detail is carried out with great care until the whole survey has been plotted. The various details are then drawn in ink and a suitable title, north point, etc., are added. Figure 3.18 shows the survey, completely plotted to a scale of 1:250 but reduced to fit the book format.

(d) Computer aided design/draughting (CAD).
The manual skills of draughtsmanship are much rarer today, and it is likely that the final survey drawing will be produced using CAD software. AutoCAD from Autodesk and Microstation from Bentley are the most widely used packages. These are expensive and complicated, having many functions which are not required for drawing a survey plot. Simpler, less expensive software such as Design CAD 3DMAX would suffice, but

because of the popularity of AutoCAD and Microstation, many survey drawings are provided to the client in one of these formats.

Instructions on the use of CAD software is beyond the scope and remit of this book. As a guide, the instructions to plot Fig. 1.6 using AutoCAD were given in Chapter 1 and the enterprising reader might wish to adapt and supplement these instructions to plot the notes of Figs 3.21 and 3.22. However, the intention here is to encourage the reader to plot these notes by hand as in Exercise 3.3 following.

Exercise 3.3

1 Using the survey notes shown in Figs 3.21 and 3.22, plot the site survey to any desired scale.

5. Answers
Exercise 3.1

108.27 m

Exercise 3.2

BD	54.28 m
BG	29.97 m
CG	30.00 m
GA	23.05 m
AD	19.33 m
FE	18.08 m
FH	15.42 m

Exercise 3.3

1 See Fig. 3.1.

Chapter summary

In this chapter, the following are the most important points:

● Every surveyor will, at some time in his or her career, have to make a linear survey using basic equipment. This type of survey requires the application of sound geometrical principles in the field. Use is made of the principles of trilateration and offsetting. In trilateration, the site is divided into well-shaped triangles, each of which is measured by tape. This forms the framework to which are added the details of the topography, trees, roads etc. These are surveyed by offsets, which are short measurements made at right angles to the framework lines.

● In linear surveying it is essential to make thorough preparations regarding the shapes of the triangles and the placing of the stations to promote easy measuring conditions and short offsets. The technicalities of measuring the lines by tape are not as onerous as in traversing surveys but the surveyor must be aware of some of the errors, gross and constant in particular, which affect the accuracy of linear surveys.

● Recording the survey is laborious and time consuming but also requires a great deal of skill to record intelligibly the large amount of data collected on such surveys. For this reason, the text shows the booking of a complete survey, which the reader is encouraged to plot in order to appreciate the necessity for clear unambiguous recording in any form of survey.

CHAPTER 4 **Levelling**

In this chapter you will learn about:

- the types and operation of levelling instruments, with emphasis on automatic and digital levels and staves
- the observation procedure in levelling, with clear definitions of the categories of sights
- the methods of booking and reducing levels, with guidance on the use of the appropriate method in differing circumstances
- the errors which arise in levelling operations and the procedures used to minimize or eliminate them
- the permanent adjustment of levels with emphasis on the two-peg test and associated calculations
- the effects of curvature and refraction on levelling operations and the method of reciprocal levelling.

Chapter 3 dealt with the simple principles of representing the earth's features in two dimensions on a plan. To be of any practical value, however, the third dimension, namely the height of the feature, must be shown by some means on the plan. In surveying, these heights are found by levelling. The heights of the points are referred to some horizontal plane of reference called the datum.

Figure 4.1 illustrates the situation where the horizontal plane through peg B is serving as datum and the heights of points A and C are required relative to this datum.

The required survey information is obtained by using an instrument, called simply a level, and a long measuring rod called a staff. The 'level' can be considered for the moment to be merely a sophisticated spirit level attached to a telescope which in turn is mounted on a tripod.

In Fig. 4.1, observations are taken to a staff held in turn on peg B (the datum point), peg A and peg C. From the diagram, it can be seen that the height of point A above datum is $1.500 - 0.850 = 0.650$ m, while the height of point C is similarly $1.500 - 1.050 = 0.450$ m above datum point B.

On large-scale Ordnance Survey maps, heights are shown by spot levels. These heights are measured relative to a datum called Ordnance Datum (OD) which is the mean level of the sea recorded at Newlyn harbour, Cornwall, over the period 1915 to 1921. From this datum levelling surveys have been made throughout the country and the levels of numerous points permanently established by bench marks (Fig. 2.6) described in Sec. 2(f) of Chapter 2.

Any levelling done on a building site may therefore be referred to OD by taking a reading to a bench mark instead of to peg B. On many sites, however, this is not

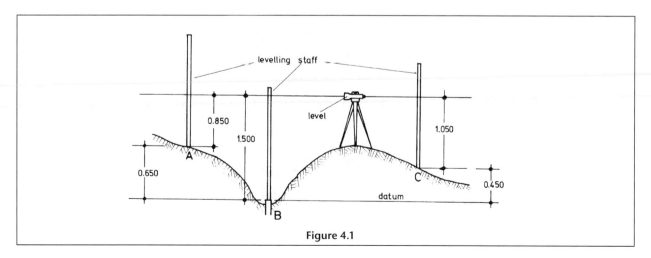

Figure 4.1

necessary and a peg concreted into the ground serves as a bench mark for the site.

1. Levelling instruments

(a) The surveying telescope

An essential part of every levelling instrument is the telescope (Fig. 4.2), which is used to sight the staff at any point of the survey. Figure 4.2(b) shows the fundamental optical arrangements of the telescope of a typical surveying instrument.

In order to use the instrument, the observer places an eye behind the eyepiece and manually aligns the telescope on to the distant levelling staff. The light rays, originating at the staff, pass into the telescope through the object glass and in the process are inverted. The rays continue through the telescope and are captured as an image on a glass disc, called the reticule.

A series of lines is etched on the reticule. The observer sees these lines and the inverted image of the staff when an eye is placed to the eyepiece. The eye-piece is really an arrangement of lenses which magnifies the image and allows the observer to read the graduations on the staff (Fig. 4.2(c)).

It is almost certain that the reticule lines, or the image of the staff, or both, will be out of focus, and in order to read the staff graduations clearly the observer must make two adjustments to the focusing arrangements:

1. The eyepiece must be focused on to the reticule until the lines of the reticule are seen clearly and sharply. This is done by slowly rotating the eyepiece in either a clockwise or anticlockwise direction. This process is known as eliminating parallax.
2. The image of the staff must be focused on to the reticule by means of the internal focusing lens. This lens is activated by the focusing screw which is situated on the right-hand side of the telescope (Fig. 4.2(a)).

When focusing has been completed the observer sees clearly the view shown in Fig. 4.2(c).

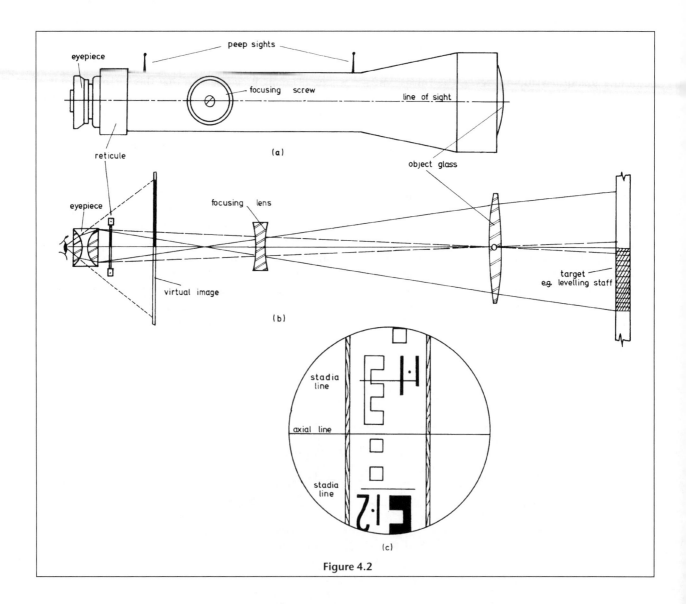

Figure 4.2

In all modern instruments, alterations are made to the optical arrangements to re-invert the image and thus present an upright view of the staff or object sighted. These alterations are usually in the form of additional lenses or prisms. Their exact form need not trouble the surveyor at this stage.

(b) Categories of levelling instrument

Levels are categorized loosely into two groups;

(i) automatic levels which, as the name suggests, set out a horizontal plane automatically, using an automatic compensator.

(ii) digital levels, also automatic, which establish and store the elevation of a point by using a bar-coded staff and indicate it digitally on an LCD screen.

(c) Automatic levels

All levelling instruments create a horizontal plane through the telescope. This horizontal plane is called the plane of collimation. Traditionally, this plane was established by using a spirit level attached to the telescope or body of the instrument. The plane was horizontal when the properly adjusted spirit level was centralized. In automatic levels, the line of collimation is established automatically by means of an optical compensator inserted into the path of the rays through the telescope. In order to allow the compensator to function correctly, the instrument is firstly levelled by means of a small circular spirit level (pond bubble) which sets the instrument to within $\frac{1}{4}$ degree of the vertical axis position.

Figure 4.3 shows the Sokkia C310 automatic level, which is a typical example of this kind of level manufactured for use by engineers and builders. It is compact, lightweight and sturdy and is an ideal basic level for use on construction sites. The compensator is a prism which acts as a pendulum to direct the incoming ray from the staff through the centre of the reticule. It is situated in front of the eyepiece and since it is light in weight, is stabilized against oscillation and vibration by four suspension wires and some form of magnetic

damping action. Figure 4.4 shows exactly how this is achieved by Sokkia.

Currently there are over fifty different automatic levels produced by the many instrument manufacturers who may use different techniques to achieve a stable horizontal line of sight in their products. These need not concern the surveyor unduly. It is sufficient to know that when the instrument is set up as in the following section a reliable horizontal line of site is established through the telescope of the instrument.

Other features of various manufacturers' automatic levels include magnification from ×20 to ×32: a horizontal circle which can read angles to one degree and an ability to measure distances with accuracies varying from 1 cm to 10 cm.

Setting up the automatic level

1. The instrument is set up with the levelling head approximately level and the instrument securely attached using the fastening screw.

2. On all automatic levels there is a small circular spirit level which is centred in exactly the same manner as a tilting level via a three-screw arrangement, a ball and socket joint or a jointed head system.

3. When the spirit level has been centred the vertical axis of rotation of the instrument is approximately vertical. The compensator automatically levels the line of sight for every subsequent pointing of the instrument.

4. Parallax is eliminated as before, the staff is sighted and brought into focus and the staff reading noted.

(d) Levelling staff

The levelling staff should conform with the British Standard Specification. A portion of such a staff is shown in Fig. 4.5. The length of the instrument is 3, 4 or 5 m, while the width of the reading face must not be less than 38 mm. Different colours must be used to show the graduation marks in alternate metres, the most common colours being black and red on a white

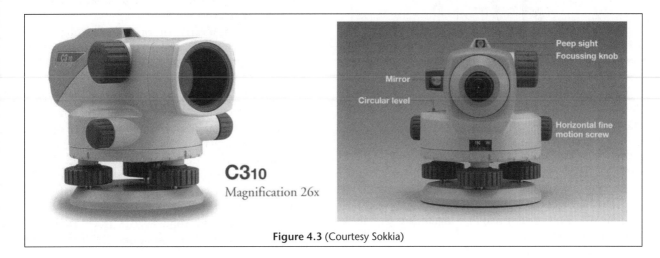

Figure 4.3 (Courtesy Sokkia)

Reliable Automatic Compensator

Even under rapidly changing atmospheric conditions, vibration and shock stability is ensured by using four suspension wires and magnetic damping action. The extended compensator range is above average at approximately 15 minutes.

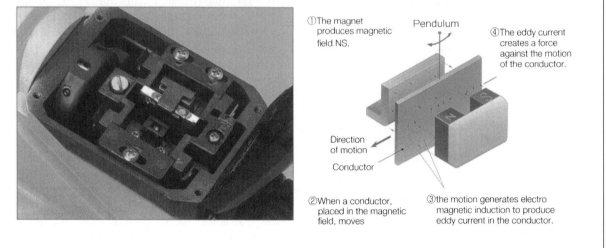

①The magnet produces magnetic field NS.

Pendulum

④The eddy current creates a force against the motion of the conductor.

Direction of motion

Conductor

②When a conductor, placed in the magnetic field, moves

③the motion generates electro magnetic induction to produce eddy current in the conductor.

Figure 4.4 (Courtesy Sokkia)

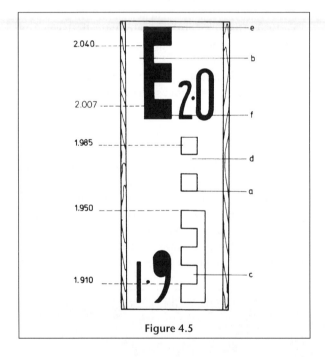

Figure 4.5

EXAMPLE

1 Make a list of the staff readings 'a' to 'f' in Fig. 4.5.

Answer

a. 1.960   b. 2.033   c. 1.915
d. 1.978   e. 2.050   f. 2.002

(e) Digital level

With the ever-increasing rate of technological advance, it is impossible for any textbook to keep abreast with the range of models available at any given time. The following description is therefore, intentionally, an overview of the main features, capabilities and applications of digital levels

Figure 4.6(a) shows the Topcon DL 102C engineer's electronic digital level. Other manufacturers produce similar models, notably the Sokkia SDL30 (Fig. 4.6(b)) and Leica DNA (Fig. 4.6(c)).

These instruments are used in conjunction with a unique patterned staff, similar to a barcode (Fig. 4.7). They are fully automatic and when properly set up, the line of sight through the telescope is horizontal, as with an automatic level. The main advantage of a digital level is that the surveyor does not need to read the staff, note the reading or calculate the results. Hence, these potential sources of error are eliminated.

The instruments are robustly constructed and are protected against water penetration. The Topcon DL

ground. Major graduations occur at 100 mm intervals, the figures denoting metres and decimal parts. Minor graduations are at 10 mm intervals, the lower three graduation marks of each 100 mm division being connected by a vertical band to form a letter E. Thus, the 'E' band covers 50 mm and its distinctive shape is a valuable aid in reading the staff. Minor graduations of 1 mm may be estimated.

In Fig 4.5, various staff readings are shown to illustrate the method of reading.

(a) **Topcon DL 102c** (Courtesy Topcon)

(b) **Sokkia SDL 30** (Courtesy Sokkia)

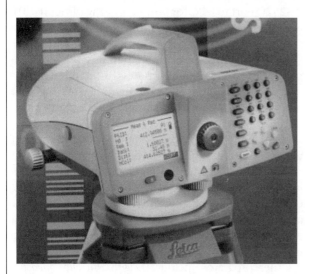

(c) **Leica DNA 10** (Courtesy Leica Geosystems)

```
————— Meas & Rec —————
Pt ID :                      A1
HO   :            285.6750 m
Rem  :            ————————
Back :              1.3341 m
Dist :               26.72 m
HCol :            287.0091 m
```

(d) Display —Leica DNA 10

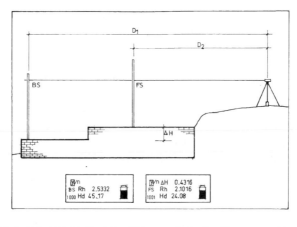

(e) **Sokkia SDL 30 "height difference" display** (Courtesy Sokkia)

Figure 4.6

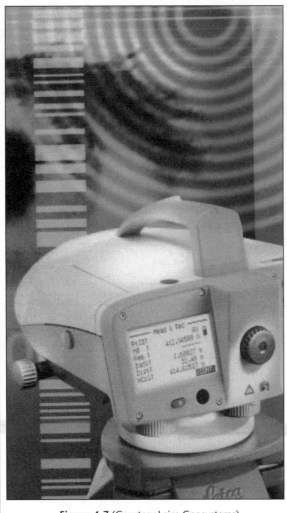

Figure 4.7 (Courtesy Leica Geosystems)

102C is powered by a Nickel Cadmiun 7.2V rechargeable battery The Sokkia SDL30 is powered by a new Lithium-ion battery 7.2V battery.

External configuration
Externally, the most obvious difference in appearance from a conventional level is the shape of the instrument (Fig. 4.6) and the most striking feature is the LCD display and keyboard with full alphanumeric capabilities.

In use, each instrument is set up approximately level using the small circular pond spirit level. Thereafter the instrument is self-levelling using a magnetically damped compressor with a working range of 1/5th of a degree.

A further innovation has been introduced with the production of auto focus. There is now no need to focus the instrument on the staff. This is done automatically with the built-in auto focus system.

Measurement times are of the order of three seconds, during which period the compensator takes about one second to set itself properly and the instrument

makes thirty to thirty-six measurements to obtain and display an average staff reading and standard deviation of the result. Other features include horizontal distance measurement with an accuracy of 10–50 mm, display of rise or fall between sights, determination of reduced levels and storage of design data for setting out purposes.

Figure 4.6(d) shows the display of the Leica DNA instrument, which is probably the most comprehensive of the measurement displays of the various instruments, having an 8-line display.

Internal arrangements
Internally, the measurement data are stored in the memory of the instrument which can accommodate up to 6000 measurements. Additionally, the data can be copied onto a world standard memory card (PCMCIA) after the measurements have been made, which allows the data to be loaded into a computer. Optionally, the data can be downloaded via an RS-232C port to a PC for post processing of results. The user can choose the format into which the data is to be converted.

Each of the instruments carries several measuring programs. Figure 4.6(d) illustrates the '*measure & record*' program of the Leica DNA where a point A1 has been measured. The display screen records point identification (A1); height (i.e. reduced level) of the observed station (HO); backsight reading (back); horizontal distance (Dist) and height of collimation (HCol).

Figure 4.6(e) illustrates the '*height difference measurement*' function of the Sokkia SDL30, where the difference in height of the ground between points is automatically calculated. The staff reading is (Rh), the difference in height is (ΔH) and the horizontal distance is (Hd).

Other programs include '*height difference measurement*' where the heights of bridges, canopies and ceilings etc. can be measured with ease; '*setting out height difference*' where the cut or fill value of excavations is calculated, and many other programs selected by the various survey companies.

Data processing
The recorded survey data can be converted into any desired user definable format. Using the Leica DNA, field-book look-alikes can be produced as in Table 4.1.

For post processing, each survey company produces its own version of software. Leica has the *LevelPak-Pro*, while Sokkia uses *Map suite plus*. However, with rapidly changing technology, that generation of software could well be outdated even before this textbook is printed.

The survey data is imported via the PCMCIA card or via the RS232-C port. It is then processed and edited to produce reports and interrogated to produce longitudinal and cross sections with their associated area and volume calculations. It is also used to find the

Table 4.1 Leica DNA: User format: Rise and Fall style

Line 1	/BF					
Date 25.12.2005 / Time 10.30.00						
Point ID	Backsight	Fore/Intm	Rise	Fall	Distance	Pt Height
Start point						
1						0.000
1	1.624 35				19.85	
1		1.529 83	+0.094 52		19.75	0.095
1	1.465 32				12.59	
2		1.659 10		−0.193 78	13.04	0.099
2	1.724 43				28.77	
3		1.612 22	+0.112 21		28.23	0.013

design levels for setting out purposes, whence the information can then be exported to the digital instrument via the data card, where it can be recalled to set out the design levels.

(f) Digital staff

Figure 4.7 is an illustration of the dedicated unique staves used in conjunction with a digital level. Aluminium or fibreglass is used in their construction to provide staves that are lightweight, strong and durable. The staves are usually dual faced with the barcode printed on one face of the staff and centimetre divisions on the other, using sophisticated high quality printing technology. The staves are produced in various lengths from 2 m to 5 m.

2. Observation procedure

At the outset, it is important to realize that all levelling surveys must be checked (i.e. closed), otherwise there can be no confidence in the results.

There are two methods of checking a levelling survey. (i) The levelling begins and finishes on the same point via the same or a different route, in which case the difference in level should be zero. (ii) The levelling begins on one point of known level and finishes on another, in which case the observed difference in level should equal the known difference.

(a) Procedure with automatic levels

1. Figure 4.8(a) illustrates a typical levelling situation, where the reduced levels of several points B, C, D, E, F and G are to be determined relative to a point A which is the bench mark. The levelling is to be closed on a second bench mark H.

The instrument has to be set up twice in this particular case, although in a practical levelling exercise there could be many more set-up points.

Every time the instrument is set up, the FIRST sight taken from that position is called a BACK sight (BS). Likewise the LAST sight taken, prior to moving the instrument, is called a FORE sight (FS). Thus, in set-up

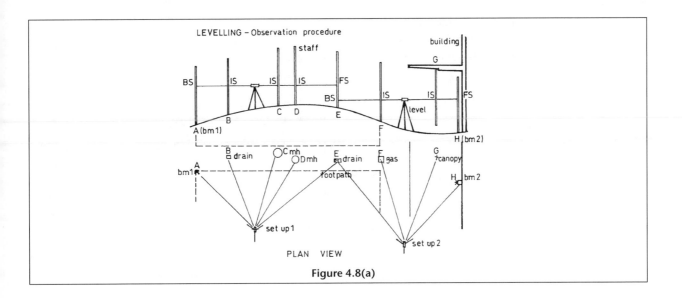

Figure 4.8(a)

number 1, point A is a backsight, while point E is a foresight. Any other sight observed between backsight and foresight is an INTERMEDIATE sight (IS). Points B, C and D are therefore intermediate sights.

At set-up number 2, the sight taken to point E is a BS; point F is an IS; point G is an IS; and finally point H is a FS. It should be noted that the sight G is taken to the underside of a beam, which is higher than the instrument. In such a case, the staff is held upside down against the point while the reading is taken. Such a sight is called an inverted staff reading. Point E, where a foresight followed by a backsight is taken, is called a change point. Both readings are entered on the same line of the field book.

Each point is given a separate line in the field book and its reading is entered on that line in its respective column, either BS, IS or FS. Numerous examples are given in Sec. 3, following.

2. At each point of the survey, the staffholder holds the staff on the mark and ensures that it is held vertically, facing towards the instrument. The observer directs the telescope towards the staff and using the focusing screw, brings the staff clearly into focus. Parallax should already have been eliminated in setting up the instrument in which case there should be no apparent movement of the cross-hairs when the head is moved up and down.

3. The observer then reads the figures on the staff and enters the reading on the appropriate line and column in the field book. The reading is taken once more and checked against the field book entry.

(b) Procedure with digital levels

The procedure is slightly different when using a digital level, the main advantage being that the surveyor does not need to read the staff, note the reading or calculate the results. These potential sources of error are therefore eliminated.

The procedure is very much the same for any digital level. This description utilizes the Sokkia SDL30 as an example. Figure 4.8(b) once again shows the instrument set up twice.

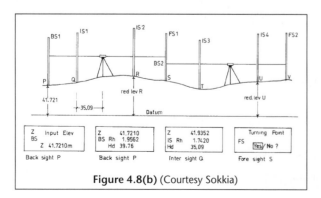

Figure 4.8(b) (Courtesy Sokkia)

(i) At the first set-up, the reduced level (or height or elevation) of the backsight station P is entered at the keyboard. This is shown as Z in the first instrument display box of Fig. 4.8(b). The backsight staff is read and the enter button is pressed to record the reading [BS Rh] in the second display box.

(ii) Intermediate sight readings are taken to points Q and R. They are recorded in the memory of the instrument, where they are used to compute the reduced levels of the stations. The data is shown on the third display box.

(iii) Point S is a foresight. The reading is taken and the reduced level is computed as above. A screen prompt asks if point S is a turning (change) point, to which question the answer is 'yes', in the fourth display box.

(iv) The procedure (i) to (iii) is then repeated for the remaining points of the survey.

3. Reduction of levels (rise and fall method)

(a) Levelling between two points

Figure 4.9(a) shows a peg A situated beside a manhole cover at the bottom of a fairly steep slope. The height of

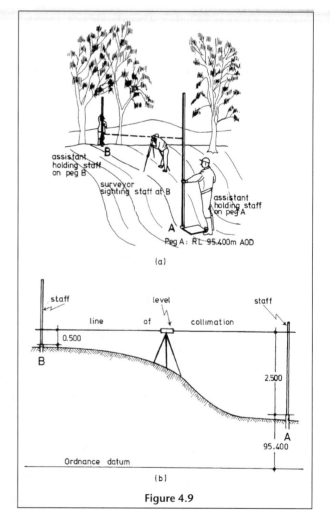

Figure 4.9

the peg A is known to be 95.400 m above Ordnance Datum (AOD), having already been surveyed from a nearby Ordnance Survey bench mark (OSBM).

RLA = 95.400 m AOD

The reduced level of peg B, at the top of the slope, is required.

Procedure
The level is set up approximately midway between the points, accurately levelled, and a first reading is taken to a staff held vertically on A. Let the reading be 2.500 m. The staff is then transferred to B, held vertically on the station and a second reading taken. Let this reading be 0.500 m.

From the sketch it is obvious that point B is higher than A by 2.500 – 0.500 = 2.000 m. In other words, the ground rises from A to B by 2.000 m (Fig. 4.9(b)).

The reduced level of point B is therefore RL A + 2.000 m, i.e. 95.400 + 2.000 = 97.400 m above datum:

RL B = 97.400 m AOD

Booking
All field notes and associated calculations must be properly recorded in a field book. The example illustrated in Fig. 4.9 is shown in Table 4.2.

Note: This is the basis for all levelling work, no matter how long or complex the particular levelling may be. It is essential therefore to understand fully the above principle.

In general terms, where the height above datum of any point is required a staff reading is taken to it and compared with a staff reading taken to a point of known height above datum. A comparison of the readings indicates a rise or fall of the ground between them. The unknown height is then determined by adding the rise to, or subtracting the fall from, the known height.

EXAMPLE

2 The following data relate to three different levelling surveys. Calculate the reduced levels of the three unknown points.

BS	IS	FS	R.L.	Remarks
3.250			135.260	Bench mark A
	1.130			Kerb B
0.752			73.270	Temporary BM
	2.896			Peg A
2.111			55.210	OSBM
		2.896		Ground level K

Answer

(a) 3.250 – 1.130 = (rise 2.120) + 135.260
      = 137.380 m (reduced level B)

(b) 0.752 – 2.896 = (fall 2.144) + 73.270
      = 71.126 m
      (reduced level peg A)

(c) 2.111 – 2.397 = (fall 0.286) + 55.210
      = 54.924 m (ground level K)

(b) Flying levelling

Figure 4.10 shows two points A and B about 150 m apart. Point A is a site bench mark (23.900 m) and point B is to be established as a new bench mark for future site works. Because of the distance and difference in elevation, more than one instrument setting is required.

Procedure
The fieldwork involves repeating, as many times as required, the exercise detailed in the previous section (a). In order to avoid collimation errors (page 68), it is important to keep the length of any foresight approximately equal to the length of its associated backsight. Figure 4.10 shows the various backsights and foresights required in this particular case, in order to level from A to B. The survey is checked by levelling from B back to A so forming a closed levelling. (Note: The check levelling is not shown on Fig. 4.10 for the sake of clarity.) The readings and relevant calculations are shown in Table 4.3. The calculations simply involve finding the difference in level between each pair of back and fore sights by always subtracting the fore from the back sight. The resultant positive or negative figures are then added successively to the known starting level to produce the reduced levels of the various station points. It should be noted that the levelling fails to close by 6 mm, which is acceptable.

Table 4.2

BS	IS	FS	Rise	Fall	Reduced level	Distance	Remarks
2.500					95.400		Peg A
		0.500	2.000		97.400		Peg B

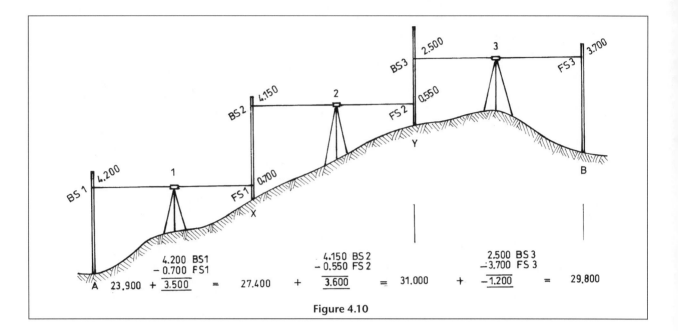

Figure 4.10

Table 4.3

BS	IS	FS	Rise	Fall	Reduced level	Remarks
4.200					23.900	Bench Mk A
4.150		0.700	3.500		27.400	Change point X
2.500		0.550	3.600		31.000	Change point Y
3.450		3.700		1.200	29.800	New Bench Mk B
0.800		2.250	1.200		31.000	X
0.750		4.398		3.598	27.402	Y
		4.246		3.496	23.906	Bench Mk A
15.850		15.844	8.300	8.294	23.906	
15.844			8.294		23.800	
0.006			0.006		0.006	

Misclosure error

As stated at the start of this section, there are two methods of checking a levelling survey. (i) The levelling begins and finishes on the same point via the same or a different route, in which case the difference in level should be zero. (ii) The levelling begins on one point of known level and finishes on another, in which case the observed difference in level should equal the known difference.

In all probability there will be an error in the closure, which should not exceed the allowance misclosure of the circuit. The allowable misclosure for normal building surveys depends upon the number of times the instrument is set up during a levelling. The generally acceptable formula for calculating the misclosure E is:

$$E = (\pm 5\sqrt{n})\ \text{mm}$$

where n is the number of times the instrument is set up during a circuit. For four set-ups the allowance is therefore $\pm(5 \times 2) = 10$ mm. Should there be an error, within the misclosure allowance, it should be distributed equally to each of the set-up points. If the error is outwith the allowance, the circuit of levelling should be repeated.

In the example of Fig. 4.10 there are in total six set-up positions. The allowable misclosure is $(\pm 5\sqrt{6})$ mm = 12 mm. The actual error of the circuit is +6 mm, so is within the allowance. Each set-up position therefore receives a correction of −1 mm cumulatively and each FS point is corrected by −1 mm. The correction to the 6th (final) foresight is therefore −6 mm.

In Fig. 4.10, the points X and Y are points where both foresights and backsights are observed. As already seen, the instrument position is moved after the foresight and reset before the backsight. The points X and Y are called change points. The letters CP are often inserted in the remarks column but are not absolutely necessary.

Arithmetic check

As in all surveying operations, a check should be provided on the arithmetic. This is shown on lines 8, 9 and 10 of Table 4.3.

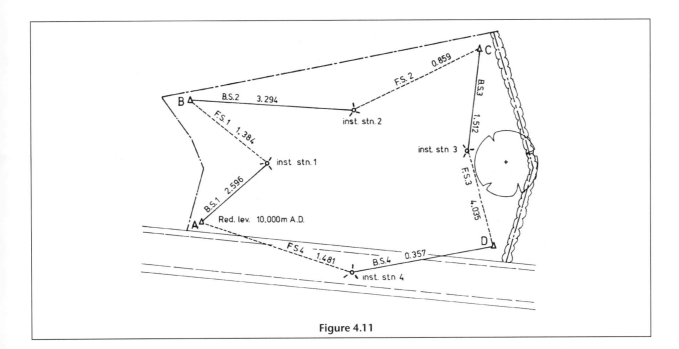

Figure 4.11

A moment's thought will show that the last reduced level is calculated as follows:

Last reduced level = first reduced level
+ all rises – all falls
Therefore last reduced level = first reduced level
+ sum rises
– sum falls

Each rise or fall, however, is the difference between its respective backsight and foresight. Therefore, the sum of the rises minus the sum of the falls must equal the sum of the backsights minus the sum of the foresights.

The complete check is therefore:

(Last reduced level – first reduced level)
= (sum rises – sum falls)
= (sum BS – sum FS)

i.e.

$(23.906 – 23.900) = (8.300 – 8.294)$
$= (15.850 – 15.844)$
$= 0.006$

This particular levelling is an example of flying levelling. The shortest route between the points A and B is chosen and as few instrument settings as possible are used.

Since the last point is actually also the first point in a closed circuit, then the last reduced level minus the first reduced level should be zero. The sum of the BS column should therefore equal the sum of the FS column.

In the field, the levelling can be very quickly verified by simply checking the sum of the BS column against that of the FS column.

While in theory the difference between the sums of the columns should be zero, in practice this is unlikely to happen, due to the minor errors that inevitably arise during a levelling. These errors are considered fully in Sec. 6

EXAMPLE

3 In Fig. 4.11, four pegs are spaced around a construction site. These pegs are to be used as temporary bench marks for the duration of the site. A flying levelling was made around the pegs as shown, in order to establish the reduced levels of the pegs. Book the readings and calculate the reduced levels of the pegs, assuming that peg A has an assumed level of 10.000 m AD.

Answer   (Table 4.4).

Exercise 4.1

1 Figure 4.12 shows the station points of the linear survey of Chapter 3 (Fig. 3.1) and Table 4.5 shows the results of a flying levelling of those stations carried out from a nearby Ordnance Survey bench mark. Calculate the reduced levels of the stations.

(c) Series levelling

When the reduced levels of many points are required, the method known as series levelling is used.

Points observed from single instrument station
In Fig. 4.13 (page 57), the reduced levels of five points B to F are required relative to a temporary bench mark A. Since all points can be observed from one instrument station, they are simply sighted in turn. A backsight is taken to TBM A, followed by intermediate sights to

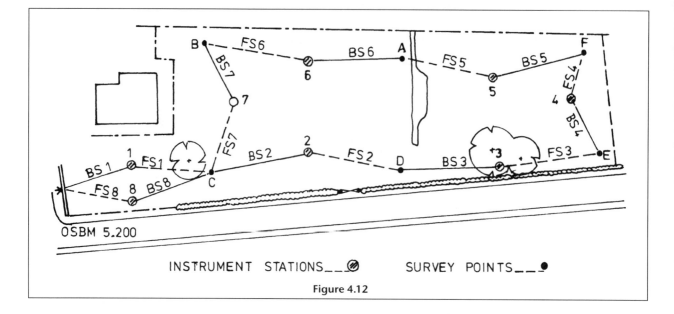

Figure 4.12

INSTRUMENT STATIONS___⊘   SURVEY POINTS___●

Table 4.4

BS	IS	FS	Rise	Fall	Reduced level	Remarks
2.596					10.000	A
3.294		1.384	1.212		11.212	B
1.512		0.059	2.435		13.647	C
0.357		4.035		2.523	11.124	D
		1.481		1.124	10.000	A
7.759		7.759	3.647	3.647	10.000	
−7.759				−3.647	−10.000	
0.000				0.000	0.000	

Table 4.5

BS	IS	FS	Rise	Fall	Reduced level	Remarks
1.955					5.200	OSBM
1.315		2.030				Station C
1.243		0.885				Station D
2.071		1.485				Station E
1.570		0.880				Station F
1.835		1.590				Station A
0.631		0.540				Station B
1.200		3.289				Station C
		1.130				OSBM

points B, C, D and E and finally a foresight to point F. The readings are entered as in Table 4.6. The reduction of levels calculation simply involves finding the difference in level between each pair of sights, A to B, B to C, C to D, D to E and E to F. The resultant positive or negative figures are then added successively to the known starting level of point A to produce the reduced levels of the various station points.

Table 4.6 shows the complete reduction. The arith-

metic check is applied in the same manner as in Sec. (b).

The arithmetical check provided in all levellings checks only that the observed readings entered in the field book have been correctly computed. It does not prove that the reduced level of any point is correct. Verification of each reduced level is required, so the fieldwork must be repeated from a different instrument station (Example 4, following) which in effect makes the total levelling a closed circuit.

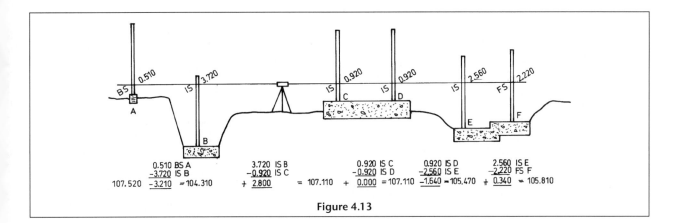

Figure 4.13

Table 4.6

BS	IS	FS	Rise	Fall	Reduced level	Distance	Remarks
0.510					107.520		A. (TBM)
	3.720			3.210	104.310		B. Foundation level 1
	0.920		2.800		107.110		C. Foundation level 2
	0.920				107.110		D. Foundation level 2
	2.560			1.640	105.470		E. Foundation level 3
		2.220	0.340		105.810		F. Foundation level 4
0.510		2.220	3.140	4.850	105.810		
2.220			−4.850		−107.520		
−1.710			−1.710		−1.710		

Table 4.7

BS	IS	FS	Rise	Fall	Reduced level	Remarks
0.240					107.520	A (TBM)
	3.450			3.210	104.310	B
	0.655		2.795		107.105	C
	0.650		0.005		107.110	D
	2.290			1.640	105.470	E
		1.955	0.335		105.805	F
0.240		1.955	3.135	4.850	105.805	
−1.955			−4.850		−107.520	
−1.715			−1.715		1.715	

The re-levelling shows that there is a 5 mm difference in the reduced levels of points C and F, which is acceptable.

Answer (Table 4.7)

Exercise 4.2

1 Figure 4.14 shows the station points of the linear survey of Fig. 3.1. Table 4.8(a) shows the results of a levelling of these points.

Calculate the reduced levels of the stations.

Table 4.8(a)

BS	IS	FS	Remarks
2.650			OSBM (5.200m)
	2.727		Station C
	2.292		Station D
	2.537		Station E
	1.346		Station F
	1.370		Station A
		0.065	Station B

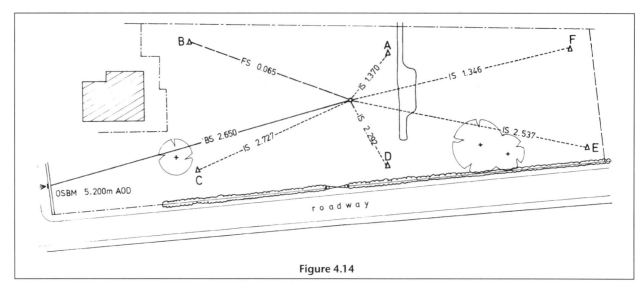

Figure 4.14

2 Table 4.8(b) shows the results of a check levelling of the stations from a different instrument set-up position.

Calculate the reduced levels of the stations, as a check on the survey, and determine which, if any, stations have been wrongly observed.

Table 4.8(b)

BS	IS	FS	Remarks
2.763			OSBM
	2.840		Stn C
	2.385		Stn D
	2.650		Stn E
	1.459		Stn F
	1.463		Stn A
		0.178	Stn B

Points observed from multiple instrument stations
Figure 4.15 shows a small site where reduced levels are required at various points of detail, fences, manholes, etc. Since there are buildings on the site, it is not possible to observe all of the points of detail from one instrument position. It is also probable that some of the points may be too high or too low or simply too far distant to permit a sight to be taken.

The first set-up station is chosen to allow the bench mark (RL 35.27 m), to be sighted as a backsight. This is followed by sights to as many points as are possible or practicable. Thus, points A, B and C become intermediate sights and point D becomes a foresight (change point I), since the next point E is too far away.

The second set-up station is chosen such that point D can be re-observed as a backsight. Points E and F are observed as intermediate sights and, since point G is

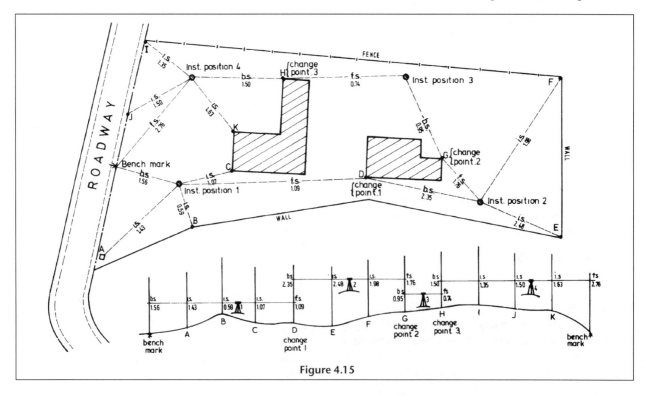

Figure 4.15

Table 4.9

BS	IS	FS	Rise	Fall	Reduced level	Distance	Remarks
1.56					35.27	Not	Bench mark
	1.43		0.13		35.40	required	A. Manhole
	0.59		0.84		36.24		B. Fence
	1.07			0.48	35.76		C. Corner of building
2.35		1.09		0.02	35.74		D. Corner of building change point 1
	2.48			0.13	35.61		E. Fence
	1.98		0.50		36.11		F. Fence
0.95		1.76	0.22		36.33		G. Corner of building change point 2
1.50		0.74	0.21		36.54		H. Corner of building change point 3
	1.35		0.15		36.69		I. Fence
	1.50			0.15	36.54		J. Fence
	1.63			0.13	36.41		K. Corner of building
		2.76		1.13	35.28		Bench mark
6.36	6.35	2.05	2.04		35.28		
−6.35		−2.04			−35.27		
=0.01		=0.01			= 0.01		

the only remaining visible station, it is observed as a foresight (change point 2).

The procedure is repeated at a third instrument station. A backsight is taken to G and a foresight to H (change point 3).

The fourth and final set-up station is chosen such that point H is visible as a backsight, points I, J and K as intermediate sights, with a final foresight observed to the BM in order to close the levelling.

The field results of this levelling are shown in Table 4.9. The reduction of levels follows the pattern detailed in the previous sections (b) and (c) *but since multiple instrument stations are the norm in levelling, the reduction is now given in detail.*

(i) Calculate the rise or fall between each pair of points in set-up no. 1.

(BS. BM) – (IS. A) = 1.56 – 1.43 = 0.13 (rise)
(IS. A) – (IS. B) = 1.43 – 0.59 = 0.84 (rise)
(IS. B) – (IS. C) = 0.59 – 1.07 = –0.48 (fall)
(IS.C) – (FS.D) = 1.07 – 1.09 = –0 02 (fall)

(ii) Calculate the rise or fall between each pair of points in set-up no. 2

(BS. D) – (IS. E) = 2.35 – 2.48 = –0.13 (fall)
(IS. E) – (IS. F) = 2.48 – 1.98 = 0.50 (rise)
(IS. F) – (FS G) = 1.98 – 1.76 = 0.22 (rise)

(iii) Calculate the rise or fall between pairs of points in set-up no. 3.

(BS. G) – (FS. H) = 0.95 – 0.74 = 0.21 (rise)

(iv) Calculate the rise or fall between each pair of points in set-up no. 4.

(BS.H) – (IS. I)   = 1.50 – 1.35 = 0.15 (rise)
(IS. I) – (IS. J)   = 1.35 – 1.50 = –0.15 (fall)
(IS. J) – (IS.K)   = 1.50 – 1.63 = –0.13 (fall)
(IS.K) – (FS.BM) = 1.63 – 2.76 = –1.13 (fall)

(v) Calculate the reduced levels of the points by successively adding the rises or falls to the BM value.

RLBM   =   = 35.27 m
RL A   = 35.27 + 0.13 = 35.40 m
RL B   = 35.40 + 0.84 = 36.24 m
RL C   = 36.24 – 0.48 = 36.76 m
RL D   = 35.76 – 0.02 = 35.74 m
RL E   = 35.74 – 0.13 = 35.61 m
RL F   = 35.61 + 0.50 = 36.11 m
RL G   = 36.11 + 0.22 = 36.33 m
RL H   = 36.33 + 0.21 = 36.54 m
RL I   = 36.54 + 0.15 = 36.69 m
RL J   = 36.69 – 0.15 = 36.54 m
RL K   = 36.54 – 0.13 = 36.41 m
RLBM   = 36.41 – 1.13 = 35.28 m

(vi) Apply the arithmetic check.

Sum BS Col – Sum FS Col = 6.36 – 6.35 = 0.01
Sum Rises – Sum Falls   = 2.05 – 2.04 = 0.01
Last RL – First RL   = 35.28 – 35.27 = 0.01

The arithmetic check shows that the levels have been correctly calculcated and also shows that an error of 10 mm has been made in the fieldwork. The normal acceptable limit of error is ±10 mm for this levelling.

It must be pointed out that it is unusual to compute the results in parallel with the fieldwork for the simple reason that if any observation is wrongly read the calculated results must also be wrong. It is therefore wise to complete the fieldwork and verify that the levelling closes before calculating any results.

The logical order in any levelling calculation is therefore:

1. Complete the fieldwork.
2. Check that the sum BS column = sum FS column within the acceptable limit of error. If the error

exceeds these limits there is no point in continuing the calculation.

3. Calculate rises and falls.
4. Check that (sum rise column – sum fall column) = (sum BS column – sum FS column).
5. Calculate reduced levels.
6. Check that (last RL – first RL) = (sum rise column – sum fall column).

It is worth noting, at this stage, a common fallacy amongst surveyors, many of whom are unaware that the arithmetic check only proves that the field book notes, as written, are correctly computed. It does not, in fact, prove that the survey is correct. In Example 5 the check proves that the change point level only is correct. All of the intermediate sights could have been observed wrongly and the check would still work. In order to fully check the level of each point, the complete levelling must be repeated.

EXAMPLE

5 Figure 4.16 shows the positions of a level and staff set up to observe levels along the time of a proposed drain. Draw up a page of a levelling book and reduce the readings, applying the appropriate arithmetic check.

Answer (Table 4.10)

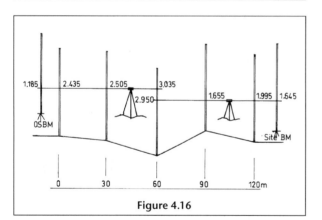

Figure 4.16

Exercise 4.3

1 Figure 4.17 shows the station points of the linear survey of Chapter 3 (Fig. 3.1) and Table 4.11 shows the results of a levelling of those stations from multiple set-up points. Calculate the reduced levels of the stations.

Table 4.11

BS	IS	FS	Rise	Fall	Reduced level	Remarks
1.256					5.200	OSBM
	1.330					C
1.100		0.906				D
	1.332					E
	0.146					F
1.875		0.166				A
0.200		0.579				B
		2.780				OSBM

4. Reduction of levels by the height of plane of collimation (HPC) method

A second method of calculating the reduced levels of points of a survey is that known as the height of the plane of collimation (HPC) method. The fieldwork of the HPC method is exactly the same as that of the rise/fall method. *The method of calculating the results is different.*

It is commonly believed that a surveyor should simply choose the reduction method, either rise/fall or HPC which he or she finds easier to use. However, this should not be the criterion in choosing. The rise/fall method should be used where a levelling involves a number of intermediate sights, because the arithmetic check does check that every reduced level has been correctly calculated, whereas the common HPC check does not. It is true that computer programs for levelling use the HPC method, but computers do not make mistakes.

Undoubtedly, the HPC method should be used when setting out levels (Chapter 12) because, in that operation, the height of collimation is always required.

Table 4.10

BS	IS	FS	Rise	Fall	Reduced level	Distance	Remarks
1.185					10.560		OSBM (10.560)
	2.435			1.250	9.310	0	Drain
	2.505			0.070	9.240	30	Drain
2.950		3.035		0.530	8.710	60	Drain
	1.655		1.295		10.005	90	Drain
	1.995			0.340	9.665	120	Drain
		1.645	0.350		10.015		Site BM (10.015)
4.135		4.680	1.645	2.190	10.015		
–4.680				–2.190	–10.560		
–0.545				–0.545	–0.545		

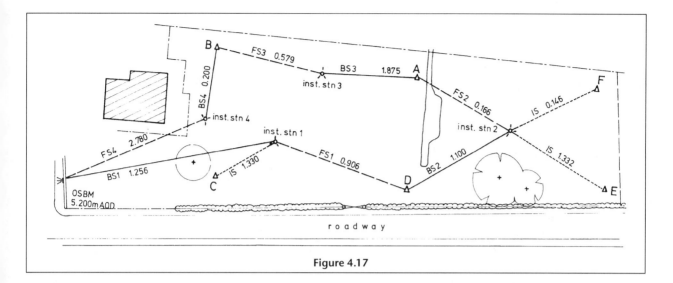

Figure 4.17

Since a great deal of setting out has to be done on construction sites, this is probably the reason for the popularity of the HPC method amongst engineers.

The overall time involved in manually calculating levels is much the same in both methods.

Line of collimation The line joining the centre of the object glass to the centre of the reticule (Fig. 4.2) is the line of collimation of the instrument. It is the line of sight along which the surveyor is looking, when using the instrument. When the telescope is rotated, it will sweep out a horizontal plane known as the plane of collimation (Fig. 4.18).

(a) Levelling between two points

The height of the plane of collimation, HPC, is the height of the line of sight above datum. It is calculated by adding the staff reading at a particular point (usually and preferably the backsight) to the reduced level of that point. The reduced level of any other point is calculated by subtracting the staff reading at the point from the HPC.

In Fig. 4.18(a) the plane of collimation cuts a levelling staff held on a point A where the reduced level is 205.500 m. The staff reading is 2.400 m.

In this method of reduction the height of the plane of collimation above datum is required for every instrument setting. Clearly, in the figure the height of the plane of collimation (HPC) above datum is the reduced level of A (205.500 m) plus the staff reading 2.400 m at that point.

i.e. HPC = reduced level A + staff reading
= 205.500 + 2.400
= 207.900 m

The plane of collimation is still at a height of 207.900 when pointing to B and since ground level B lies 1.800 m below the HPC the reduced level of B is HPC minus staff reading B.

i.e. reduced level B = HPC – staff reading
= 207.900 – 1.800
= 206.100 m

The field book is ruled differently, of course, and Table 4.12 shows the book ruled for the collimation system of reduction. Since the actual observations are in no way altered, the sighting to A is a backsight while B is a foresight. The HPC 207.900 is entered on line 1. The HPC is written only once, opposite the backsight, and is understood to refer to the whole of that instrument setting. The reduced level of B, 206.100, is entered on line 2 in its appropriate column.

(b) Series levelling

Points observed from single instrument station
In Fig. 4.18 (b) the reduced levels of six points are required. (The figure and the readings used are those of Fig. 4.13, where the levels were calculated by the rise and fall method.)

Table 4.12

	BS	IS	FS	HPC	Reduced level	Distance	Remarks
Line 1	2.400			207.900	205.500		A. Ground level
Line 2			1.800		206.100		B. Ground level

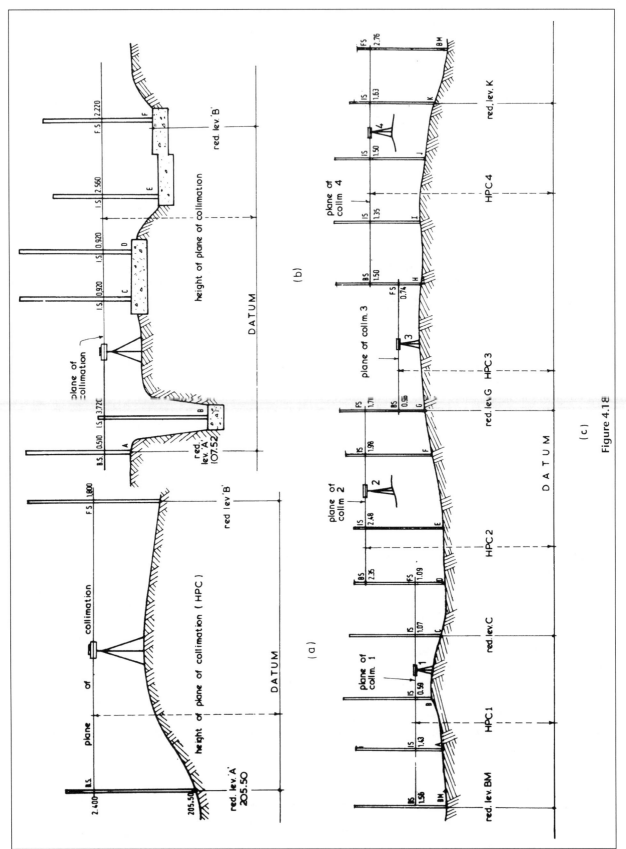

Figure 4.18

(a) The height of collimation is found as before.

$$HPC = \text{reduced level A} + \text{staff reading (BS)}$$
$$= 107.520 + 0.510$$
$$= 108.030 \text{ m}$$

The HPC is entered in the field book on line 1 (Table 4.13). It is then held to apply to the whole of that instrument setting and need not be repeated down the whole HPC column as it is in the table.

(b) The reduced level of every other point is found thus:

$$RL \text{ (any point)} = HPC - \text{staff reading at that point}$$

Therefore

$$RLB = 108.030 - 3.720 = 104.310$$
$$RLC = 108.030 - 0.920 = 107.110$$
$$RLD = 108.030 - 0.920 = 107.110$$
$$RLE = 108.030 - 2.560 = 105.470$$
$$RLF = 108.030 - 2.220 = 105.810$$

The reduced levels are entered in the RL column (Table 4.13) opposite their respective staff readings. They agree with the levels calculated in Table 4.6 by the rise and fall method.

EXAMPLE

6 The levelling shown in Fig. 4.18 (b) was checked from a second instrument station. The observations were as follows:

A. 0.240	B. 3.450	C. 0.655
D. 0.650	E. 2.290	F. 1.955

Enter these results in a levelling table and then calculate the reduced levels of the stations from the temporary bench mark A (RL 107.520 m).

Answer   (Table 4.14)

Note.   The reduced levels agree with those calculated in Table 4.7, using the rise and fall method.

Multiple instrument settings

Figure 4.18(c) shows a whole series of points where the reduced levels are required. (The figure and the readings are those of Fig. 4.15 where the levels were reduced by the rise and fall method.)

(a) The height of collimation of the first instrument setting is found by adding the reduced level of

Table 4.13

BS	IS	FS	HPC	Reduced level	Distance	Remarks
0.510			108.030	107.520		A. (TBM)
	3.720		108.030	104.310		B. Foundation level 1
	0.920		108.030	107.110		C. Foundation level 2
	0.920		108.030	107.110		D. Foundation level 2
	2.560		108.030	105.470		E. Foundation level 3
		2.220	108.030	105.810		F. Foundation level 4
0.510		2.220		105.810		
−2.220				−107.520		Check
−1.710				−1.710		

Table 4.14

BS	IS	FS	HPC	Reduced level	Distance	Remarks
0.240			107.760	107.520		A. (TBM)
	3.450		107.760	104.310		B. Foundation level 1
	0.655		107.760	107.105		C. Foundation level 2
	0.650		107.760	107.110		D. Foundation level 2
	2.290		107.760	105.470		E. Foundation level 3
		1.955	107.760	105.805		F. Foundation level 4
0.240		1.955		105.805		
−1.955				−107.520		Check
−1.715				−1.715		

Table 4.15

Line	BS	IS	FS	HPC	Reduced level	Remarks
1	1.56			36.83	35.27	BM
2		1.43			35.40	A
3		0.59			36.24	B
4		1.07			35.76	C
5	2.35		1.09	38.09	35.74	D
6		2.48			35.61	E
7		1.98			36.11	F
8	0.95		1.76	37.28	36.33	G
9	1.50		0.74	38.04	36.54	H
10		1.35			36.69	I
11		1.50			36.54	J
12		1.63			36.41	K
13			2.76		35.28	BM
14	6.36	12.03	6.35		35.28	
15	−6.35				−35.27	
16	0.01				0.01	

the bench mark and the backsight reading observed to it.

HPC (1) = reduced level BM + staff reading (BS)
$$= 35.27 + 1.56$$
$$= 36.83 \text{ m}$$

The HPC is entered in the field book (Table 4.15 line 1).

(b) The reduced levels of stations A, B, C and D, observed from instrument position 1, are obtained by subtracting their staff readings from the HPC.

Therefore

Reduced level A = 36.83 − 1.43 = 35.40 m
Reduced level B = 36.83 − 0.59 = 36.24 m
Reduced level C = 36.83 − 1.07 = 35.76 m
Reduced level D = 36.83 − 1.09 = 35.74 m

These levels are entered in the field book (Table 4.15 lines 2–5) opposite their respective staff readings.

(c) The fieldwork and calculations finished with a foresight to point D, which is a change point. The instrument is removed to set-up point 2 and a backsight taken to station D. The instrument therefore has a new height of collimation which is found as before:

HPC (2) = reduced level (change point)
         + staff reading (BS)
$$= 35.74 + 2.35$$
$$= 38.09 \text{ m}$$

The HPC (2) is entered in the field book (Table 4.15 line 5) opposite the BS, to which it refers.

(d) The reduced levels of the second set of points are obtained as before by subtracting their staff readings from the HPC.

Therefore

Reduced level E = 38.09 − 2.48 = 35.61 m
Reduced level F = 38.09 − 1.98 = 36.11 m
Reduced level G = 38.09 − 1.76 = 36.33 m

These levels are entered in the field book (Table 4.15 lines 6–8) opposite their respective staff readings.

(e) The same calculation procedure is carried out for set-up points 3 and 4.

At (3), HPC (3) = RL G + BS G
$$= 36.33 + 0.95 = 37.28 \text{ m}$$
and RL H = HPC (3) − FS
$$= 37.28 − 0.74 = 36.54 \text{ m}$$
At (4), HPC (4) = RL H + BS H
$$= 36.54 + 1.50 = 38.04 \text{ m}$$
RLI   = HPC (4) − IS I
$$= 38.04 − 1.35 = 36.69 \text{ m}$$
and RL J = 38.04 − 1.50 = 36.54 m
RL K   = 38.04 − 1.63 = 36.41 m
RLBM   = 38.04 − 2.76 = 35.28 m

The results are entered in the field book (Table 4.15, lines 8 to 13).

Arithmetic check The commonly applied check is shown on lines 14, 15 and 16. It mirrors the rise and fall system check, where it was shown that the difference between the first and last reduced levels equals the difference between the sum of the BS column and the sum of the FS column, namely 0.01 m.

While this check is very widely used, it does *not* completely check the levelling. For example, if reduced level B had been calculated wrongly as 37.24 m instead of 36.24 m, the simple check shown above would still work indeed, if every intermediate reduced level were wrong, the check would not detect the errors.

The complete arithmetical check shown below is complex and is almost certainly the reason for the adoption of the simple check shown in Table 4.15. Complete check:

Sum of reduced
levels (except first) = sum (each height of collimation
× number of IS and FS
observed from each)
− sum (IS column + FS column)

In Table 4.15,

Sum of RLs except first = 432.65 m
Sum of HPC × number of IS and FS

$$= 36.83 \times 4 = 147.32$$
$$+ 38.09 \times 3 = +114.27$$
$$+ 37.28 \times 1 = + 37.28$$
$$+ 38.04 \times 4 = +152.16$$
$$\underline{451.03}$$

Sum of IS column = 12.03

Sum of FS column = $\underline{6.35}$
$\underline{18.38}$
$(451.03 - 18.38) = 432.65$ m

EXAMPLE

7 The level booking sheet (Table 4.16) shows the ground levels through which it is proposed to run a drain, rising from manhole 1 to manhole 5. Calculate the reduced levels of each staff position and apply a full arithmetic check.

Answers (Table 4.17)

Exercise 4.4

1 Figure 4.19 shows levelling information observed as part of a proposed road. The readings 1.727 and 0.573

Table 4.16

Backsight	Intermediate sight	Foresight	Height of collimation	Reduced level	Distance	Remarks
1.579				100.000	0.0	BM 1 on cover plate; manhole 1
	1.295				20.0	
	1.873				40.0	
	2.018				60.0	
	1.884				80.0	Manhole 2; cover level
	1.625				100.0	
2.441		1.000			105.0	Start of steps
	1.807				118.5	Top of steps
	1.495				122.3	Manhole 3; cover level
	1.807				129.6	Manhole 4; cover level
		1.020		102.000	135.0	BM 2 on cover plate; manhole 5

Table 4.17

Backsight	Intermediate sight	Foresight	Height of collimation	Reduced level	Distance	Remarks
1.579			101.579	100.000	0.0	BM 1 on cover plate; manhole 1
	1.295			100.284	20.0	
	1.873			99.706	40.0	
	2.018			99.561	60.0	
	1.884			99.695	80.0	Manhole 2; cover level
	1.625			99.954	100.0	
2.441		1.000	103.020	100.579	105.0	Start of steps
	1.807			101.213	118.5	Top of steps
	1.495			101.525	122.3	Manhole 3; cover level
	1.807			101.213	129.6	Manhole 4; cover level
		1.020		102.000	135.0	BM 2 on cover plate; manhole 5
4.020	13.804	2.020		102.000		
−2.020				−100.000		
= 2.000				= 2.000		

Check:
Sum of RL (except first) = 1005.73
$$= (101.579 \times 6) + (103.02 \times 4) - (13.804 + 2.020)$$
$$= 1005.73$$

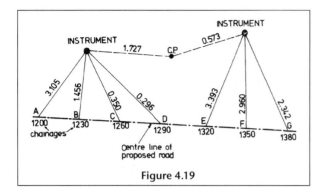

Figure 4.19

were on a change point while the remainder of the readings at points A, B, C, D, E, F and G were taken at 30 m intervals along the line of the proposed road. Given that the reduced level of point A is 13.273 m above datum, calculate the reduced levels of all points along the road.

(City and Guilds of London Institute Surveying and Levelling Examination)

5. Inverted staff readings

In all of the previous examples on levelling, the points observed all lay below the line of sight. Frequently on building sites, the reduced levels of points above the height of the instrument are required, e.g. the soffit level of a bridge or underpass, the underside of a canopy, the level of roofs, eaves, etc., of buildings. Figure 4.20 illustrates a typical case.

The reduced levels of points A, B, C and D on the frame of a multi-storey building require checking. The staff is simply held upside down on the points A and C and booked with a *negative sign* in front of the reading, e.g. –1.520. Alternatively, the reading may be put in brackets, e.g. (1.520), or an asterisk may be put alongside the figure, e.g. 1.520*. Such staff readings are called inverted staff readings.

Table 4.18 shows the readings observed to the points A, B, C and D on the multi-storey building of Fig. 4.20. The levels may be reduced by either (a) the rise and fall method or (b) the HPC method.

(a) Reduction by the rise and fall method

The rise or fall is required between successive points as before. The second reading is subtracted from the first as follows:

$$
\begin{array}{rl}
\text{BS to bench mark} = & (1.750) \\
\text{IS to frame A} = - & (-3.100) \\
\text{Difference BM to A} = & +4.850 \text{ rise} \\
\text{IS to frame A} = & (-3.100) \\
\text{IS to floor level B} = & -(1.490) \\
\text{Difference A to B} = & -4.590 \text{ fall} \\
\text{IS to floor level B} = & 1.490 \\
\text{FS to canopy C} = - & (-2.560) \\
\text{Difference B to C} = & +4.050 \text{ rise} \\
\text{BS to canopy C} = & (-4.210) \\
\text{FS to kerb D} = & -(4.200) \\
\text{Difference C to D} = & -8.410 \text{ fall}
\end{array}
$$

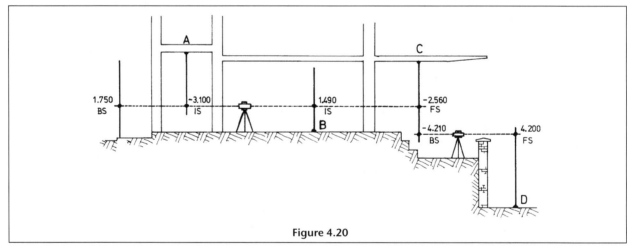

Figure 4.20

Table 4.18

BS	IS	FS	Rise	Fall	Reduced level	Distance	Remarks
1.750					72.300		Bench mark (72.300)
	–3.100		4.850		77.150		A Frame (lift shaft)
	1.490			4.590	72.560		B Floor level
–4.210		–2.560	4.050		76.610		C Canopy
		4.200		8.410	68.200		D Bench mark (68.195m)
–2.460		+1.640	8.900	13.000	68.200		
–1.640				–13.000	–72.300		
–4.100				–4.100	–4.100		

The reduced levels are obtained by the algebraic addition of the rises and falls as before (Table 4.18).

The arithmetic check is applied in the usual manner. The BS and FS columns are added algebraically, the inverted staff readings being regarded as negative.

$$(\text{Last RL} - \text{first RL}) = (\text{sum rises} - \text{sum falls})$$
$$= (\text{sum BS} - \text{sum FS})$$
$$= 68.200 - 72.300$$
$$= 8.900 - 13.000$$
$$= -2.460 - 1.640$$
$$= -4.100 \text{ m}$$

(b) Reduction by the HPC method

The usual HPC rules are followed, viz.

$$\text{HPC} = \text{reduced level of point}$$
$$= + \text{staff reading at point}$$
$$\text{RL point} = \text{HPC} - \text{staff reading at point}$$
$$\text{HPC (1)} = \text{reduced level BM} + \text{BS staff reading}$$
$$= 72.300 + 1.750$$
$$= 74.050$$
$$\text{RL A} = \text{HPC(1)} - \text{IS staff reading A}$$
$$= 74.050 - (-3.100)$$
$$= 77.150 \text{ m}$$
$$\text{RL B} = \text{HPC(1)} - \text{IS staff reading B}$$
$$= 74.050 - 1.490$$
$$= 72.560 \text{ m}$$
$$\text{RL C} = \text{HPC(1)} - \text{FS staff reading C}$$
$$= 74.050 - (-2.560)$$
$$= 76.610 \text{ m}$$
$$\text{HPC (2)} = \text{reduced level change point C}$$
$$= + \text{BS staff reading C}$$
$$= 76.610 + (-4.210)$$
$$= 2.400 \text{ m}$$
$$\text{RL D} = \text{HPC(2)} - \text{FS staff reading D}$$
$$= 72.400 - 4.200$$
$$= 68.200 \text{ m}$$

The complete reduction is shown in Table 4.19.

Exercise 4.5

1 A tunnel, being driven as part of a water supply scheme, is to be checked for possible subsidence of the roof line. Levels were taken as shown in Fig. 4.21 and Table 4.20. Reduce the levels by (a) the rise and fall method and (b) the height of collimation method.

Table 4.20

BS	IS	FS	Reduced level	Remarks
1.10			10.000	A. (TBM) floor level
	−2.05			B. Roof level
−2.68		−1.45		C. Roof level
	1.05			D. Floor level
	−1.30			E. Roof level
	−0.46			F. Roof level
		2.78		G (TBM) floor level

6. Errors in levelling

As in all surveying operations, the sources and effects of errors must be recognized and steps taken to eliminate or minimize them. Errors in levelling can be classified under several headings.

(a) Gross errors

These are mistakes arising in the mind of the observer. They are usually due to carelessness, inexperience or fatigue.

1. *Wrong staff readings* This is probably the most common error of all in levelling. Examples of wrong staff readings are: misplacing the decimal point, read-

Table 4.19

BS	IS	FS	HPC	Reduced level	Remarks
1.750			74.050	72.300	Bench mark (72.300)
	−3.100			77.150	A Frame (lift shaft)
	1.490			72.560	B Floor level
−4.210		−2.560	72.400	76.610	C Canopy
		4.200		68.200	D Bench mark (68.195m)
−2.460	−1.610	1.640		68.200	
−1.640				−72.300	
−4.100				−4.100	

Check:
Sum of RLs except first = 294.52
$$= (74.050 \times 3) + 72.400 - (-1.61 + 1.64)$$
$$= 222.15 + 72.40 - 0.03$$
$$= 294.52$$

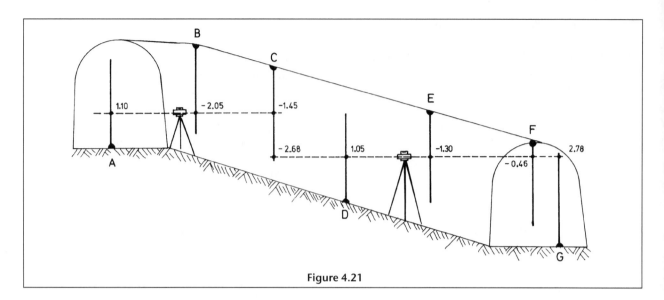

Figure 4.21

ing the wrong metre value and reading the staff wrong way up, with inverted staff readings.

2. *Using the wrong cross-hair* Instead of reading the staff against the axial line, the observer reads against one of the stadia lines. This error is common in poor visibility.

3. *Wrong booking* The reading is noted with the figures interchanged e.g. 3.020 instead of 3.002.

4. *Omission or wrong entry* A staff reading can easily be written in the wrong column or even omitted entirely.

5. *Spirit level not centred* When using automatic levels, the small circular spirit level (pond bubble) must be accurately centred. If not, the compensator may become jammed, since it has only a limited amount of movement.

All of these errors can be small or very large and every effort must be made to eliminate them. The only way to eliminate gross errors is to make a double levelling, i.e. to level from A to B then back from B to A. Theoretically the levelling should close without any error but this will very seldom happen. However, the error should lie within the limits already stated at the start of Sec. 2, 'Observation procedure'.

(b) Constant errors

These errors are due to instrumental defects and will always be of the same sign.

1. *Non-verticality of the staff* This is a serious source of error. Instead of being held vertically the staff may be leaning forward or backward. In Fig. 4.22, the staff is 3° out of vertical. If a reading of 4.000 m is observed, it will be in error by 5 mm. The correct reading is

$4 \times \cos 3° = 3.995$ m

The error can be eliminated by fitting the staff with a

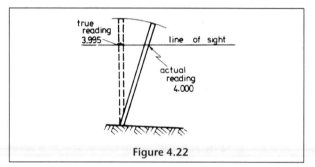

Figure 4.22

circular spirit level. The staffholder must ensure that the bubble is centred when the staff is being read. A second method of eliminating the error is for the staff-holder to swing the staff slowly backwards and forwards across the vertical position during the observation. The observer then reads the lowest reading.

2. *Collimation error in the instrument* In a properly adjusted level, the line of sight must be perfectly horizontal, when the instrument has been set up, ready for use. If not, there will be an error in the staff reading. In Fig. 4.23, the line of sight is inclined and the resultant error is e, increasing with length of sight. The error can be entirely eliminated by making backsights and foresights equal in length. The error e will be the same for each sight and the true difference in level will be the difference in the readings.

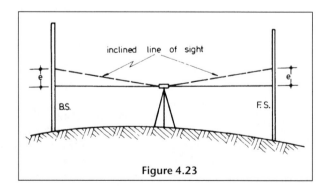

Figure 4.23

3. *Staff graduation errors* With the improvement of quality in the manufacture of levelling staves, particularly in the printing process, staff graduation errors are very uncommon. However, care should be taken when using a telescopic staff, to ensure that the staff is fully extended.

(c) Random errors

These errors are due to physical and climatic conditions. The resulting errors are small and are likely to be compensatory.

1. *Effect of wind and temperature* The stability of the instrument may be affected, causing the height of collimation to change slightly.

2. *Soft and hard ground* When the instrument is set on soft ground it is likely to sink slightly as the observer moves around it. When set in frosty earth, the instrument tends to rise out of the ground. Again the height of collimation changes slightly.

3. *Change points* At any change point the staff must be held on exactly the same spot for both foresight and backsight. A firm spot must be chosen and marked by chalk. If the ground is soft a change plate (Fig. 4.24) must be used. The plate is simply a triangular piece of metal bent at the apices to form spikes. A dome of metal is welded to the plate and the staff is held on the dome at each sighting.

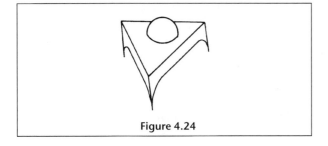

Figure 4.24

4. *Human deficiencies* Errors arise in estimating the millimetre readings, particularly when visibility is bad or sights are long.

All of the errors in this class tend to be small and compensating and are of minor importance only, in building surveying.

7. Permanent adjustments

In the preceding errors section, it was pointed out that the line of collimation might not be horizontal. The levelling will still be correct provided the sights are of equal length. This is not always possible, however, particularly when many intermediate sights are to be taken. The only way in which these errors can be eliminated is to ensure that the instrument is in good adjustment.

(a) Automatic and digital levels

1. The vertical axis of both types of level must be within ¼ degree of the true plumb line. This verticality is indicated by the small circular spirit level (Fig. 4.3) or pond level as it is frequently called. When set up, the bubble should remain central when the instrument is turned to any position. If this does not happen, adjustments can be made to the level vial by means of two small screws, using the tools supplied by the manufacturer

2. The line of sight must be perfectly horizontal, when the bubble of the pond level is central. This will only happen when the compensator and reticule are set correctly, otherwise a collimation error is present. In order to detect a collimation error, a so-called 'two-peg' test must be carried out.

(b) Two-peg test

Test (Fig. 4.25)

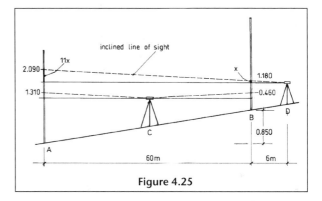

Figure 4.25

1. Hammer two pegs A and B firmly into the ground, 60 metres apart.
2. Set up a level exactly midway between the pegs at a point C and read a staff held on each peg in turn.

Reading on peg A = 1.310 m
Reading on peg B = 0.460 m
True difference in level A to B = 0.850 m

This is the true difference in level between the pegs irrespective of whether the line of collimation is inclined or not. If there were a collimation error it would be equal in both directions CA and CB since the distances are equal.

3. Remove the level to a point D some distance beyond peg B. The best distance is 1/10th of distance AB, which is 6 metres. Take a second set of readings on pegs A and B.

Reading on peg A = 2.090 m
Reading on peg B = 1.180 m
Apparent difference in level A to B = 0.910 m

4. Since the true difference in level does not equal the apparent difference, there is a collimation error in the line of sight of the instrument. Because the

apparent difference is greater than the true difference, the line of collimation error points upwards in this case.

5. The correct staff readings required to adjust the level are now required. They are calculated by one of the following methods.

Calculation of adjustment

(i) Let x be the collimation error at B

Therefore the true staff reading at B $= (1.180 - x)$

By proportion, the collimation error at A is

$$\frac{\text{Error(A)}}{x} = \frac{66}{6}$$

Therefore error A $= 11x$

So the true staff reading at A $= (2.090 - 11x)$

True difference in level $= 0.850$ m

Also, true difference in level $= (2.090 - 11x)$

$$- (1.180 - x)$$

$$= 2.090 - 1.180$$

$$- 10x$$

$$= 0.910 - 10x$$

Therefore $0.850 = 0.910 - 10x$

$$10x = 0.060$$

$$x = 0.006 \text{ m}$$

True reading B $= 1.180 - x$

$$= 1.180 - 0.006$$

$$= \underline{1.174 \text{ m}}$$

True reading A $= 2.090 - 10x$

$$= 2.090 - 0.066$$

$$= 2.090 - 0.066$$

$$= \underline{2.024 \text{ m}}$$

Check:

Difference in level AB $\quad = 2.024 - 1.174$

$$= \underline{0.085 \text{ m}}$$

(ii) True difference in level AB $\quad = 1.310 - 0.460$

$$= \underline{+0.850 \text{ m}}$$

Apparent difference in level AB $= 2.090 - 1.180$

$$= \underline{+0.910 \text{ m}}$$

The line of collimation is elevated by $(0.910 - 0.850 = 0.060)$ m over 60 m

i.e. $+0.060$ m in 60 m $= \quad 0.1\%$

Therefore true reading at B $= \quad 1.180 - 0.1\%$ of 6 m

$$= (1.10 - 0.006)$$

$$= \underline{1.174 \text{ m}}$$

and true reading at A $= \quad 2.090 - 0.1\%$ of 66 m

$$= (2.090 - 0.066)$$

$$= \underline{2.024 \text{ m}}$$

Check:

Difference in level AB $\quad = (2.024 - 1.174)$

$$= \underline{0.850 \text{ m}}$$

Adjustment The line of sight through the instrument is adjusted by means of the reticule (eyepiece). On most instruments there is one screw which is moved slowly up or down using a special tool supplied by the

manufacturer, until the correct reading on the staff held on point A is obtained. A check is then made to peg B.

The instrument manual will give all the information on a particular instrument. It may show that the adjustment method requires an adjustment to be made to the compensator. This is definitely not a job for the amateur and the instrument should be returned to the manufacturer where it will be adjusted under laboratory conditions.

Exercise 4.6

1 In a two-peg test, a level was set up exactly midway between two pegs X and Y which are 50 metres apart. The readings to the pegs were:

Peg X   2.394
Peg Y   1.971

The level was then set up at a point Z, 5 metres beyond peg Y on the line XY and further readings taken as follows:

Peg X   2.723
Peg Y   2.330

Calculate:

(a)   the collimation error in the level as a percentage,
(b)   the staff readings required to correctly adjust the instrument.

2 The following list of readings was taken in sequence during a levelling survey.

Reading (m)	Remarks
1.250	BM, 1.435 m AOD
1.285	Peg A
1.125	Peg B
0.810	Change point
1.555	
−1.400	Inverted staff reading taken to the underside of a bridge
1.235	Peg C
0.665	Change point
1.905	
0.070	BM, 4.600 m AOD

Adopt a standard form of booking and reduce the levels to Ordnance Datum.
(OND, Building)

3 The table below shows the results of a levelling along the centreline of a roadway where settlement has taken place. The road was initially constructed at a uniform gradient rising at 1 in 75 from A to B. Assuming no settlement has taken place at station A:

(a)   Reduce the levels.
(b)   Calculate the maximum settlement along the line AB.

BS	IS	FS	Remarks	Chainage (m)
3.540			OBM 78.675	
0.410		3.665		
0.525		2.245		
	2.840		Station A	0
	2.440			30
	2.045			60
2.475		1.655		90
	2.090			120
	1.700			150
	1.315			180
	0.900			210
2.465		0.485		240
	2.055			270
	1.645		Station B	300
		2.040	TBM 78.000	

(IOB, Site Surveying)

4 Table 4.21 shows the readings taken to determine the clearance between the river level and the soffit of a road bridge. Reduce the levels and determine the clearance between the river level and the soffit of the bridge.

5 Table 4.22 shows the staff readings obtained from a survey along the line of a proposed roadway from A to B.

(a) On the table, reduce the levels from A to B applying all the appropriate checks.
(b) Given that the finished roadway has to be evenly graded from A to B, calculate the depths of cut or fill at 20 metre intervals between A and B.

8. Curvature and refraction

Throughout this chapter, reference has been made to a horizontal line as distinct from a level line. If the Earth

Table 4.21

BS	IS	FS	RL	Remarks
0.872			21.460	OBM
0.665		3.980		
	2.920			River level at A
	−1.332			Soffit of bridge at A
	−1.312			Soffit of bridge at B
	−1.294			Soffit of bridge at C
	−1.280			Soffit of bridge at D
	2.920			River level at D
4.216		0.597		
		1.155		OBM

(City and Guilds of London Institute)

is considered to be a perfect sphere (Fig. 4.26), a level line would be, at all points, equidistant from the centre. However, the line of sight through a levelling instrument is a horizontal line tangential to the level line. If a staff were held on B, the staff reading, observed from A, would be too great by the amount BB_1. This is the curvature correction c, which is calculated as follows:

In the triangle L is the length of sight in kilometres and R is the mean radius of the Earth (6370 km). By Pythagoras's theorem:

$$(R + c)^2 = R^2 + L^2$$

i.e.

$$R^2 + c^2 + 2Rc = R^2 + L^2$$

Therefore

$$c(c + 2R) = L^2$$
$$c = L^2/(c + 2R)$$

Since c is so small compared with R, it can be ignored. Therefore,

$$c = (L^2/2R) \text{ kilometres}$$

Table 4.22

BS	IS	FS	Rise Fall or HPC	Surface reduced level	Grade reduced level	Fill	Cut	Remarks
0.824				39.220				TBM 1
	1.628							A
	0.790							20 m from A
	0.383							40 m from A
2.154		1.224						60 m from A
	2.336							80 m from A
	2.757							100 m from A
2.555		0.461						Change point
	2.275							120 m from A
	0.436							140 m from A
	0.227							160 m from A
	0.716							180 m from A
	0.652							B, 200 m from A
		0.233						TBM 2

(SCOTVEC, Ordinary National Diploma in Building)

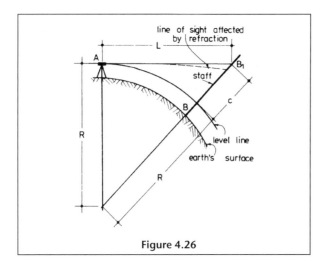

Figure 4.26

8 Calculate the corrected staff reading for a sight of 1500 metres if the observed reading is 3.250.

Answer

$$\text{Length of sight} = 1.5 \text{ km}$$
$$\text{Correction for curvature} = (0.0673 \times 1.5^2) \text{ m}$$
$$\text{and refraction} = (0.0673 \times 2.25) \text{ m}$$
$$= 0.151 \text{ m}$$
$$\text{Observed staff reading} = 3.250$$
$$\text{Correction} = -0.151$$
$$\text{Correct staff reading} = 3.099 \text{ m}$$

9 Calculate the correction due to curvature and refraction over a length of sight of 120 metres.

Answer

$$c = (0.0673 \times 0.12^2) \text{ m}$$
$$= (0.0673 \times 0.0144) \text{ m}$$
$$= 0.001 \text{ m}$$

Since 0.001 m is negligible, the correction can be neglected for lengths of sight less than 120 metres. It is good practice, when levelling, to restrict the length of sight to about 50 metres. Furthermore, at this distance the staff reading should never be allowed to fall below 0.5 m because of the variation in refraction caused by fluctuations in the density of the air close to the ground.

i.e.

$$c = \left(\frac{L^2}{12\,740}\right) \text{km}$$

However, c is required in metres while L remains in kilometres:

$$c = \left(\frac{L^2 \times 1000}{12\,740}\right) \text{metres}$$
$$c = 0.0785L^2 \text{ metres (where } L \text{ is in kilometres)}$$

The line of sight is not really horizontal. It is affected by refraction in such a manner that the line of sight is bent downwards towards the earth. Refraction is affected by pressure, temperature, latitude, humidity, etc., and its value is not constant. Its value is taken as 1/7th of the curvature and is opposite in effect to that of curvature. Thus,

$$\text{Combined correction} = 0.0785L^2 - \tfrac{1}{7}(0.0785L^2)$$
$$= \tfrac{6}{7}(0.0785L^2)$$
$$= 0.0673L^2 \text{ metres}$$

(where L is in kilometres).

9. Reciprocal levelling

The importance of having sights of equal length has already been stressed. Briefly, collimation errors are eliminated by this technique. If the length of sight does not exceed 120 metres, curvature and refraction errors are negligible. However, there are occasions when a long sight must be taken and collimation and curvature errors are present. For example, in Fig. 4.27, the

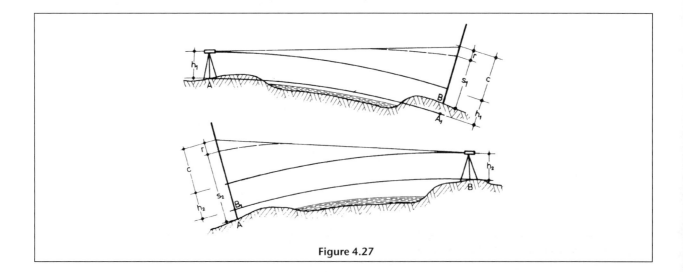

Figure 4.27

difference in level between two points A and B on opposite banks of a wide river is required.

The instrument is set at A and the instrument height h_1 is measured. The staff is held on B and the staff reading s_1 recorded. In Fig. 4.27, r, the refraction error, and c, the curvature error, are clearly shown. Since AA_1 is the level surface through A the difference in level between A and B is the distance A_1B. Now

$$A_1B = h_1 + c - r - s_1$$

The instrument is then removed to B, the height h_2 measured and the staff reading s_2 recorded: r and c are the refraction and curvature errors as before. The difference in level between B and A is the distance B_1A:

$$B_1A = s_2 + r - c - h_2$$

Note that since the sights are of equal length, the collimation error is the same and is therefore cancelled.

The difference in level is the mean of $A_1B + B_1A$

$$= \tfrac{1}{2}(h_1 + c - r - s_1 + s_2 + r - c - h_2)$$
$$= \tfrac{1}{2}(h_1 - h_2 + s_2 - s_1)$$

Strictly, this is not the true difference of level since the value of r will be slightly different in the two equations. However, it will be sufficiently close for building

surveys provided the observations are made as soon after each other as possible.

EXAMPLE

10 Observations were made between points X and Y on opposite sides of a wide water-filled quarry as follows:

Level at X	Instrument height = 1.350
	Staff reading Y = 1.725
Level at Y	Instrument height = 1.410
	Staff reading X = 1.055

Calculate the true difference in level between the stations and the reduced level of Y if X is 352.710 AOD.

Solution

$$\begin{aligned}
\text{True difference in level} &= \tfrac{1}{2}(h_1 - h_2 + s_2 - s_1) \\
&= \tfrac{1}{2}(1.350 - 1.725 \\
&\quad + 1.055 - 1.410)\ \text{m} \\
&= \tfrac{1}{2}(2.405 - 3.135)\ \text{m} \\
&= \tfrac{1}{2}(-0.730)\ \text{m} \\
&= -0.365\ \text{m}
\end{aligned}$$

Reduced level X = 352.710 m AOD
Therefore fall X − Y = −0.365 m
and reduced level Y = 352.345 m AOD

---

10. Answers

Exercise 4.1

1 Table 4.23

Table 4.23

BS	IS	FS	Rise	Fall	Reduced level	Remarks
1.955					5.200	OSBM
1.315		2.030		0.075	5.125	C
1.243		0.885	0.430		5.555	D
2.071		1.485		0.242	5.313	E
1.570		0.880	1.191		6.504	F
1.835		1.590		0.020	6.484	A
0.631		0.540	1.295		7.779	B
1.200		3.289		2.658	5.121	C
		1.130	0.070		5.191	OSBM
11.820		11.829	2.986	2.995	5.191	
−11.829			−2.995		−5.200	
−0.009			−0.009		−0.009	

Checks indicate that (a) there is a survey error of 9 mm and (b) there is no arithmetic error.

Exercise 4.2

1 Table 4.24 (a)

Table 4.24(a)

BS	IS	FS	Rise	Fall	Reduced level	Remarks
2.650					5.200	OSBM
	2.727			0.077	5.123	C
	2.292		0.435		5.558	D
	2.537			0.245	5.313	E
	1.346		1.191		6.504	F
	1.370			0.024	6.480	A
		0.065	1.305		7.785	B
2.650			2.931	0.346	7.785	
−0.065			−0.346		−5.200	
2.585			2.585		2.585	

2 Table 4.24 (b)

Table 4.24(b)

BS	IS	FS	Rise	Fall	Reduced level	Remarks
2.763					5.200	OSBM
	2.840			0.077	5.123	C
	2.385		0.455		5.578	D
	2.650			0.265	5.313	E
	1.459		1.191		6.504	F
	1.463			0.004	6.500	A
		0.178	1.285		7.785	B
2.763		0.178	2.931	0.346	7.785	
−0.178			−0.346		−5.200	
2.585			2.585		2.585	

Stations A and D show errors.

Exercise 4.3

1 Table 4.25

Table 4.25

BS	IS	FS	Rise	Fall	Reduced level	Remarks
1.256					5.200	BM
	1.330			0.074	5.126	C
1.100		0.906	0.424		5.550	D
	1.332			0.232	5.318	E
	0.146		1.186		6.504	F
1.875		0.166		0.020	6.484	A
0.200		0.579	1.296		7.780	B
		2.780		2.580	5.200	BM
4.431		4.431	2.906	2.906		
−4.431			−2.906			
0.000			0.000			

Exercise 4.4

1 Table 4.26

Table 4.26

BS	IS	FS	Rise	Fall	Reduced level	Distance	Remarks
3.105					13.273	1200	TBM A
	1.456		1.649		14.922	1230	B
	0.350		1.106		16.028	1260	C
	0.296		0.054		16.082	1290	D
0.573		1.727		1.431	14.651		Change point
	3.393			2.820	11.831	1320	E
	2.960		0.433		12.264	1350	F
		2.342	0.618		12.882	1380	G
3.678	4.069	3.860	4.251		12.882		
−4.069		−4.251			−13.273		
−0.391		−0.391			−0.391		

Exercise 4.5

1 (a) Using the rise and fall method (Table 4.27)

Table 4.27

BS	IS	FS	Rise	Fall	Reduced level	Remarks
1.10					10.00	A. (TBM) floor level
	−2.05		3.15		13.15	B. Roof level
−2.68		−1.45		0.60	12.55	C. Roof level
	1.05			3.73	8.82	D. Floor level
	−1.30		2.35		11.17	E. Roof level
	−0.46			0.84	10.33	F. Roof level
		2.78		3.24	7.09	G. Floor level
−1.58		1.33	5.50	8.41	7.09	
−1.33			−8.41		−10.00	
−2.91			−2.91		−2.91	

(b) Using the HPC method (Table 4.28)

Table 4.28

BS	IS	FS	HPC	Reduced level	Remarks
1.10			11.10	10.00	A. (TBM) floor level
	−2.05			13.15	B. Roof level
−2.68		−1.45	9.87	12.55	C. Roof level
	1.05			8.82	D. Floor level
	−1.30			11.17	E. Roof level
	−0.46			10.33	F. Roof level
		2.78		7.09	G. Floor level
	−2.76	1.33			

Sum RLs (except first) = 63.11
(11.10 × 2) + (9.87 × 4) = 61.68
61.68 − (−2.76) − 1.33 = 63.11 check

Exercise 4.6

1 (a) 0.06%
(b) X 2.756 m
Y 2.333 m

2 Table 4.29

Table 4.29

BS	IS	FS	HPC	RL	Remarks
1.250			2.685	1.435	BM
	1.285			1.400	Peg A
	1.125			1.560	Peg B
1.555		0.810	3.430	1.875	CP
	−1.400			4.830	Inverted staff
	1.235			2.195	Peg C
1.905		0.665	4.670	2.765	CP
		0.070		4.600	BM
4.710	2.245	1.545		4.600	
−1.545				−1.435	
3.165				3.165	Simple partial check

Sum RLs (except first)   = 19.225 m
                          2.685 × 3 =   8.055
                        + 3.430 × 3 = 10.290
                        + 4.670 × 1 =   4.670
                                      23.015
Sum IS column   2.245
Sum FS column   1.545
                                      3.790
                   23.015 − 3.790 = 19.225 m

3 Table 4.30

Table 4.30

BS	IS	FS	Rise	Fall	Reduced level	Remarks	Chainage	Formation level	Settlement
3.540					78.675	OBM			
0.410		3.665		0.125	78.550				
0.525		2.245		1.835	76.715				
	2.840			2.315	74.400	Station A	0	74.400	0.000
	2.440		0.400		74.800		30	74.800	0.000
	2.045		0.395		75.195		60	75.200	0.005
2.475		1.655	0.390		75.585		90	75.600	0.015
	2.090		0.385		75.970		120	76.000	0.030
	1.700		0.390		76.360		150	76.400	0.040
	1.315		0.385		76.745		180	76.800	0.055
	0.900		0.415		77.160		210	77.200	0.040
2.465		0.485	0.415		77.575		240	77.600	0.025
	2.055		0.410		77.985		270	78.000	0.015
	1.645		0.410		78.395	Station B	300	78.400	0.005
		2.040		0.395	78.000	TBM			
9.415	10.090	3.995	4.670		78.000				
−10.090		−4.670			−78.675				
−0.675		−0.675			−0.675				

Original level station A = 74.400. Since the construction gradient of the roadway is 1 in 75 rising from A, the formation levels of the chainage points must increase uniformly by (30/75) = 0.400 m, giving the formation levels in the table above. The maximum settlement occurs at chainage 180 m.

4 Table 4.31

Table 4.31

BS	IS	FS	HPC	Reduced level	Remarks
0.872			22.332	21.460	OBM
0.665		3.980	19.017	18.352	
	2.920			16.097	River level at A
	−1.332			20.349	Soffit of bridge at A
	−1.312			20.329	Soffit of bridge at B
	−1.294			20.311	Soffit of bridge at C
	−1.280			20.297	Soffit of bridge at D
	2.920			16.097	River level at D
4.216		0.597	22.636	18.420	
		1.155		21.481	OBM
5.753	0.622	5.732		21.481	
−5.732				−21.460	
0.021				0.021	

Sum RLs (except first) = 171.733 = [22.332 + (19.017 × 7) + 22.636]
$$-[0.622 + 5.732]$$
$$= 171.733$$

5 Table 4.32

Table 4.32

BS	IS	FS	Height of collimation	Surface reduced level	Grade reduced level	Fill	Cut	Remarks
0.824			40.044	39.220				TBM 1
	1.628			38.416	38.416	—	—	A
	0.790			39.254	38.816		0.438	20 m from A
	0.383			39.661	39.216		0.445	40 m from A
2.154		1.224	40.974	38.820	39.616	0.796		60 m from A
	2.336			38.638	40.016	1.378		80 m from A
	2.757			38.217	40.416	2.199		100 m from A
2.555		0.461	43.068	40.513				Change point
	2.275			40.793	40.816	0.023		120 m from A
	0.436			42.632	41.216		1.416	140 m from A
	0.227			42.841	41.616		1.225	160 m from A
	0.716			42.352	42.016		0.336	180 m from A
	0.652			42.416	42.416		—	B, 200 m from A
		0.233		42.835				TBM 2
5.533	12.200	1.918		42.835				
−1.918				−39.220				
3.615				3.615				

Sum of RLs (except first) = 527.388 = [(40.044 × 4) + (40.974 × 3) + (43.068 × 6)] − [12.200 + 1.918]
$$= 527.388$$
Gradient A to B = (42.416 − 38.416) in 200
$$= 1 \text{ in } 50$$
Rise per 20 m = (1/50) × 20
$$= 0.400 \text{ m}$$

Chapter summary

In this chapter, the following are the most important points:

- In a levelling survey, the surveyor is concerned with finding the third dimension of a surveyed point, namely the height. Use is made of an instrument called a level, which sets out a horizontal plane from which measurements are made downwards to the ground, by means of a staff, read through the telescope of the level. The most common modern instruments are the automatic and digital levels, which, as their names suggest, set out the horizontal plane automatically.

- The readings are converted to reduced levels, which is a surveying term meaning the elevation or height of the ground above a datum, which may be sea level if the survey is connected to the Ordnance Survey Datum, via a bench mark.

  The calculation of the levels is made either by the rise and fall method or the HPC method. In the first method, the rise or fall of the ground between a known and an unknown point is calculated by comparing their two staff readings. The rise is then added to the height of the known point or the fall is subtracted from the height of the known point to produce the height of the unknown point.

  In the height of the plane of collimation (HPC) method the staff reading at the known point is added to the reduced level of the point to produce the height of the telescope of the level, known as the height of the plane of collimation. Subsequent staff readings are subtracted from that HPC to produce the level of the unknown point. Both methods have advantages and disadvantages which are fully covered at various points throughout the text.

- A surveyor must know if and when his or her instrument may be malfunctioning. A simple 'two peg' test is carried out which reveals when the instrument is in need of adjustment. The computation and adjustment are given on pages 69 to 70.

- It is good practice in levelling surveys to make back and forward sights equal in length, to negate errors of collimation. Sometimes, though very infrequently, this cannot be achieved and recourse has to be made to a form of levelling known as reciprocal levelling. This form of levelling nullifies the effects of the Earth's curvature and minimizes the effect of atmospheric refraction. This levelling is of an advanced nature and is fully explained with examples on pages 71 to 73.

CHAPTER 5 Contouring

In this chapter you will learn about:
• the meaning, significance and characteristics of contour lines on construction site drawings
• the methods of surveying used to find contour lines in the field and the methods of interpolating contour lines from the surveyed data
• the various uses of contour lines in construction site planning and earthwork calculations.

Chapter 3 dealt with the principles of making a linear survey and producing a two-dimensional plan from the results. Figure 3.1 is the end product and shows clearly the surveyed parcel of ground, bordered by hedges, fences, road, etc. The plan would be greatly enhanced and much more useful if the third dimension, namely height, could be represented in some way.

On large-scale plans of civil engineering and building projects, contouring is the commonly used method of showing height.

Contoured plans can be produced from a variety of surveyed data. The most cost-effective and accurate method of collecting the data is automatically by total station instruments from which the data is post processed and contoured plans produced digitally by computer. The database can then be interrogated to produce sections from which volumes of earthworks can be computed easily (Chapters 6 and 10).

The main aim of this textbook, however, is to provide the aspiring young surveyor with a background of basic knowledge which, with experience, will be supplemented but never superseded by modern methods.

This chapter therefore concentrates on the basic methods of producing contoured plans from levelling data.

1. Contour characteristics

(a) Contour line

A contour line is a line drawn on a plan joining all points of the same height above or below some datum. The concept of a contour line can readily be grasped if a reservoir is imagined. If the water is perfectly calm, the edge of the water will be at the same level all the way round the reservoir forming a contour line. If the water level is lowered by, say, five metres the water's edge will form a second contour.

Further lowering of the water will result in the formation of more contour lines (Fig. 5.1). Contour lines are continuous lines and cannot meet or cross any other contour line, nor can any one line split or join any other line, except in the case of a cliff or overhang.

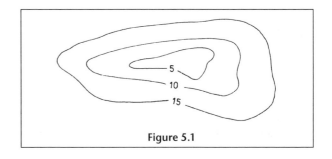

Figure 5.1

Figure 5.2 shows the contour plan and section of an island. The tide mark left by the sea is the contour line of zero metres value. If it were possible to pass a series of equidistant horizontal planes 10 metres apart through the island, their points of contact with the island would form contours with values of 10, 20, 30 and 40 m.

(b) Gradients

The height between successive contours is called the vertical interval or contour interval and is always constant over a map or plan. On the section the vertical interval is represented by AB. The horizontal distance between the same two contours is the distance BC. This is called the horizontal equivalent.

The gradient of the ground between the points A and C is found from

$$\text{Gradient} = \frac{AB}{BC} = \frac{\text{vertical interval}}{\text{horizontal equivalent}}$$

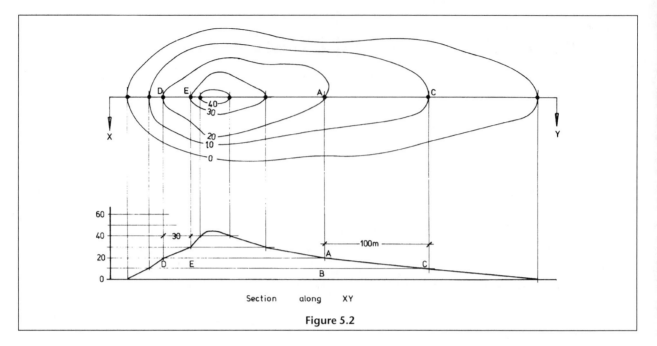

Figure 5.2

Since the vertical interval is constant throughout any plan, the gradient varies with the horizontal equivalent, for example

$$\text{Gradient along AC} = \frac{10}{100} = \frac{1}{10} = 1 \text{ in } 10$$

$$\text{Gradient along DE} = \frac{10}{30} = \frac{1}{3} = 1 \text{ in } 3$$

(c) Reading contours

It should be clear from the above examples that the gradient is steep where the contours are close together and conversely flat where the contours are far apart.

In Fig. 5.3 three different slopes are shown. The contours are equally spaced in Fig. 5.3(a), indicating that the slope has a regular gradient. In Fig. 5.3(b) the contours are closer at the top of the slope than at the bottom. The slope is therefore steeper at the top than at the bottom and such a slope is called a concave slope. Conversely, Fig. 5.3(c) portrays a convex slope.

A river valley has a characteristic 'V-shape' formed by the contours (Fig. 5.3(d)). The V always points towards the source of the river, i.e. uphill. In contrast, the contours of Fig. 5.3(e) also form a V-shape but point downhill forming a 'nose' or spur.

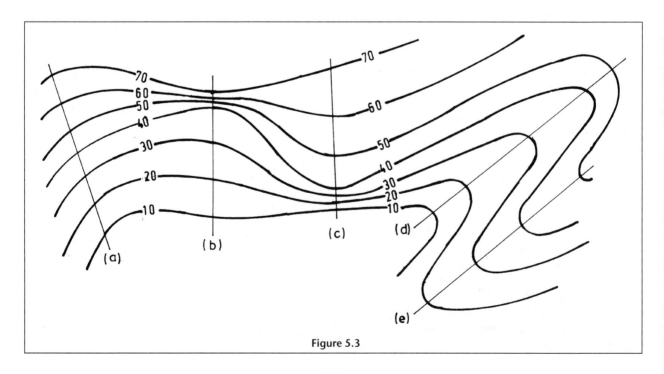

Figure 5.3

EXAMPLE

1 Figure 5.4 shows contours at vertical intervals of 0.5 m over a building site, drawn at scale 1:500.
(a) Using a scale rule, measure the length and breadth of the site.
(b) Describe briefly the terrain along the lines AB and AC.
(c) Calculate the gradient between:
   (i) points D and E,
   (ii) points F and G.

Answer

(a) 80 m × 51 m
(b) From point A, the ground falls on a fairly regular slope, towards B. Line AB is, in fact, the line of a ridge, where the ground falls, both northwards and southwards, from the ridge line. On the line AC, the ground falls in the form of a concave slope, the slope being steeper towards the north end of the line.

   The whole area is the side of a hill, probably formed by bulldozing the area during the construction of the building and car park.
(c) Line DE

   Fall = 2 m, horizontal distance = 8.0 m
   Therefore gradient = 1 in 4

Line FG
   Fall = 2 m, horizontal distance = 25 m
   Therefore gradient = 1 in 12.5

Exercise 5.1

1 Figure 5.5 shows a contoured plan of two car parks, divided by a high retaining wall and bounded by various other walls and fences.

(a) Calculate the average gradient across the lines AB, CD and EF.
(b) Estimate, from the values of the contour lines, the height of the retaining wall dividing the car parks.

2. Methods of contouring

Once an accurate survey has been completed the surveyor knows the planimetric position of all points on the site relative to each other. The second task is to make a levelling to enable accurate positions of the contour lines over the site to be drawn.

(a) Choice of vertical interval

The vertical interval of the contour lines on any plan depends on several factors, namely:

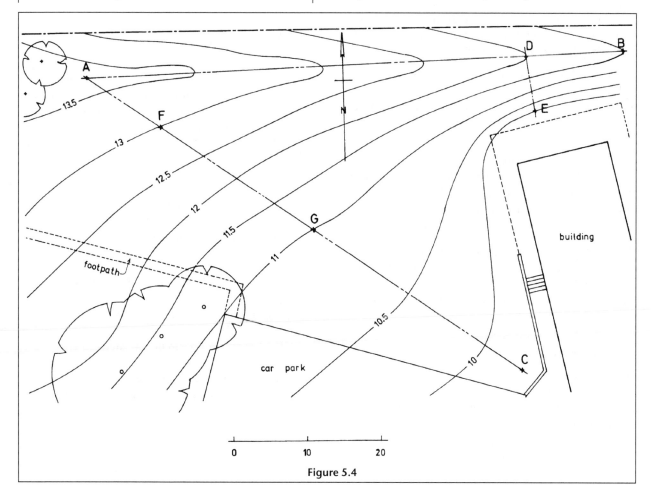

Figure 5.4

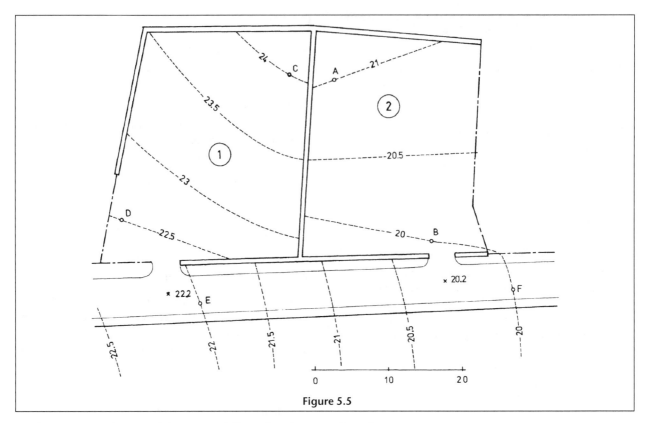

Figure 5.5

1. *The purpose and extent of the survey* Where the plan is required for estimating earthwork quantities or for detailed design of works a small vertical interval will be required. The interval may be as small as 0.5 metre over a small site but 1 to 2 metres is more common, particularly where the site is fairly large.

2. *The scale of the map or plan* Generally, on small-scale maps the vertical interval has to be fairly large. If not, some essential details might be obscured by the large number of contour lines produced by a small vertical interval.

3. *The nature of the terrain* In surveys of small sites, this is probably the deciding factor. A close vertical interval is required to portray small undulations on relatively flat ground. Where the terrain is steep, however, a wider interval would be chosen.

The methods of contouring can be divided into (a) direct and (b) indirect methods. Before studying the direct method, it is necessary to understand how a point is physically set out, on the ground, at a predetermined height.

(b) Setting out a point of known level

The basic principle of setting out a point on the ground at a predetermined level is illustrated in Fig. 5.6. Point A is a temporary bench mark (RL 8.55 m AD). Markers, in the form of pegs or arrows, are to be placed in the ground, in positions where the ground has, firstly, a level of 9.00 metres and, secondly, a level of 8.00 metres.

Procedure

1. The observer sets up the level at a height convenient for observing the TBM (RL 8.55 m) and takes a backsight staff reading. The reading is 1.25 m.

$$\text{HPC} = 8.55 + 1.25$$
$$= 9.80 \text{ m AD}$$

2. The staffman walks to an area where it is estimated that the ground has an approximate level of 9.00 m and turns the staff towards the instrument. The staff is then moved slowly up or down the ground slope until the base of the staff is at a height of 9.00 m exactly. This will occur when the observer reads 0.80 m on the staff, since

$$\text{HPC} = 9.80 \text{ m}$$
$$\text{Required level} = 9.00 \text{ m}$$
Therefore staff reading = 0.80 m (to achieve the required level)

3. The ground position (B), is marked with an arrow or peg and the staff is removed to another location. The operation is repeated and a second peg is placed at C, when the observer again reads 0.80 m on the staff. This operation can be repeated any number of times.

4. When a different ground level, say 8.00 m, is to be marked, a new calculation is made, as follows:

$$\text{HPC} = 9.80 \text{ m}$$
$$\text{Required level} = 8.00 \text{ m}$$
Therefore staff reading = 1.80 m (to achieve the required level)

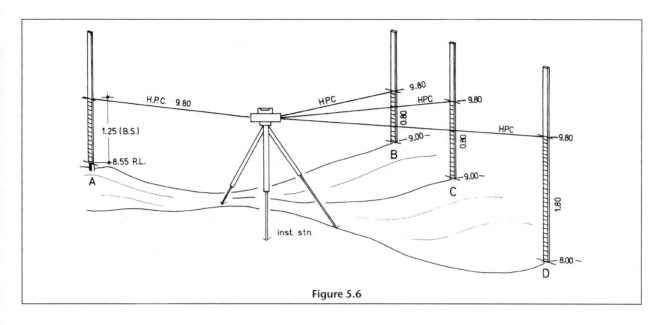

Figure 5.6

5. The observer directs the staffman downhill and directs him or her to move the staff until the reading becomes 1.80 m, whereupon the staffman marks the position (D) with a peg of a different colour.

(c) Direct method of contouring

Using this method the contour lines are physically followed on the ground. The work is really the reverse of ordinary levelling, for whereas, by the latter operation, the levels of known positions are found, in contouring it is necessary to establish the positions of known levels, using the techniques described in Sec. 2(b).

Figure 5.7 shows the plan of the building plot of Fig. 3.1. Contour lines are required over the site at vertical intervals of 1 metre. Temporary bench marks have been established on the survey points A to F by flying levelling.

In order to find the contour positions on the ground, two distinct surveying operations are necessary.

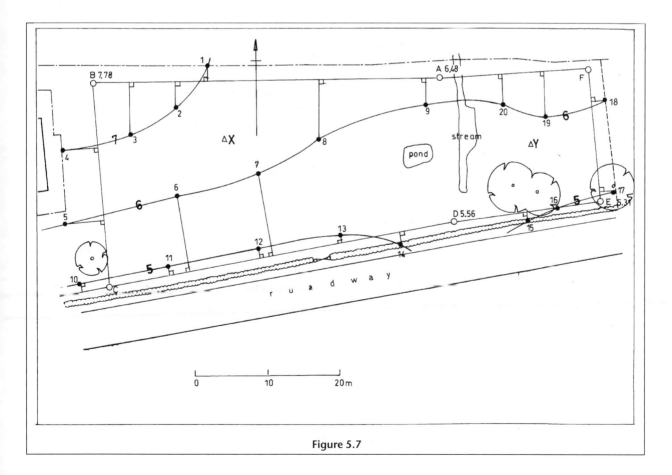

Figure 5.7

Levelling

A level is set on site at some position from which a bench mark can be observed comfortably. In Fig. 5.7, the level is set at position X and a backsight of 0.92 m observed on the BM (peg B, RL 7.78 m). The height of collimation = 7.78 + 0.92 = 8.70 m. From this instrument position, the contours at 7.00 m and 6.00 m can be observed. The staffman holds the staff facing the instrument and backs slowly downhill. When the observer reads 1.700 m, the bottom of the staff is at 7.000 m, since

$$\text{Height of collimation} = 8.70 \text{ m}$$
$$7 \text{ metre contour} = 7.00 \text{ m}$$
$$\text{Staff reading} = 1.70 \text{ m}$$

The staffman marks the staff position by knocking a peg into the ground. He or she then proceeds roughly along a level course, stopping at frequent intervals, while being directed to move the staff slowly up or downhill until the 1.700 m reading is observed from the instrument. A peg is inserted at each correct staff position. In Fig. 5.7, pegs 1 to 4 have been established on the 7.00 m contour. The 6.00 m contour is similarly established at pegs 5 to 9 using a staff reading of 2.70 m.

In order to set out the 5.00 m contour, a staff reading of 3.70 m is required. The contour line is denoted by pegs 10 to 14. Table 5.1 shows the appropriate booking. In order to check the levelling, a FS reading of 3.14 m has been taken to TBM peg D.

The eastern end of the site is too far removed from instrument station X to allow accurate contouring to take place. Consequently, a further instrument set-up is established at point Y and the levelling operation is continued from there.

Survey of the pegs

Figure 5.7 shows the various peg positions denoting contour lines 7.0, 6.0 and 5.0 m, on both sides of the small stream. The pegs are surveyed by offsets from the existing survey framework, all of which are clearly shown on the figure.

On a larger site, instrumental methods would be employed, e.g. electromagnetic radial positioning. The plan positions of the contours are plotted directly on to the site plan and smooth curves drawn through them.

EXAMPLE

2 The contour survey of the proposed building plot was completed on the east side of the stream by removing the instrument to set-up point Y. Table 5.2 shows the relevant readings.

Complete the table by calculating the staff readings required to set out the 6.0 and 5.0 m contour lines.

Answer
$$\text{HPC} = 6.48 + 1.05 = 7.53$$
$$\text{Therefore staff reading} = 7.53 - 6.00$$
$$= 1.53 \text{ (for 6 m contour)}$$
$$\text{and} = 7.53 - 5.00$$
$$= 2.53 \text{ (for 5 m contour)}$$
$$\text{RLE} = 7.53 - 2.23$$
$$= 5.30 \text{ (which checks to 1 cm)}$$

The 5 and 6 m contour lines are set out on the ground and marked by pegs 15 to 20 (Fig. 5.7).

Table 5.2

BS	IS	FS	HPC	Reduced level	Remarks
1.05				6.48	TBM A
	?			6.00	6 m contour line
	?			5.00	5 m contour line
		2.23		?	TBM E (RL 5.31 m)

(d) Indirect method of contouring

When using this method, no attempt is made to follow the contour lines. Instead a series of spot levels is taken at readily identifiable locations, e.g. at trees, gateposts, manholes and at intersections of walls and fences. Contour positions are then interpolated between them.

On open areas, where there are no easily identifiable features, a grid of squares or rectangles is set out on the ground and spot levels are taken to each point of the grid.

In theory, when contouring by this method, spot levels are required only at the tops and bottoms of all

Table 5.1

BS	IS	FS	HPC	Reduced level	Distance	Remarks
0.920			8.700	7.780		TBM peg B
	1.700		8.700	7.000		7 m contour line
	2.700		8.700	6.000		6 m contour line
	3.700		8.700	5.000		5 m contour line
		3.140	8.700	5.560		TBM peg D (RL 5.56 m)

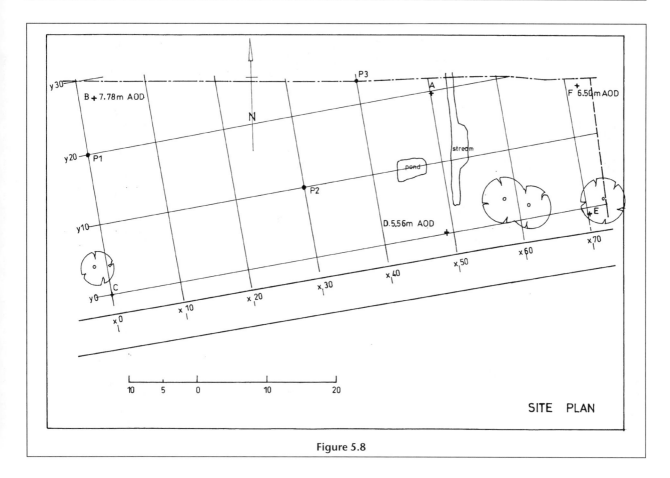

Figure 5.8

slopes. The ground is then treated as a plane surface between them. In practice, many more levels are taken. Mistakes are usually easy to identify.

Three distinct operations are involved in the indirect method of contouring.

Setting out a grid
1. On any site, a long side is chosen as the baseline and ranging rods or small wooden pegs are set out along the line at regular intervals. The interval depends on the contour vertical interval; the factors affecting this interval have been previously discussed. Briefly, the baseline intervals should be 5 or 10 metres where the vertical interval is to be close (say 0.5 or 1 m) and should be a maximum of 20 metres in all other cases.
2. At 30 m intervals, or more closely if deemed necessary, lines are set out at right angles to the baseline, using the horizontal circle of the level or by tape, using the theorem of Pythagoras. Further pegs are inserted at 10 m intervals along these latter lines. The other pegs on the grid are established by simple taped measurements.

The numerous pegs must be easily identified at a later stage so each is given a grid reference. There are many reference systems in common use, each with its own advantages and disadvantages. Though slightly cumbersome, an excellent system is formed by giving all distances along and parallel to the baseline an x chainage and all distances at right angles to the baseline a y chainage.

In Fig. 5.8, line CD is chosen as baseline and extended eastwards. Pegs are then established at 10 m intervals along this line and at right angles to it. Peg P1 therefore has a grid reference of ($x0$, $y20$) while the reference of peg P2 is ($x30$, $y10$).

The grid values of any point on the boundary can be obtained by measuring along the appropriate line to the boundary. The point P3 would have the identification of ($x40$, $y23$). This is the main advantage of the identification system.

Levelling
A levelling survey is made over the site to every point on the grid and to every subsidiary point. The reduced levels are then calculated and added to the plan as in Fig. 5.9.

Interpolating the contours
Figure 5.10(a) shows a part of the grid, with its appropriate reduced levels. Contour lines may be interpolated on the grid either mathematically or graphically.

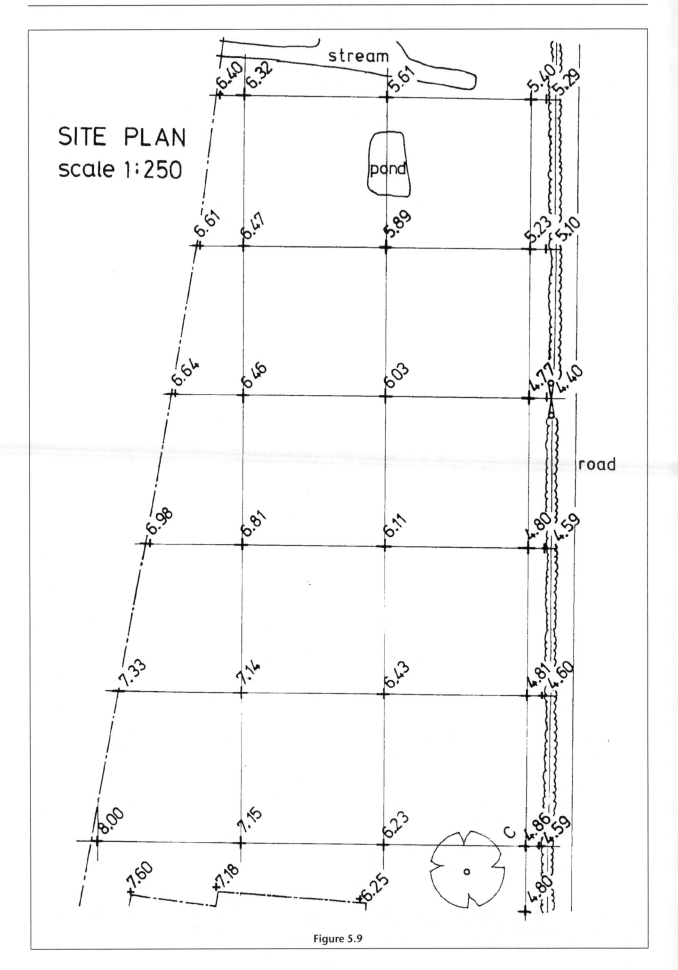

Figure 5.9

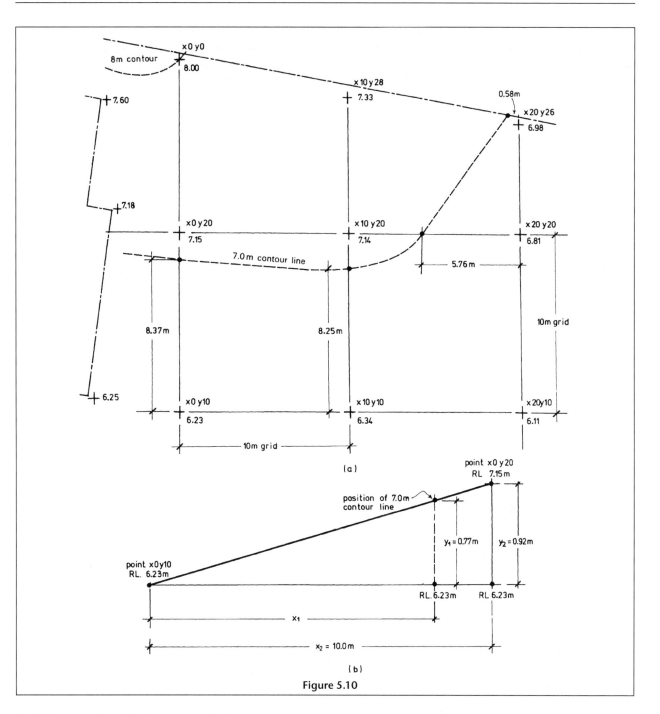

Figure 5.10

1. *Mathematically* The positions of the contours are interpolated mathematically from the reduced levels by simple proportion; e.g. the 7 m contour passes somewhere between the point $(x0, y10)$ and $(x0, y20)$ (Fig. 5.10(a)).

Referring to Fig. 5.10(b), the plan position of the contour line is calculated as follows:

Horizontal distance between $(x0, y10)$
and $(x0, y20) = x_2 = 10$ m

Difference in level between $(x0, y10)$
and $(x0, y20) = y_2 = (7.15 - 6.23) = 0.92$ m

Difference in level between $(x0, y10)$ and the
7.0 m contour line $= y_1 = (7.0 - 6.23) = 0.77$ m

Using the geometric principles of similar triangles,

$$\frac{x_1}{x_2} = \frac{y_1}{y_2}$$

Therefore $x_1 = \dfrac{y_1 \times x_2}{y_2}$

$$= \frac{0.77 \times 10}{0.92}$$

$$= 8.37 \text{ m}$$

In other words, the horizontal distance from point $(x0, y10)$ to the 7.0 m contour line = 8.37 m.

When a great many points are to be interpolated, a hand calculator is used, particularly when levels are taken to two decimal places. Use is made of a calculation table (Table 5.3). The interpolation of the 7 m contour, shown above, appears in the first line of the table while the remainder of the table is devoted to the calculations necessary for plotting the 7 m contour line completely.

The contour positions are plotted on the plan (Fig. 5.10(a)) and joined by a smooth curve. Similar calculations are carried out for other contour lines, which are then added to the plan.

Table 5.3

1	2	3	4	5	6	7
Lower grid point 1	Higher grid point 2	Contour line value	Horizontal distance from point 1 to 2	Difference in level (high point 2 – low point 1) (column 2 – column 1)	Difference in level (contour – low point 1) (column 3 – column 1)	Horizontal distance from low point to contour line $\left(\dfrac{column\ 6}{column\ 5}\right) \times column\ 4$
x0, y10	x0, y20	7.00	10.00	0.92	0.77	8.37
x10, y10	x10, y20	7.00	10.00	0.80	0.66	8.25
x20, y20	x10, y20	7.00	10.00	0.33	0.19	5.76
x20, y26	x10, y28	7.00	10.20	0.35	0.02	0.57

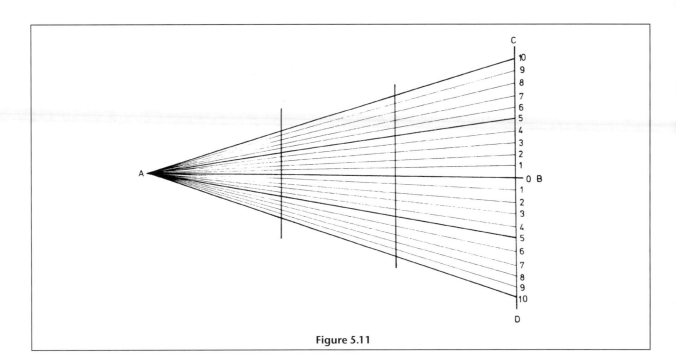

Figure 5.11

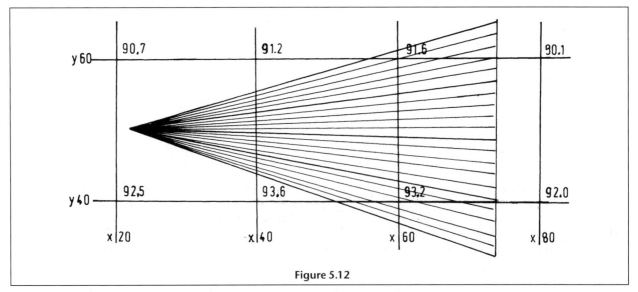

Figure 5.12

2. *Graphically* An alternative to the somewhat tedious mathematical method is a graphical interpolator. The graph is drawn on tracing paper and is constructed as in Fig. 5.11. Two mutually perpendicular lines AB and CD are drawn. The shorter line CD is divided into twenty equal parts and radial lines are drawn from A to each of the divisions.

In Fig. 5.12, the graph is shown in use. The 92 m contour position is required between the stations ($x60$, $y40$) and ($x60$, $y60$) where the reduced levels are 93.2 and 91.6 m respectively. The difference in level between the points is therefore 1.6 m. The overlay is laid on the grid with AB and CD parallel to the grid lines, until 16 divisions are intercepted between the two grid stations.

Each division represents 0.1 m in this case and the 92 m contour will therefore be four divisions from station ($x60$, $y60$). A pin-hole is made through the overlay at this point and the mark denoted by a small circle on the site plan. It should be noted that the overlay can be manoeuvred until any convenient number of divisions is intercepted by the grid stations. If, for example, eight divisions were chosen, each would represent 0.2 metre.

Since the overlay is not drawn to any scale, it can be kept and used for any contour interpolation. It is probably the fastest method of interpolating and it is therefore worth while spending a few minutes' time in preparing a really good overlay.

Exercise 5.2

1 Figure 5.9 shows the network of spot levels on the western part of the building site plan, used throughout the book, plotted to a scale of 1:500. Using either the mathematical or graphical method, interpolate on the plan the positions of the 7, 6 and 5 m contour lines.

3. Uses of contour plans

The following are the most relevant uses of contour lines in the construction industry.

(a) Vertical sections

A vertical section is really the profile of the ground that would be obtained by cutting the ground surface along any chosen line. Figure 5.13 shows the $x0$ line of the contoured plan of Fig 5.9. A vertical section is to be

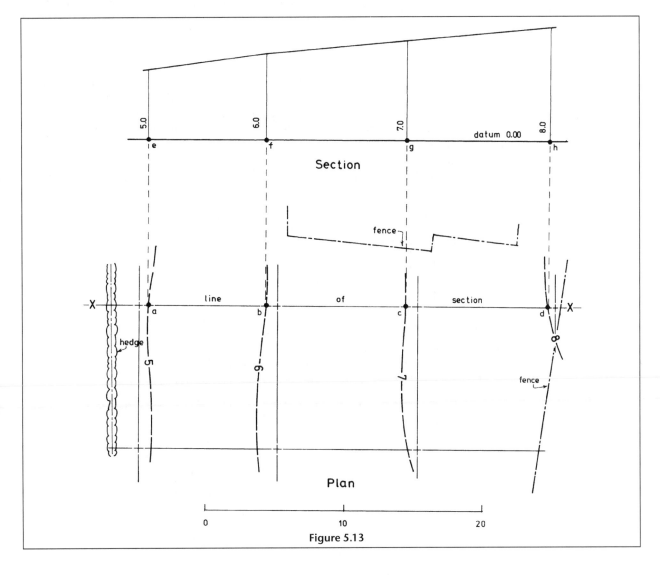

Figure 5.13

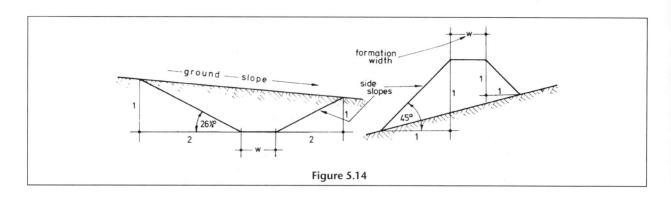

Figure 5.14

drawn along this line. (Fig 5.9 is the subject of Exercise 5.2, which the student should have finished. If not, the solution is given in the Answers (Fig. 5.25).)

1. A datum line is drawn parallel to the line of the section, clear of the plan, and a height is selected for the datum. In this case, the chosen value is zero (0) metres.
2. The contour positions along the section line (points a, b, c, d) are projected at right angles, to reach the datum line (points e, f, g, h).
3. The contour values of 5, 6, 7 and 8 m are scaled at right angles to the datum, to form the ground surface points.
4. The ground surface points are joined to form the ground profile.

(b) Intersection of surfaces

Whenever earth is deposited it will adopt a natural angle of repose. The angle will vary according to whether the material is clayey or sandy, wet or dry, but will almost certainly be between 45° and 26½°; i.e. the

material will form side slopes of 1 in 1 to 1 in 2. In Fig. 5.14 the cutting has side slopes that batter at 1 in 2 while the side slopes of the embankment batter at 1 in 1.

Basic principles
New construction work is often shown on plans by a system of contour lines. The portion of the embankment shown in Fig. 5.15 has a formation width of 10 metres, is 10 metres high and is to accommodate a level roadway 100 metres long. The sides of the embankment slope at 1 unit vertically to 1½ units horizontally and the embankment is being constructed on absolutely flat ground.

Contour lines at 1 metre vertical intervals are drawn parallel to the roadway at equivalent 1½ metre horizontal intervals. Since the embankment is formed on level ground the outer limits, denoting the bottom of the embankment, will be perfectly straight lines parallel to the top of the embankment. These lines will be 15 metres on either side of the formation width.

Very seldom will the ground surface be perfectly

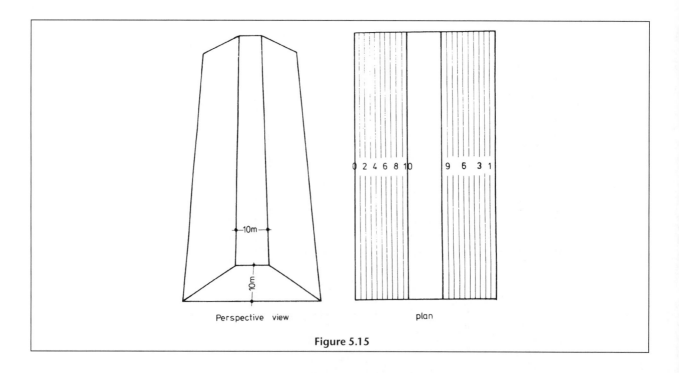

Perspective view   plan

Figure 5.15

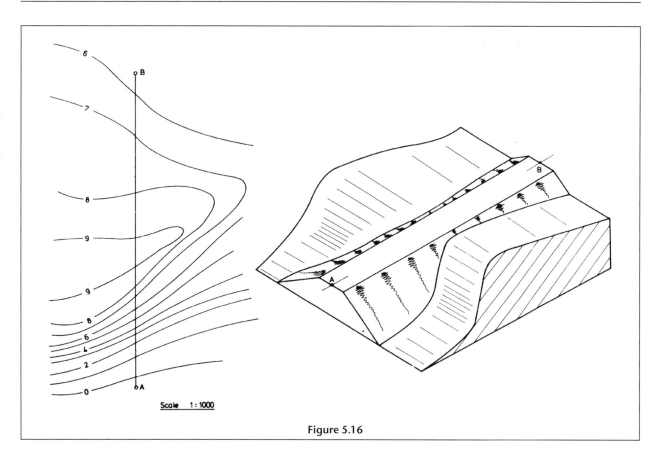

Figure 5.16

level. Figure 5.16 shows a series of contour lines portraying an escarpment, the southern scarp slope of which rises steeply from 0 to 9 m, while the dip slope slopes gently northwards from 9 to 6 m. The embankment already described is to be laid along the line AB. The perspective view in Fig. 5.16 illustrates the situation.

Approximately halfway along the embankment, the ground surface is at a height of 9 m, i.e. 1 m below the formation level of the embankment. It follows therefore that the width of the embankment will be very much narrower than at the commencement of the embankment, i.e. at point A. The plan position of the outer limits will no longer be a straight line. The actual position is determined by superimposing the embankment contours on the surface contours.

The intersection of similar values forms a point on the tail of the embankment (Fig. 5.17). When all of the intersection points are joined a plan position of the outer limits is obtained.

When cuttings are to be constructed, the procedure for drawing the outline is identical to that described above. However, when the contours along the cutting slopes are drawn, values increase with distance from the formation, whereas the contour values decrease on embankments.

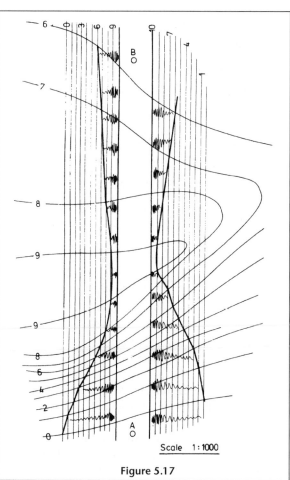

Figure 5.17

EXAMPLE

3 Figure 5.18 shows the position of a proposed factory building, together with ground contours at 1 metre vertical intervals. The formation level over the whole site is to be 0.5 m below the finished level, i.e. 23.000 m AOD, and any cuttings or embankments are to have side slopes of 1 unit vertically to 1 unit horizontally.

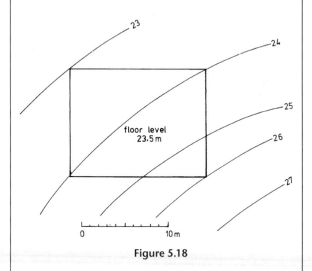

Figure 5.18

Draw on the plan the outline of the required earthworks.

Answer (Fig. 5.19)

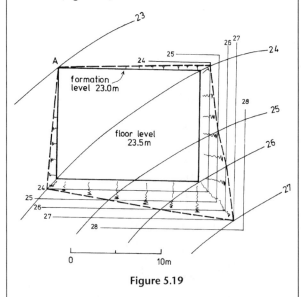

Figure 5.19

(a) There is no excavation or filling at point A, since the ground level and formation level are the same.
(b) All earthworks southwards are in cutting. Contours are drawn parallel to the building at 1 metre horizontal intervals and numbered from 24 to 28 m.

(c) The intersections of ground contours and earthwork contours of similar value form the edge of the embankments or cuttings.

Sloping earthworks
When earthworks are to be formed to some specific gradient, the drawing of the contours is not as straightforward as with level works. Figure 5.20 shows an embankment being formed on flat ground, the level of which is 67.000 m. The gradient of the embankment is 1 in 20 rising from a formation level of 70.000 m at A towards B. The length of the embankment is 60 m and the sides slope at 1 in 2.

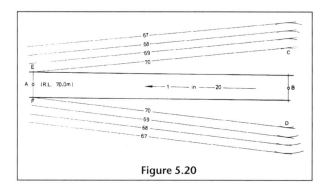

Figure 5.20

The embankment contours are formed by joining two points of the same level on the embankment sides.

The reduced level of point A is 70.000 m; therefore the reduced level of point B is

70.000 + (60/20) m = 73.000 m

A point of 70.000 m level will therefore be 3 metres vertically below point B. The horizontal equivalent of 3 m vertical at a gradient of 1 in 2 is 6 m. Points C and D are drawn 6 m from the formation edge and joined to points E and F respectively to form the 70 m embankment contours.

All other contours at vertical intervals of 1 metre are parallel to these two lines, and are spaced at 2 metre horizontal intervals.

EXAMPLE

4 Figure 5.21 shows ground contours at vertical intervals of 1 metre. AB is the centre line of a proposed roadway, which is to be constructed to the following specification:

(i) Formation width, 5 metres
(ii) Formation level A, 60.00 m AD
(iii) Gradient AB, 1 in 25 rising
(iv) Side slopes, 1 in 3

(a) Show clearly the outline of any earthworks formed between A and B.

(b) Draw a cross section of the embankment on the line indicated.

Answer (Fig. 5.22)

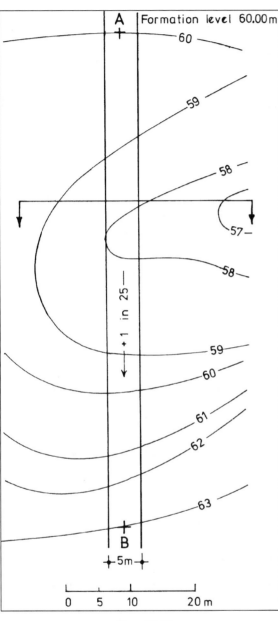

Figure 5.21

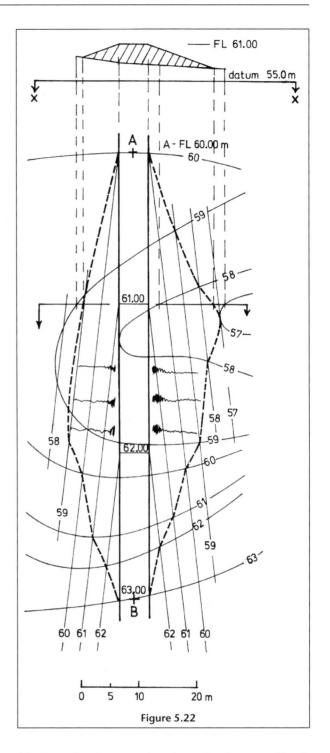

Figure 5.22

Exercise 5.3

1 Figure 5.23 shows ground contours at 1 metre vertical intervals. ABC is the line of a proposed level roadway having the following specification:

(i)   Formation level, 69.0 m
(ii)   Formation width, 5.0 m
(iii)   Side slopes, 1 in 1 (45°)
(a)   Show on the plan the outline of the earthworks required to form the roadway.
(b)   Draw natural cross-sections at A, B and C to show the roadway embankment and the ground surface.

(c)   From the plan or section, determine the ground levels at the base of the side slopes and on the centre line of the embankment across the sections at A, B and C.

Note: This exercise will be used as an example in the 'Volumes' section of this book (Chapter 15).

2 Figure 5.24 shows the positions, and Table 5.4 shows the reduced levels, of twelve points on an area of land that is to be developed as part of a new road scheme. The new road is to be constructed on a centre line XYZ to the following specification:

(i)   Formation level, 20.00 m
(ii)   Road width, 5.00 m

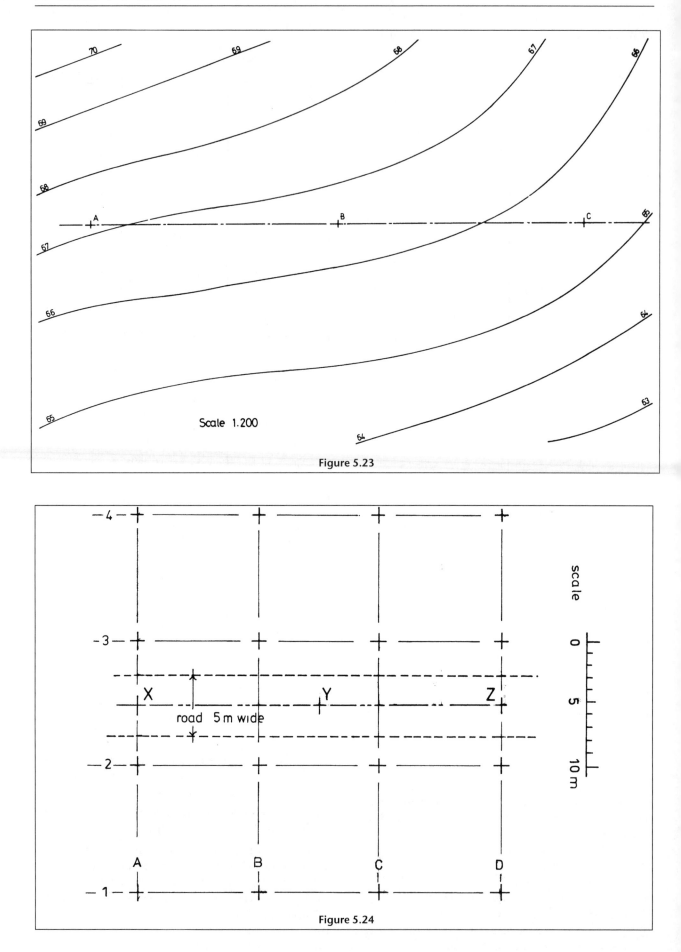

Scale 1.200

Figure 5.23

Figure 5.24

Table 5.4

Point	Red.Lev.	Point	Red.lev.
A4	27.370	A3	25.740
B4	26.750	B3	24.650
C4	25.490	C3	23.700
D4	24.040	D3	22.830

Point	Red.Lev.	Point	Red.lev.
A2	23.550	A1	21.010
B2	23.000	B1	21.010
C2	22.240	C1	21.000
D2	21.280	D1	19.530

(iii)   Road gradient, level
(iv)   Side slopes, 1 in 1 (45°)
(a) Print the levels on the plan, in their appropriate locations.
(b) Interpolate the ground contours, at 1 metre vertical intervals.

(c) Show on the plan the outline of the cutting required to accommodate the roadway.
(d) Draw natural cross-sections, at points X, Y and Z, to show the roadway cutting and the ground surface.
(e) From either the plan or sections, determine the reduced levels at the tops of the cutting side slopes and on the centre line of the cutting, across each of the three sections at X, Y and Z.

Note: This exercise will be used as an example in the 'Volumes' section of this book (Chapter 14).

4. Answers

Exercise 5.1

1 (a) Gradient AB = 1 in 25
Gradient CD = 1 in 20
Gradient EF = 1 in 21
(b) Height of wall = 3 m

Exercise 5.2

1. Figure 5.25

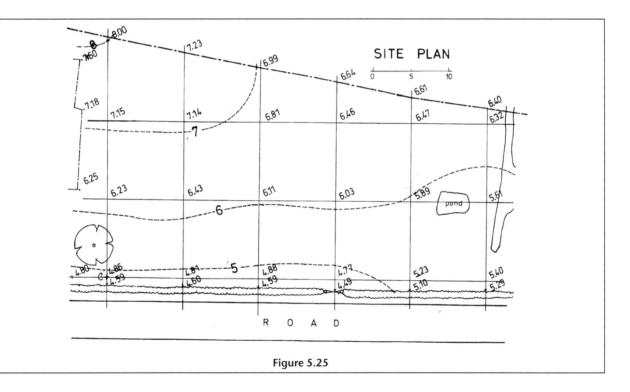

Figure 5.25

Exercise 5.3

1 (a) Figure 5.26

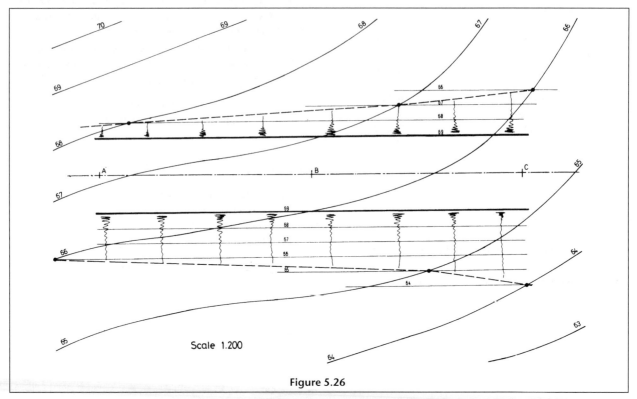

Figure 5.26

(b) (c) Figure 5.27

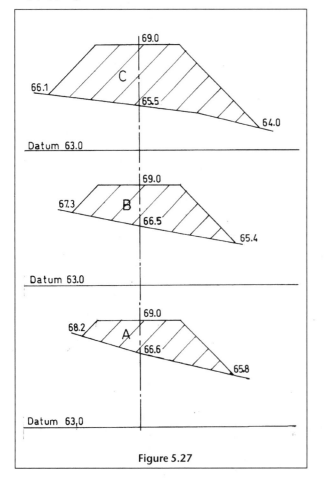

Figure 5.27

2 Figure 5.28

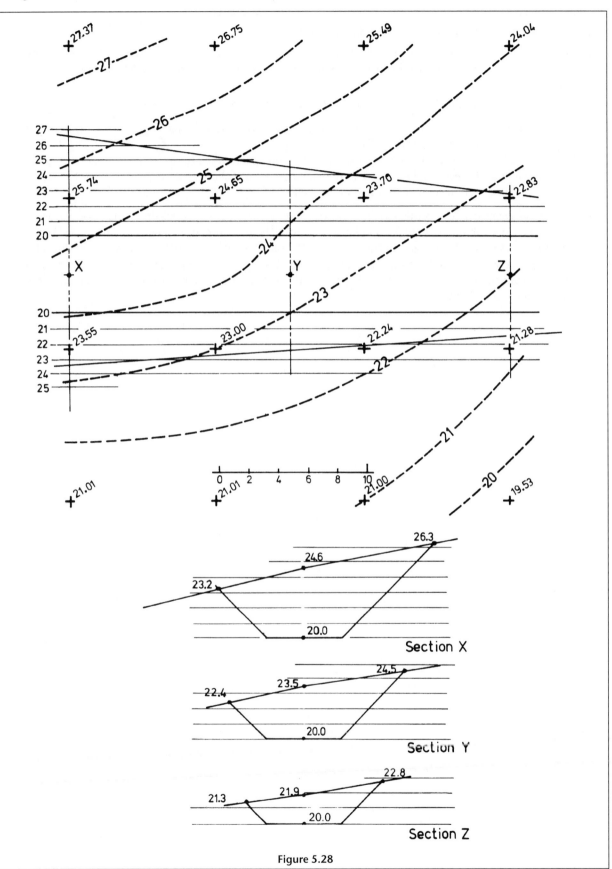

Figure 5.28

Chapter summary

In this chapter, the following are the most important points:

- A contour line is a line drawn on a map or plan to indicate the height of the ground all along the line. Successive lines close to one another indicate steep terrain and conversely wider spacing between lines indicates less steep areas.

- Two methods are common in construction circles for finding the ground positions of contour lines.

  1. EDM equipment will record, compute and store the heights of a large number of ground points from which computer generated contour lines will be drawn on a plan.
  2. The contour positions can be found by using a levelling procedure whereby points are set out on the ground in a grid pattern. The levels of the ground at those points are calculated and plotted on a plan and, from the levels, existing ground contour lines are interpolated.

- The positions of proposed construction site earthworks are shown on the site plan for the purposes of calculating areas and volumes of rock and sub-soil and for correct positioning of other nearby construction works. The contour lines of the earthworks are superimposed on the plan contour lines and the intersections of similarly valued existing and proposed contours form the outline of the proposed construction works. The principles are not easy to understand in theory, so as reinforcement, the text includes a variety of progressive examples and exercises on pages 90 to 97.

CHAPTER 6 Vertical sections

In this chapter you will learn about:

- the nature of longitudinal and cross sections
- the fieldwork involved in obtaining data to plot the longitudinal and cross sections of a variety of site earthworks
- the methods of plotting sections and adding construction details to the plot
- the method of generating sections using computer software packages

Longitudinal and cross sections can be produced from a variety of surveyed data. The most cost-effective and probably most accurate method of collecting the data is automatically by total station instruments, from which the data is post processed and contoured plans produced digitally by computer. The database can then be interrogated to produce longitudinal and cross sections from which areas and volumes of earthworks are readily computed. This method will be covered in Chapter 10.

The main aim of this textbook, however, is to provide the aspiring young surveyor with a background of basic knowledge which, with experience, will be supplemented but never superseded by modern methods.

This chapter therefore concentrates on the basic methods of producing longitudinal and cross sections from levelling data.

Whenever narrow works of long length, e.g. roads, sewers, drains, etc., are to be constructed, it is necessary to draw vertical sections showing clearly the profile of the ground. Two kinds are necessary:

(a) longitudinal sections, i.e. a vertical section along the centre line of the complete length of the works,

(b) cross-sections, i.e. vertical sections drawn at right angles to the centre line of the works.

The information provided by the sections provides data for:

(a) determining suitable gradients for the construction works,
(b) calculating the volume of the earthworks,
(c) supplying details of depth of cutting or height of filling required.

1. Development plan

Figure 6.1 is the contoured plan of the proposed building site shown in Fig. 3.1. Contours have been added from Fig. 5.25 to produce this development plan. The plan shows the positions of two houses; the centre line of a service roadway and the lines of two sewers which will serve the houses.

Roadway specification

Width	4.00 m
Gradient	1 in 50 rising from R1 to R4.
Side slopes	1 in 1 (45°)

Drainage specification

Storm water	Manhole S1 to S4 rising at 1 in 40
	Invert level manhole S1 – 2.197 m
Foul water	Manhole F1 to F4 rising at 1 in 40
	Invert level manhole F1 – 2.950 m

In dealing with drains and sewers, reference is made to the invert level rather than the formation level. The invert level is the inside of the bottom of the pipe, but for practical purposes it may be considered to be the bottom of the excavation.

2. Longitudinal sections

In construction work, the sewers and roadways are usually constructed first of all. They must be set out, on the ground, in their correct locations. Additional plans, in the form of vertical sections, are required.

In order to cost the development accurately, the volume of material required to construct the earthworks (cuttings or embankments) has to be calculated, generally by the quantity surveyor.

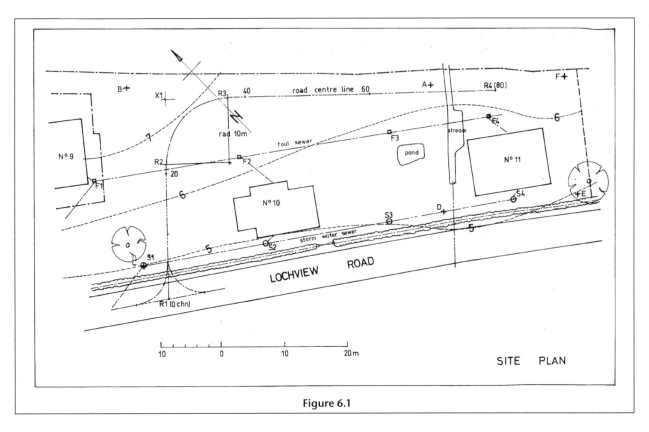

Figure 6.1

The first step in calculating the earthworks quantities and in preparing the setting-out information is to make longitudinal sections and cross sections along the lines of the proposed works.

(a) Fieldwork

1. The centre lines of the road and sewers are set out, on the ground, by a series of stakes. The setting out of the centre lines is usually done using a theodolite. (Setting out is the subject of Chapter 12. It is sufficient for the moment to assume that the stakes have been set out in their correct positions.)
2. A levelling is made along the centre line with levels taken at all changes of gradient. A level is also taken at every tape length whether or not it signifies a change in gradient.
3. Horizontal measurements are made between all the points at which levels were taken. The measurements are accumulated from the first point such that all points have a running chainage. Pegs are left at every tape length to enable cross sections to be taken later (Fig. 6.2).

Procedure Generally a surveyor and three assistants are required if the section is long. The surveyor takes the readings and does the booking; one assistant acts as staffman, while the other two act as chainmen taking all measurements, lining-in ranging poles along the previously established centre line and leaving pegs at all tape lengths.

A flying levelling is conducted from some nearby bench mark to the peg denoting zero chainage. Thereafter the levelling is in series form with intermediate sights taken as necessary.

The tape is held at peg zero chainage and stretched out along the line of the section. The chainmen and staffman work together and, while the latter holds the staff as a backsight at peg zero, the chainman marks the changes in gradient and calls out the chainages of these points to the observer. The staffman follows up and holds the staff at all of the changes of gradient. When one tape length has been completed, a peg is left and the next length is observed. The procedure is repeated until the complete section has been levelled.

As in all surveying work, a check must be provided. In sectioning this can be done by flying levelling from

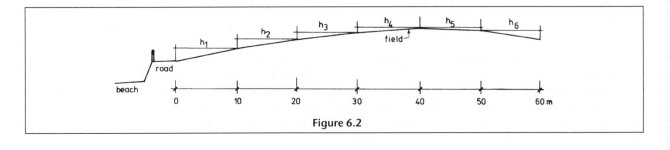

Figure 6.2

Table 6.1

BS	IS	FS	Rise	Fall	Reduced level	Chainage (m)	Remarks
1.350					5.200		OSBM (5.200)
	2.150			0.800	4.400	0	R1 centre line of road
	1.130		1.020		5.420	10	Centre line of road
1.650		0.285	0.845		6.265	20	Centre line of road
	1.020		0.630		6.895	30	Centre line of road
	1.735			0.715	6.180	40	Centre line of road
	1.605		0.130		6.310	50	Centre line of road
	1.765			0.160	6.150	60	Centre line of road
	1.465		0.300		6.450	70	Centre line of road
	1.515			0.050	6.400	80	R4 Centre line of road
		0.125	1.390		7.790		Peg B (7.780)
3.000		0.410	4.315	1.725	7.790		
−0.410			−1.725		−5.200		
2.590			2.590		2.590		

the last point of the section to the commencing bench mark, or to some other, closer, bench mark.

(b) Plotting longitudinal sections

1. The fieldwork results are computed and all checks applied. Table 6.1 shows the series of levels taken along the centre line of the proposed roadway of Fig. 6.1 from point R1 (chainage 0 m) to point R4 (chainage 80 m). The levels have been fully calculated, ready for plotting.

It will be noticed that the levelling does not close exactly on to bench mark B, the discrepancy being 0.010 m. This closure error is acceptable and the reduced levels are considered to be satisfactory.

2. The scales are chosen for the section drawing, such that the horizontal scale is the same as the scale of the plan view of the site. (*In this book, however, the scale has been chosen to fit the width of the page.*) Compared to the length of the section, the differences in elevation of the section points will always be comparatively small: consequently the vertical scale of the section is exaggerated to enable the differences in elevation to be readily seen. The horizontal scale is usually enlarged two to ten times, producing the following scales in Fig. 6.3:

Horizontal scale   1:500
Vertical scale   1:200

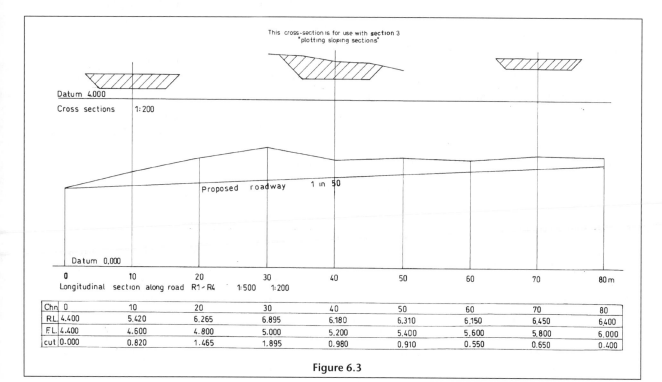

Figure 6.3

3. The reduced levels are scrutinized and the lowest point is found. A horizontal line is drawn to represent a datum, some way below the lowest point of the section. In this case (Fig. 6.3), Ordnance Datum has been chosen, because the levels are all less than 10 m. In other circumstances, a multiple of 10 m would be selected as datum for the section.

4. The horizontal chainages are carefully measured along the datum line and perpendiculars are erected at each chainage point.

5. The reduced level of each station is carefully scaled off along the perpendicular. Each station so established is then joined to the next by a straight line to produce an exaggerated profile of the ground along the section. The lines joining the stations must *not* be drawn as curves since the levels have been taken at all changes of slope and the gradient is therefore constant between any two points.

6. The proposed works are added to the drawing. The formation level of any construction works is the level to which the earth is excavated or deposited to accommodate the works. In Fig. 6.3 the formation level of the new roadway at R1 is 4.400 m. The roadway is to rise at a gradient of 1 in 50 from R1 at chainage 0 m to R4 at chainage 80 m. Since the profile is exaggerated true gradients cannot be shown in the section.

Formation level R1 is plotted. The formation level of a second point, R4 (chainage 80 m), is calculated and also plotted:

Rise from R1 to R4 = 1/50th of 80 m = 1.600 m

Formation. level R4 = FL.R1 + 1.600
$$= 4.400 + 1.600$$
$$= 6.000 \text{ m}$$

When joined, the line between the points represents the gradient. The depth of the excavation required to accommodate the roadway is scaled from the section, the depths being the distance between the surface and the line representing the roadway formation level. Since all scaling and calculations are performed on this section, it is called the working drawing.

7. The presentation drawing is then compiled by tracing the working drawing on to a plastic sheet or piece of tracing paper in black ink only. From this tracing any number of prints can be obtained.

(c) Calculation of cut and fill

In addition to being scaled from the section, the depths of cut or height to fill are also calculated. The method of calculation is similar for all vertical sections. Generally, it consists of:

1. Calculating the reduced level at each chainage point.
2. Calculating the proposed level of the new works at each chainage point.

3. Subtracting one from the other. Where the surface level is higher than the proposed level, there must be cutting and where the proposed level exceeds the surface level, filling will be required.

The complete calculation of cut/fill along the centre line of the proposed roadway R1 to R4 is as follows:

1. Calculate the surface levels along the centre line of the roadway. These levels are shown in Table 6.1.
2. Calculate the formation levels of the proposed roadway, given that the road is to rise from point R1 to point R4 at a gradient of 1 in 50.

Roadway gradient R1–R4	= 1 in 50 rising
Rise over any length	= 1/50 × length
Therefore rise over 10 m	= 1/50 × 10 m
	= 0.200 m

The various formation levels are therefore:

FL 0m	= 4.400 m
FL 10 m = 4.400 + 0.200	= 4.600 m
FL 20 m = 4.600 + 0.200	= 4.800 m
FL 30 m = 4.800 + 0.200	= 5.000 m
FL 40 m = 5.000 + 0.200	= 5.200 m
FL 50 m = 5.200 + 0.200	= 5.400 m
FL 60 m = 5.400 + 0.200	= 5.600 m
FL 70 m = 5.600 + 0.200	= 5.800 m
FL 80 m = 5.800 + 0.200	= 6.000 m

3. Calculate the depth of cut or height of fill at each chainage point. In this case, the proposed level is less than the existing ground level at each point, therefore the roadway will be in cutting for the whole of its length.

Depth of cut	= (ground level – formation level)
Depth at 20m	= (6.265 – 4.800) m
	= 1.465 m

The depth of cut at each chainage point along the roadway is similarly computed. Table 6.2 shows the complete calculation.

All calculations are tabulated as part of the section drawing (Fig. 6.3), or as a separate table (Table 6.2).

Table 6.2

Chainage (m)	Reduced level	Formation level	Cut (+) Fill (−)
0.0	4.400	4.400	0.000
10.0	5.420	4.600	+0.820
20.0	6.265	4.800	+1.465
30.0	6.895	5.000	+1.895
40.0	6.180	5.200	+0.980
50.0	6.310	5.400	+0.910
60.0	6.150	5.600	+0.550
70.0	6.450	5.800	+0.650
80.0	6.400	6.000	+0.400

EXAMPLE

1 Table 6.3 shows the chainages and reduced levels obtained during a levelling survey along the line of sewer F1–F4 (Fig 6.1).

Draw a longitudinal section along the line of the sewer, showing clearly:
(i) the ground surface,
(ii) the proposed sewer rising from existing invert level 2.950 m at manhole F1 to manhole F4 at a gradient of 1 in 40.

Table 6.3

Reduced level	Chainage (m)	Remarks
7.780		Peg B (7.780m AOD)
6.660	0.0	Manhole F1 (existing)
6.590	8.0	Centre line of proposed sewer
6.750	16.0	Centre line of proposed sewer
6.410	24.0	Proposed manhole F2
6.270	32.0	Centre line of proposed sewer
6.200	40.0	Centre line of proposed sewer
5.870	48.0	Proposed manhole F3
5.800	56.0	Centre line of proposed sewer
6.020	64.0	Proposed manhole F4
5.550		Peg D (5.555m AOD)

Answer Figure 6.4

3. Cross sections

(a) Level cross sections

No definition of a level cross section actually exists but it is generally taken to mean that the top and bottom of the earthwork (cutting or embankment) are parallel, with the result that the plotted earthwork has the shape of a trapezium or rectangle.

Fieldwork
It is not necessary to observe levels in the field. The ground across the centre line at any chainage point is taken to be level or nearly so, in which case the centre line level is assumed to apply across the line of the section.

Plotting
The plotting is similar to the plotting of longitudinal sections. Generally, however, the cross section is plotted to a natural scale, i.e. the horizontal and vertical scales of the plot are the same.

The proposed roadway R1–R4 (Fig. 6.3) is to be 4 metres wide and any earthworks are to be formed with sides sloping at 45° (i.e. 1 unit vertically to 1 unit horizontally). The cross sections are drawn as follows:

1. A line, representing the section datum (in this case, 4.00 m AOD), is drawn above the longitudinal section.
2. The construction lines of each chainage point of the longitudinal section are extended vertically to cut

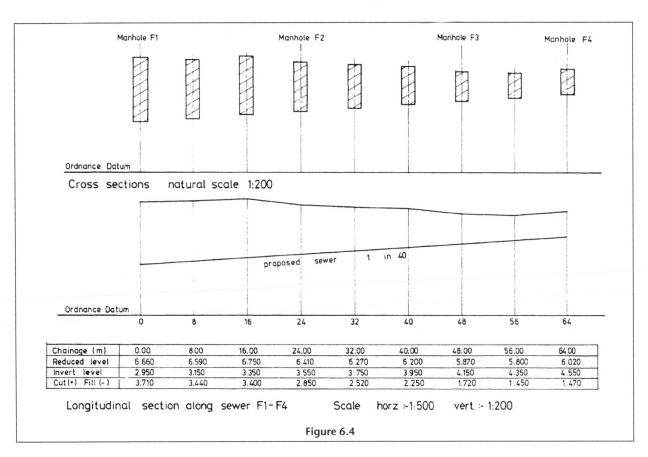

Chainage (m)	0.00	8.00	16.00	24.00	32.00	40.00	48.00	56.00	64.00
Reduced level	6.660	6.590	6.750	6.410	6.270	6.200	5.870	5.800	6.020
Invert level	2.950	3.150	3.350	3.550	3.750	3.950	4.150	4.350	4.550
Cut (+) Fill (−)	3.710	3.440	3.400	2.850	2.520	2.250	1.720	1.450	1.470

Longitudinal section along sewer F1–F4   Scale   horz :-1:500   vert. :- 1:200

Figure 6.4

this datum and act as centre lines for the cross sections.

3. The ground level and the formation level of the roadway are plotted from the datum on the centre lines and horizontal lines drawn lightly through both points.

4. The width of the road is marked on the line representing the formation level and the sloping sides of the cutting drawn at 45° to cut the line representing ground level.

5. The resulting trapezia representing the road cutting are drawn boldly (Fig. 6.3).

In example 1, (Fig. 6.4), the sewer is to be 0.8 m wide and the excavation is to have vertical sides. The sections are plotted as follows (Fig 6.4):

1. A line representing the section datum of 0.00 m is drawn lightly above the longitudinal section.

2. The construction lines of the chainage points of the longitudinal section are extended vertically to cut this datum and act as centre lines for the cross sections.

3. The ground level and the sewer invert level are plotted from the datum on the centre line and horizontal lines drawn lightly through both points.

4. The width of the sewer (0.80 m) is marked and the vertical sides of the drain drawn to cut the horizontals in (3) above.

5. The rectangle denoting the excavation is drawn boldly.

✎ Exercise 6.1

1 Table 6.4 shows the chainages and reduced levels obtained on a levelling survey along the line of the storm water sewer S1–S4 (Fig. 6.1).

Table 6.4

Chainage	Reduced Level	Remarks
0.000	4.900	Manhole S1
10.000	4.850	Centre line of sewer
20.000	4.900	Proposed manhole S2
30.000	4.775	Centre line of sewer
40.000	5.300	Proposed manhole S3
50.000	5.455	Centre line of sewer
60.000	5.230	Proposed manhole S4

(a) Draw a longitudinal section along the line S1–S4 showing clearly:
  (i) the ground surface and
  (ii) the proposed sewer, rising from the existing invert level (2.197 m AOD) of manhole S1, to manhole S3 at a gradient of 1 in 30 and then at such a gradient that the invert level of the pipe in manhole S4 is 1.200 m below the ground level at the manhole.

(b) Calculate the gradient of the pipe between manholes S3 and S4.

(c) Draw cross sections at every chainage point, assuming that the ground is level across the section and given that:
  (i) the width of the sewer track is 2.0 m,
  (ii) the trench has vertical sides.

(b) Sloping cross sections

In cases where the ground across the centre line at any chainage point is obviously not flat, the following fieldwork is required to obtain levels for plotting the cross sections.

Fieldwork

Cross sections are taken at right angles to the longitudinal section at every point observed on the latter. Generally, this rule is not strictly observed and cross sections are taken usually at every tape length. The following fieldwork is necessary:

1. Right angles are set out, using the horizontal circle of the level or simply by eye. A ranging rod is inserted at either side of the centre line of the cross section.

2. A levelling must be made from the peg previously established on the centre line of the longitudinal section, to every point where the gradient changes on the line of the cross section (Fig. 6.5). In addition, levels are always taken on either side of the centre line at points C_1 and C_2 denoting the formation width.

    Where the ground is relatively flat, one instrument setting is usually sufficient (Fig. 6.5(a)). If the cross gradient is steep, a short series levelling is required (Fig. 6.5(b)).

    Each cross section is independent of every other. The peg on the centre line at each tape length is a temporary bench mark for its particular cross section.

3. Horizontal measurements must be made between all points at which levels are taken to cover the total width of the proposed works.

4. *Procedure* The procedure is much the same as that for longitudinal sections. Since the distances involved are short, the surveyor and one assistant take the levels and measurements without further assistance.

Plotting

In Fig. 6.1, the site has a cross slope from north to south, therefore additional levels are taken at chainages 40 m, 50 m and 60 m to determine accurately the nature of the slope. Table 6.5 shows these levels fully reduced for plotting the cross sections.

The plotting is similar to the plotting of longitudinal sections. One essential difference, however, is that the

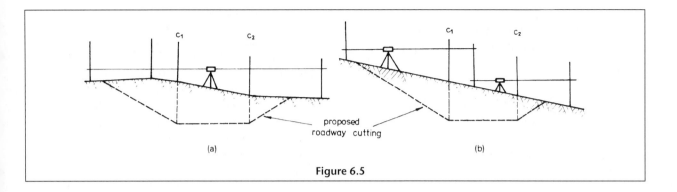

(a)   (b)

proposed
roadway cutting

Figure 6.5

Table 6.5

BS	IS	FS	HPC	Reduced level	Distance	Remarks
1.630		7.185		5.555		Peg D (5.555 m AOD)
1.230		1.005	7.410	6.180	40.000	Centre line of proposed road
	0.850			6.560		4 m left of centre line
	0.910			6.500		2 m left of centre line
	1.310			6.100		2 m right of centre line
	1.710			5.700		4 m right of centre line
	1.100			6.310	50.000	Centre line of proposed road
	0.890			6.520		4 m left of centre line
	1.110			6.300		2 m left of centre line
	1.390			6.020		2 m right of centre line
	1.600			5.810		4 m right of centre line
1.390		1.260	7.540	6.150	60.000	Centre line of proposed road
	1.025			6.515		4 m left of centre line
	0.960			6.580		2 m left of centre line
	1.490			6.050		2 m right of centre line
	1.590			5.950		4 m right of centre line
	2.020			5.520	70.000	Centre line of proposed road
		1.980		5.560		Peg D (5.555 m AOD)

sections are plotted to a natural scale, i.e. the horizontal and vertical scales are the same.

Figure 6.3 shows the longitudinal section of roadway R1–R4. The description of plotting a cross section in the next paragraphs will also utilize Fig. 6.3, as will the tutorial Exercise 6.2 which follows that description.

Procedure   Plotting a cross section is carried out in the following manner:

1. The levels are reduced (Table 6.5).
2. The longitudinal section is plotted (Fig. 6.3).
3. A line representing some chosen height above datum (in this case 4.000 m) is drawn for each cross section and the measurements to left and right of the centre line are scaled accurately (Fig. 6.3).
4. Perpendiculars are erected at each point so scaled and the reduced levels of each point plotted.
5. The points are joined to form a natural profile of the ground.
6. The formation level of the proposed works is obtained from the table accompanying the longi-

tudinal section. In the data table of Fig. 6.3, the formation level at chainage 40 m is 5.20 m. This level is accurately plotted on the cross section and a horizontal line, representing the roadway, is drawn through the point.

7. The finished width of the road, called the formation width, is marked on the line and the side slopes of the cutting are added. In Fig. 6.3 the slope is 45°. The finished shape of the cutting is then drawn boldly.

(c) Embankments

The earlier parts of this section dealt with the plotting of cross sections. All of the examples and exercises have concentrated on the plotting of cuttings, in order to preserve continuity of theme. However, earthworks equally involve the plotting of embankments and are exemplified in Examples 2 and 3.

When earthworks involve embankments, the formation level of the proposed works is higher than the actual ground level, so the side slopes of the works must slope downwards from the formation level to the ground level. In all other aspects, the plotting procedure is as already described in the earlier parts of this section.

EXAMPLES

2 The following reduced levels were obtained along the centre line of a proposed road between two points A and B.

Chainage (m)	Reduced level (m)	Chainage (m)	Reduced level (m)
A 0	83.50	50	82.45
10	83.84	60	82.20
20	84.06	70	82.41
30	83.66	80	82.70
40	83.30	B 90	83.05

The roadway is to be constructed so that there is one regular gradient between points A and B.
(a) Draw a longitudinal section along the centre line at a horizontal scale of 1:1000 and a vertical scale of 1:100.
(b) Determine from the section the depth of cutting or height of fill required at each chainage point to form a new roadway.
(c) Check the answers by calculation.

Answer

(a) The longitudinal section is shown in Fig. 6.6.
(b) *Gradient of AB*

$$Reduced\ level\ of\ A = 83.50\ m$$
$$Reduced\ level\ of\ B = 83.05\ m$$
$$Difference = 0.45\ m$$
$$Distance\ A\ to\ B = 90\ m$$
$$Therefore\ gradient\ A\ to\ B\ falling = 0.45\ in\ 90$$
$$= 1\ in\ 200$$

Proposed levels (formation levels):

$$Fall\ over\ 10\ metres = 0.05\ m$$
Therefore proposed level at chainage
$$10\ m = 83.50 - 0.05 = 83.45\ m$$

Proposed level at chainage
$$20\ m = 83.45 - 0.05 = 83.40\ m\ etc.$$

Chainage (m)	Reduced level (m)	Proposed level (m)	Cut (m)	Fill (m)
A 0	83.50	83.50		
10	83.84	83.45	0.39	
20	84.06	83.40	0.66	
30	83.66	83.35	0.31	
40	83.30	83.30	—	—
50	82.45	83.25		0.80
60	82.20	83.20		1.00
70	82.41	83.15		0.74
80	82.70	83.10		0.40
90 B	83.05	83.05		—

3 Cross sections are required at chainages 20, 40 and 60 m in Example 2 above. Levellings were made, and the results tabulated in the field book are shown in Table 6.6.

Using a natural scale of 1:100, draw cross sections at 20, 40 and 60 m chainages, given that the formation width of the roadway is 5 m and that the sides of any cuttings or embankments slope at 30° to the horizontal.

Answer

The cross sections are shown in Fig. 6.7. The method of plotting is briefly as follows:
(i)   Draw arbitrary datum lines of 82.00 m.
(ii) Plot centre line, 5 m left and 5 m right along the datum lines.
(iii) Plot the reduced levels of these points and join them to produce the surface profiles.
(iv) Plot the formation levels at 20, 40 and 60 m. These are 83.40, 83.30 and 83.20 respectively from the calculations of Example 2.
(v)   Plot the formation widths, 2.5 m on both sides of the centre line.
(vi) Draw the side slopes at 30° to the horizontal. Where the reduced levels exceed the formation level, a cutting is produced as in Fig. 6.7(a). In Fig. 6.7(c) the formation level exceeds the

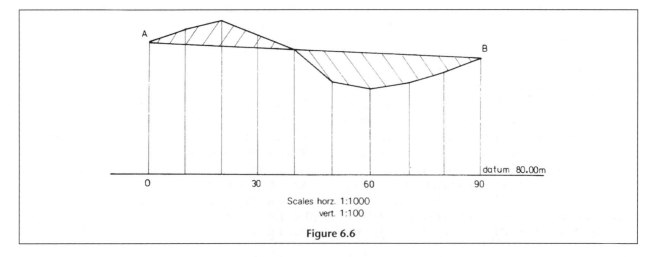

Scales horz. 1:1000
vert. 1:100

Figure 6.6

Table 6.6

BS	IS	FS	HP collimation	Reduced level	Distance	Remarks
1.52			85.58	84.06	20	Peg on centre line
	0.88			84.70	20	5 m left of centre line
		2.02		83.56	20	5 m right of centre line
1.61			84.91	83.30	40	Peg on centre line
	0.56			84.35	40	5 m left of centre line
		2.34		82.57	40	5 m right of centre line
1.47			83.67	82.20	60	Peg on centre line
	0.67			83.00	60	5 m left of centre line
		1.66		82.01	60	5 m right of centre line

reduced level producing an embankment, while in Fig. 6.7(b), the cross section is part cutting and part embankment.

Exercise 6.2

1 Table 6.5 (page 105) shows cross sectional levels taken at 40, 50, and 60 m chainages along the proposed roadway R1–R4.

(a) Plot cross sections 50 m and 60 m to a natural scale of 1:100. (Note – cross section 40 m is plotted on Fig. 6.3 as a guide.)
(b) Add the construction details to the drawing given that:

(i) road width = 4 m
(ii) side slopes are at 45°
(iii) FL 50 m = 5.400 m and FL 60 m = 5.600 m

2 Table 6.7 shows levels taken along the centre line of a proposed sewer. The sewer at chainage 0 m has an invert level of 88.900 and is to fall towards chainage 50 m at a gradient of 1 in 100.

(a) Reduce the levels and apply the appropriate checks.
(b) Draw a vertical section along the centre line of the sewer on a horizontal scale of 1:500 and vertical scale of 1:100.
(c) From the section determine the depth of cover at each chainage point.
(d) A mechanical excavator being used to form the sewer

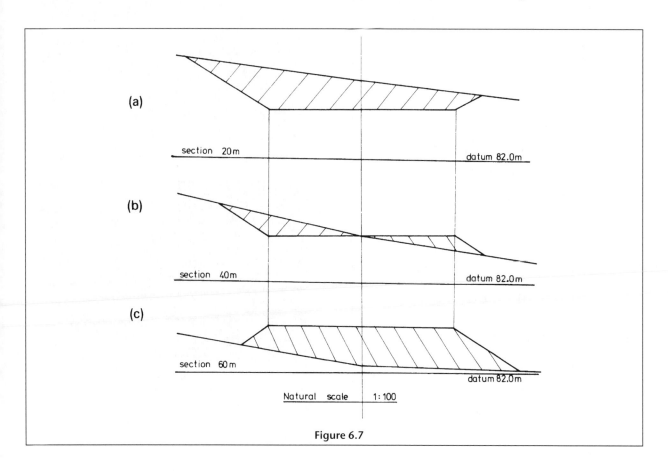

Figure 6.7

Table 6.7

BS	IS	FS	Rise	Fall	Reduced level	Distance	Remarks
1.670					92.550		BM
1.520		3.870				0	Ground level
	0.910					10	Ground level
	1.590					20	Ground level
	1.770					30	Ground level
	1.660					40	Ground level
	−4.200					50	Underside of bridge
		0.720				50	Ground level

requires a minimum working height of 4.50 m. What clearance (if any) will it have when working below the bridge?

(e) Given that the sewer track is 0.60 m wide with vertical sides, draw cross sections to a scale of 1:100 to show the excavation at each chainage point.

4. Computer generated sections

The formation of contour plans and digital terrain models is fully described in Chapter 10 on detail surveying. At the moment it is sufficient to know that the models form the basis for the design of construction details of roads, railways, pipelines etc.

Figure 10.16 is the gridded contour plan of a simple actual survey to which a road alignment has been added and is here shown in Fig. 6.8.

From this plan a longitudinal section is generated with specified scales and labelling (Fig. 6.9) and the

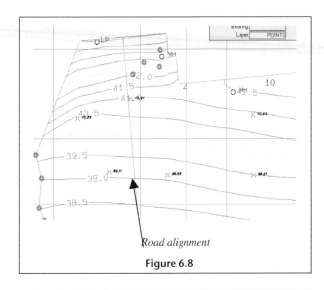

Road alignment

Figure 6.8

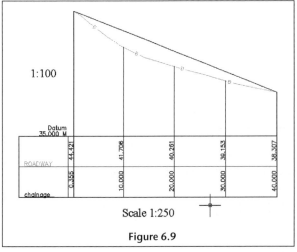

Scale 1:250

Figure 6.9

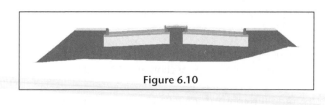

Figure 6.10

Figure 6.11 3D model with layer colouring Autodesk Civil 3D< (courtesy Autodesk<) *Autodesk, the Autodesk logo, Autodesk Map 3D and Autodesk Civil 3D are registered trademarks of Autodesk, Inc. in the USA and/or other countries*

specified gradient of a proposed roadway is added to the section.

Figure 6.10 is a cross section generated from the road alignment and shows standard cambers, kerb details, road make-up and batters. Schedule tables can also be generated to allow material quantities and costs to be computed.

More complicated examples, as shown in Fig. 6.11, would require alignment curve parameters to be set. Feasibility studies of alternative routes could be carried out with mass haul calculations to equalize cut and fill. The 3D terrain model is vital for environmental impact consideration when planning applications are made.

Most survey software has a modelling option, but for more sophisticated applications, the survey data can be exported to packages such as MX (formerly MOSS) from Bentley and Autodesk Map 3D< or Autodesk Civils 3D<. Rendering packages such as Bryce 3D offer a cheaper solution if visualization only is required.

5. Answers

Exercise 6.1

See Figure 6.12

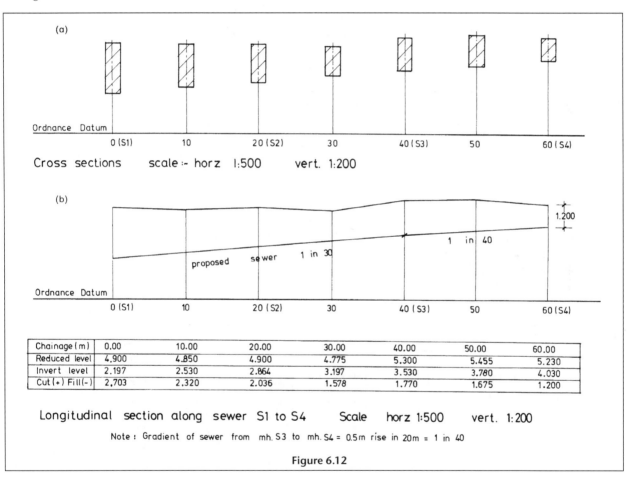

Chainage (m)	0.00	10.00	20.00	30.00	40.00	50.00	60.00
Reduced level	4.900	4.850	4.900	4.775	5.300	5.455	5.230
Invert level	2.197	2.530	2.864	3.197	3.530	3.780	4.030
Cut (+) Fill (−)	2.703	2.320	2.036	1.578	1.770	1.675	1.200

Longitudinal section along sewer S1 to S4   Scale   horz 1:500   vert. 1:200

Note : Gradient of sewer from mh. S3 to mh. S4 = 0.5m rise in 20m = 1 in 40

Figure 6.12

Exercise 6.2

1 See Figure 6.13

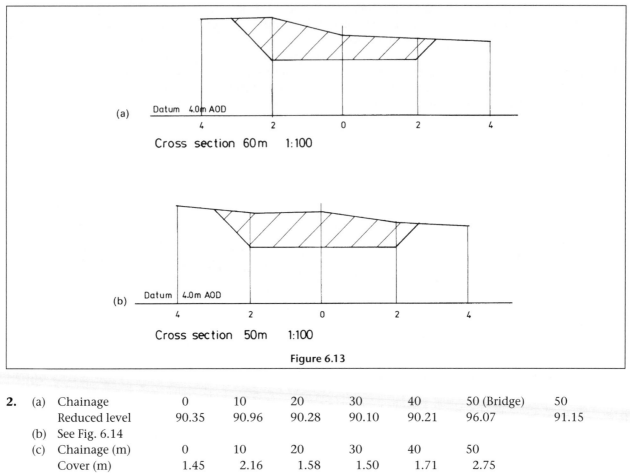

(a)

Datum 4.0m AOD

Cross section 60m   1:100

(b)

Datum 4.0m AOD

Cross section 50m   1:100

Figure 6.13

2. (a)

Chainage	0	10	20	30	40	50 (Bridge)	50
Reduced level	90.35	90.96	90.28	90.10	90.21	96.07	91.15

(b) See Fig. 6.14

(c)

Chainage (m)	0	10	20	30	40	50
Cover (m)	1.45	2.16	1.58	1.50	1.71	2.75

(d) Clearance 0.42 m

(e) See Figure 6.14.

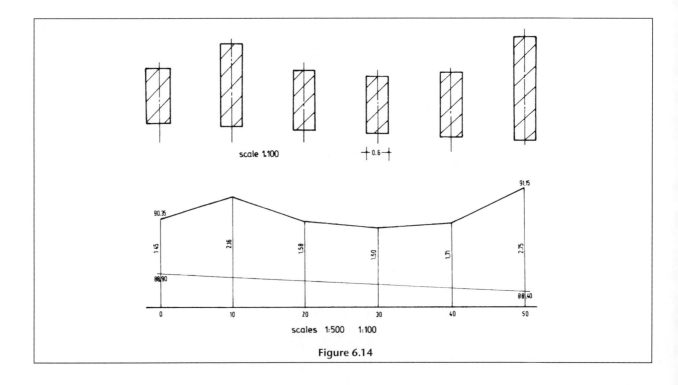

scale 1:100   ⊢0.6⊣

scales 1:500   1:100

Figure 6.14

Chapter 6 summary

In this chapter, the following are the most important points:

- Longitudinal and cross sections are profiles, drawn of the existing and proposed ground surfaces along and across construction site earthworks. They can be produced from a variety of surveyed data. The fundamental source of such data is a levelling survey from which sections are produced manually, or the data is produced by EDM survey for computer generation of sections.

- In manual plotting of longitudinal sections of long ribbon-like earthworks (e.g. roadways or sewers), chainages are set out at regular intervals along the proposed works and levels are taken at each of these chainage points. Each chainage point together with its associated reduced level is plotted to some suitable scale. (Since the overall height or depth of a longitudinal section will be only a fraction of the length, the vertical scale is usually exaggerated to present a clearer picture of the section). The points indicating the reduced levels are joined to produce a profile of the ground. The proposed levels of the earthworks are added and depths or heights of the earthworks are scaled and/or computed from the section.

- Cross sections at any or all of the chainage points along the earthworks are drawn in identical manner but since the overall height or depth of the cross section will not vary greatly from the width, the horizontal and vertical scales are usually chosen to be the same, thus giving a more realistic picture of the section.

- The sections can be computer generated from EDM data much more quickly but the basic principles are still fundamental to the operation. An introduction to computer generation is given on page 108 and is expanded in a following chapter.

- In later chapters, dealing with areas and volumes of earthworks, examples from this chapter will be used as a continuing theme. The reader is therefore advised to familiarize himself or herself with the methods of sectioning.

CHAPTER 7 Theodolites and total stations

In this chapter you will learn about:

- the component parts of a theodolite
- the categories of instruments
- the procedure to measure horizontal and vertical angles
- the possible sources of error and how to minimize them
- the component parts of a total station
- the principles of electronic distance measurement (EDM)
- the factors affecting the accuracy of measurement
- the categories of total station instruments and their features
- the common applications of total stations.

In the succeeding chapters of this book, the topics of traversing, radial positioning and setting out will be considered in detail. In these topics, it is essential that the surveyor has a sound knowledge of the instruments and methods used in the measurement of horizontal and vertical angles.

For two centuries the theodolite has been the workhorse of surveyors for measuring and setting out angles. In 1791 Ramsden's 24 inch brass circle vernier theodolite was used for the first national triangulation network in the UK. Sadly, British manufacturers such as Watts and Cooke Troughton and Simms had closed down by the 1960s. In Europe the first optical theodolite using a glass circle was developed by Heinrich Wild in Switzerland in the 1920s. Manufacturers such as Wild and Kern in Switzerland and Zeiss in Germany dominated the market.

In the late 1950s a South African company developed electronic distance measurement (EDM) with the Tellurometer, which used microwaves. This was followed in the 1960s by the Swedish company AGA, which developed the Geodimeter, the first EDM to use light waves. These were very bulky instruments which were powered by car batteries. More compact EDM units using light waves were introduced which attached on top of the theodolite. Topcon developed the EDM unit integrated within the telescope.

Electronic theodolites were developed in the late 1970s by Kern and Wild.

Today Leica, which absorbed Wild and Kern, and

Trimble, a US company, which has incorporated Geodimeter, dominate the high-end market. Japanese manufacturers, who originally offered low-cost solutions, are catching up and companies such as Topcon and Sokkia offer increasingly sophisticated total stations as well as optical and electronic theodolites. No doubt Chinese companies will eventually appear on the European market.

THEODOLITES

1. Classification

A theodolite is generally classified according to the method used to read the circles. There are two methods commonly in use today:

(a) optical micrometer,
(b) opto-electronic.

Figure 7.1 shows some examples of these two types of instrument.

Both systems are based on a graduated glass protractor.

(a) In optical micrometer instruments the circle is graduated clockwise from 0° to 360°. It is read by projecting daylight reflected by a mirror internally through the glass scale via a system of prisms and lenses to a reading telescope.

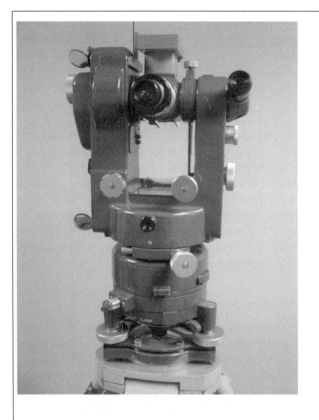

Hilger & Watts Microptic No. 2 1″ double circle reading

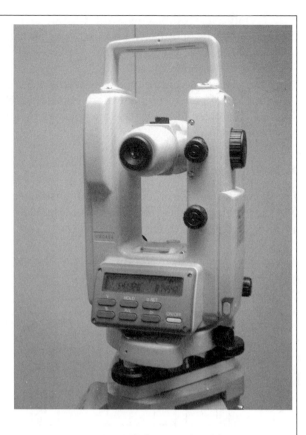

Topcon DT 20. 20″ Electronic theodolite

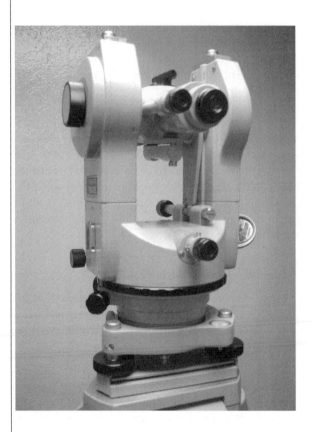

Sokkia TM20E. 5″ optical micrometer theodolite

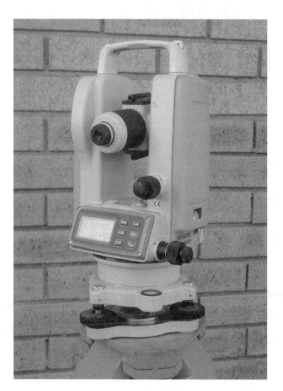

Pentax ETH 300 series electronic theodolite 2″, 5″ or 10″

Figure 7.1

Only a few instruments of this type are manufactured now, but many organizations, particularly educational establishments, still use this type of instrument.

(b) The glass circle of this type of instrument has a fine radial pattern of graduations etched round its circumference and these are counted by an optical sensor.

2. Principles of construction

The principles of construction of both instrument types are similar. The main components of a theodolite are illustrated in Fig. 7.2. These are as follows:

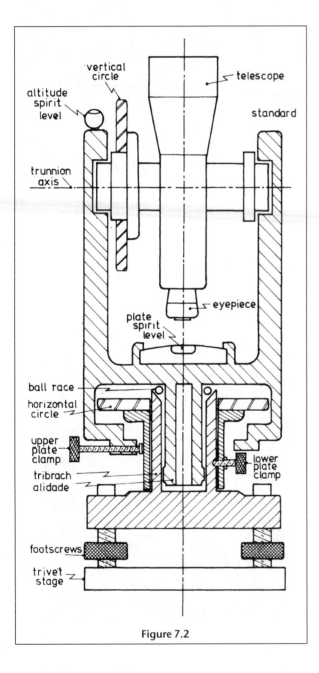

Figure 7.2

(a) Trivet stage

The trivet stage is the flat base of the instrument which screws on to the tripod and carries the feet of the levelling screws.

(b) Tribrach

The tribrach is the lower body of the instrument carrying the upper parts. In older theodolites the tribrach has a hollow, cylindrical, socket into which fits the remainder of the instrument (Fig. 7.2) being enabled to turn freely on a cylindrical ball race. In modern instruments the term tribrach refers to the whole of the base of the instrument including the trivet stage and footscrews. The upper part of the instrument is detachable and incorporates the cylindrical ball race.

(c) Levelling arrangement

The instrument is levelled by means of two spirit levels. A circular sprit level is attached to the tribrach and is set approximately level by adjusting the footscrews.

The main spirit bubble, the plate bubble, is usually situated on the cover plate of the horizontal circle. The sensitivity of the spirit level is of the order of 2 mm = 40 seconds of arc.

(d) Horizontal circle (lower plate)

The horizontal circle is mounted on a cylindrical axis which is free to rotate around the vertical axis through the centre of the instrument. The circle can be locked in position. The glass plate is about 80 mm in diameter.

(e) Alidade (upper plate)

The alidade is the remainder of the theodolite comprising the uprights or standards, which support the telescope and vertical circle. The alidade is coaxial with the lower plate and carries the index mark for reading the horizontal circle. It can be clamped for sighting and a slow motion or tangent screw allows fine rotation for interesecting the target.

(f) Controls for measuring horizontal angles

As mentioned previously, the horizontal motion of the theodolite can be controlled by a clamp and slow motion screw. There are three possible configurations.

(i) *The double centre system* uses upper and lower plate clamps. Figure 7.2 shows the double centre axis system and Fig. 7.3 shows the two clamps. It is essential that the function of these clamps be understood since they control the entire operation of measuring a horizontal angle.

When both clamps are open the lower plate (horizontal circle) and upper plate (alidade) are free

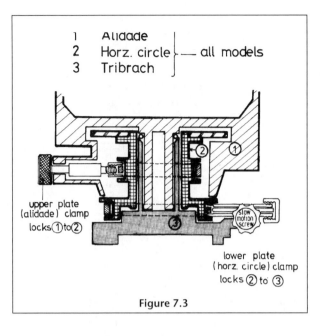

Figure 7.3

to rotate relative to the tribrach and to each other (Fig. 7.3).

When the lower plate is clamped, the horizontal circle is locked in position and as the alidade rotates, the horizontal circle reading changes.

When both clamps are locked, neither plate can rotate. If the lower plate clamp is released the upper and lower plates will move together and there will be no change in the horizontal circle reading.

(ii) *The Circle-setting screw is used* in more accurate optical micrometer theodolites, which are generally used for control survey rather than setting out.

This type of instrument has no lower plate clamping system as shown in Fig. 7.4. The horizontal circle can be rotated by a continuous-drive circle-setting screw. It may prove difficult to set a particular scale reading such as zero as no fine motion is available.

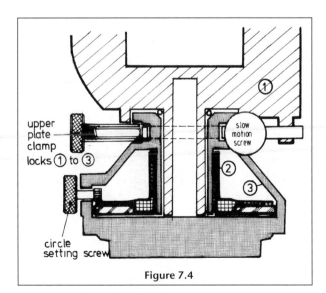

Figure 7.4

(iii) *The zero setting key* on an electronic theodolite fulfills, with great ease, the same function as the circle setting screw in (ii) above. With one press of the key, the reading is set to zero. In addition there is a key which locks any particular reading for an instrument pointing.

(g) Index marks

In order to read the circle for any pointing of the telescope it is convenient to imagine an index mark mounted on the alidade directly below the telescope. As the alidade is rotated the index mark moves over the horizontal circle. When the alidade is locked the index mark is read against the circle. Actually the index mark is a line etched on a glass plate somewhere in the optical train of the theodolite. In an electronic theodolite the index mark is the optical sensor beneath the horizontal circle.

When measuring a horizontal angle, e.g. angle PQR in Fig. 7.5, the theodolite is set over point Q and the horizontal scale reading is taken with the telescope pointing at P and R. The horizontal angle is the difference of the two readings. To simplify the arithmetic the first reading can be set at zero by one of the three methods in sec. (f) above.

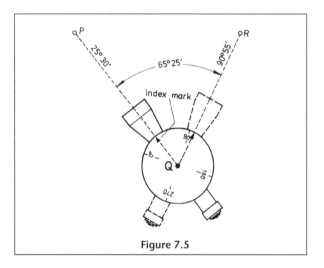

Figure 7.5

(h) Transit axis or trunion axis

The transit axis rests on the limbs of the standards and is securely held in position by a locknut. Attached to the transit axis are the telescope and vertical circle. They are free to rotate in the vertical plane about the transit axis. This rotation is controlled by the vertical clamp and slow motion screw.

A typical specification for a theodolite telescope is:

(i) internal focusing, damp and rust resistant;
(ii) shortest focusing distance: 2 m;
(iii) magnification: 30×;
(iv) object lens diameter: 42 mm;
(v) field of view: 1° 12′.

The vertical circle is attached to the telescope. Modern theodolites are graduated with the 0° reading in the zenith position. When the telescope is horizontal, the reading is either 90° (face left) or 270° (face right).

The vertical is established automatically by either a pendulum device or a liquid compensator (Fig. 7.17). In older instruments this was achieved by a separate altitude spirit level.

(i) Optical plummet/laser plummet

This device allows the instrument to be accurately centred over the survey mark rather than by use of a plumb-bob.

Figure 7.6 is a section through the optical system of an optical plummet. When the theodolite is levelled the view through the eyepiece is deflected vertically downwards through the centre of the instrument. In a laser plummet the eyepiece is replaced with a laser diode which emits a narrow beam of red laser light. The red laser spot is directed onto the point vertically beneath the centre of the instrument. The plummet is built into either the tribrach or the base of the alidade. To centre the instrument, the screw holding the trivet plate to the tripod is slightly slackened and the instrument is slid in one direction until it is exactly over the survey station. Care should be taken not to rotate the tribrach as it is being moved as this will alter the levelling.

3. Reading the circles

(a) Optical micrometer reading

Single reading micrometer
Figure 7.7 illustrates the optical train through a higher order theodolite, namely Watts 1. The optics are simple and the illumination is very clear. In the figure, the double open line follows the optical path through the

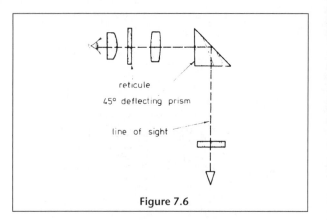

Figure 7.6

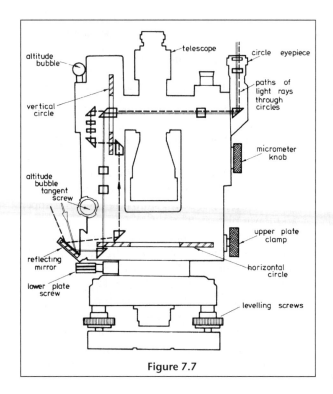

Figure 7.7

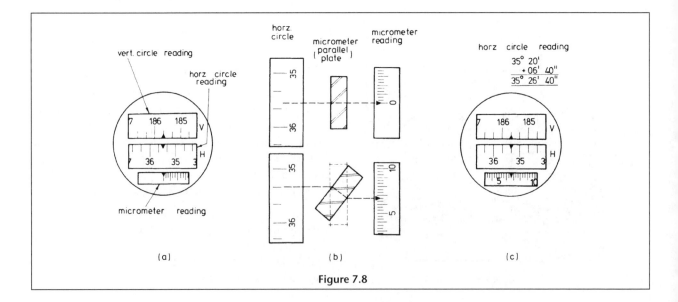

Figure 7.8

horizontal circle while the dashed line follows that of the vertical circle.

The readings of both circles are seen through the circle eyepiece mounted on the outside of the standard upright. It can be rotated from one side to the other for comfortable viewing.

The eyepiece contains three apertures, the horizontal and vertical circle graduations appearing in those marked H and V respectively. The circles are divided into 20-minute divisions. The horizontal circle, as read against the index arrow in Fig. 7.8(a), is therefore

$$35° \ 20' + x$$

The fractional part x is read in the third aperture by means of a parallel plate optical micrometer inserted into the light path of the instrument.

A parallel plate micrometer is simply a glass block with parallel sides. In physics, the law of refraction states that a ray of light striking a parallel-sided glass block at right angles will pass through the block without being refracted. If the block is tilted, however, the ray of light will be refracted but the emergent (i.e. exiting) ray will be parallel to the incident (i.e. entering) ray.

In Fig. 7.8(b) the ray of light from the horizontal circle is passing through the parallel plate when in the vertical position. The plate is directly geared to a drum mounted on the standard upright. Rotation of the drum causes the plate to tilt and the main scale reading of 35° 20′ is thereby made to coincide with the index mark. The resultant displacement, x, is read in minutes and seconds on the micrometer scale. The horizontal circle reading shown in Fig. 7.8(c) is therefore

$$\begin{aligned} &35° \ 20' \\ &\underline{+ \ 06' \ 40''} \\ = \ &\underline{35° \ 26' \ 40''} \end{aligned}$$

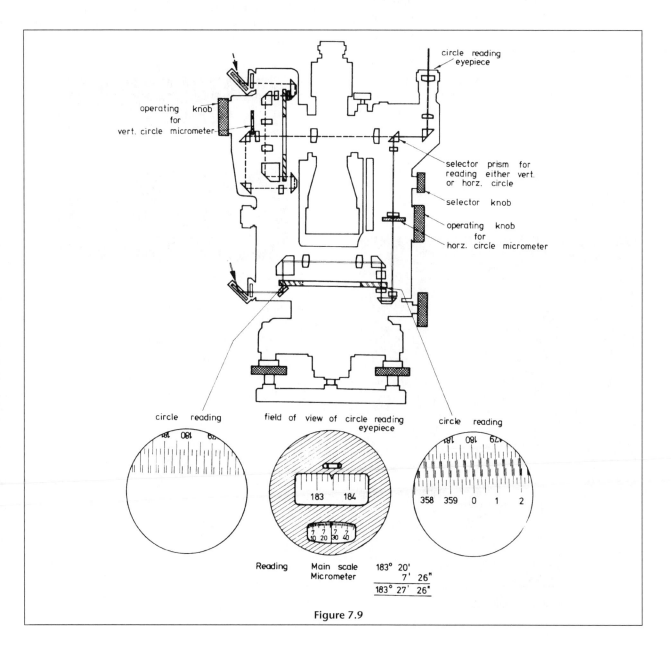

Figure 7.9

Double reading micrometer

It is frequently necessary in engineering surveying to measure angles with a higher degree of accuracy than can be obtained from a 20-second theodolite. In such cases a theodolite reading directly to 1 second is used.

It is possible to show that if the spindles of the upper and lower plates are eccentric, the measurement of the horizontal angles will be in error. The effects of eccentricity are entirely eliminated if two readings, 180° apart, are obtained and the mean of them taken.

On a 1-second theodolite, the reading eyepiece is again mounted on the upright. In most cases the horizontal and vertical circle readings are not shown simultaneously. On the Hilger and Watts 2 microptic theodolite, a knob is situated below the reading eyepiece which, when turned to either H or V, brings into view the horizontal or vertical circle scale.

The images of divisions diametrically opposed to each other are automatically averaged when setting the micrometer (Fig. 7.9 [see page 117]) and the reading direct to 1 second is free from circle eccentricity. The observer is actually viewing both sides of the circle simultaneously. In order to obtain a reading, the observer turns the micrometer drum until in the smallest aperture of the viewing eyepiece the graduations from one side of the circle are seen to be correctly superimposed on those from the other side, as in Fig. 7.9.

Both circles are read to 1 second, each circle having its own light path and separate micrometer.

The reading of the horizontal circle in Fig. 7.9 is

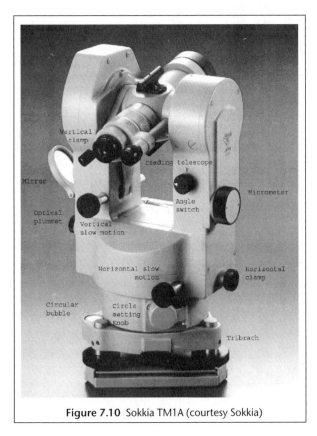

Figure 7.10 Sokkia TM1A (courtesy Sokkia)

$$183° \ 20' + x \ \text{(main scale)}$$
$$\underline{ 7' \ 26''} \ \text{(micrometer scale)}$$
$$183° \ 27' \ 26''$$

Figure 7.10 shows a Sokkia TM1A double reading optical micrometer theodolite, which is one of the few optical instruments still available.

The instrument has a detachable tribrach which contains a circular level.

A circle setting knob allows a particular reading to be set on the horizontal circle. Consequently there is only one horizontal clamp and slow motion.

The vertical circle compensator is a pendulum system with magnetic damping similar to an automatic level.

Reading the circles is achieved by directing daylight through a mirror and the image is projected through an optical system, containing a micrometer, to the reading telescope adjacent to the main telescope. It is a double reading system with the image of both sides of the circle being adjusted for coincidence by the micrometer as shown in Fig. 7.11.

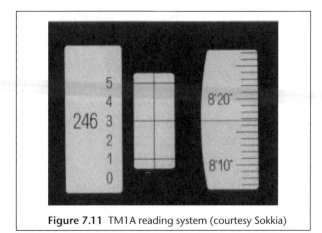

Figure 7.11 TM1A reading system (courtesy Sokkia)

An angle switch flicks between horizontal or vertical angle readings.

The circle reading illustrated in Fig. 7.11 is

$$246° \ 30' \qquad \text{on the main scale plus}$$
$$\underline{ 8' \ 16.7''} \ \text{on the micrometer scale}$$
$$246° \ 38' \ 16.7'' \ \text{is therefore the reading.}$$

E X A M P L E

1 Figure 7.12 shows the circle readings of three types of theodolite.

Determine the readings shown.

Answers

(a) Sokkia TM6 micrometer   H = 263° 15′ 24″
(b) Pentax TH06D micrometer   V = 58° 25′ 48″
(c) Sokkia DT5 electronic   V = 67° 05′ 10″
                                        H = 137° 08′ 00″

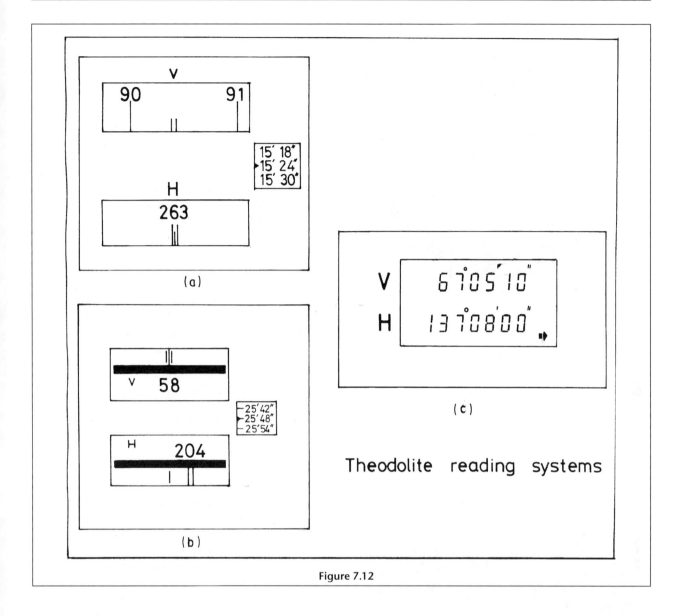

Figure 7.12

(b) Opto-electronic

The angle sensor of this type of instrument consists of a glass disc, which has a pattern of graduation code lines etched on its surface. Some sensors have coarse and fine graduations, others only a single pattern.

Automatic counting of the graduations is achieved by reflecting a spot of light from a diode off the graduated disc and analysing the light on a CCD sensor. In more accurate instruments two sensors are used on opposite sides of the disc.

Figure 7.13 shows a typical sensor and its mounting on the horizontal circle. Figure 7.14 shows the dual sensor of the Trimble S6 Total Station. Here a laser light source is used with a metal oxide semiconductor image sensor. This allows very high turning speeds of up to 115 degrees per second, which speeds up robotic operations.

The horizontal and vertical readings are displayed on a LCD screen, which may be on one or both sides of the instrument. Modern angle sensors are called absolute encoders. There is no zero index mark. Setting the zero on the horizontal circle is done with the press of a button.

A lower plate clamp and slow motion screw are not required. A particular horizontal reading can be locked if bearings require to be set out. The horizontal circle readings can be set to increase clockwise or anti-clockwise. Vertical readings are automatically set at zero with the telescope pointing vertically upwards.

Figure 7.15 shows the components and keyboard of a Pentax electronic theodolite.

4. Setting up the theodolite (temporary adjustment)

The sequence of operations required to prepare the instrument for measuring an angle is as follows.

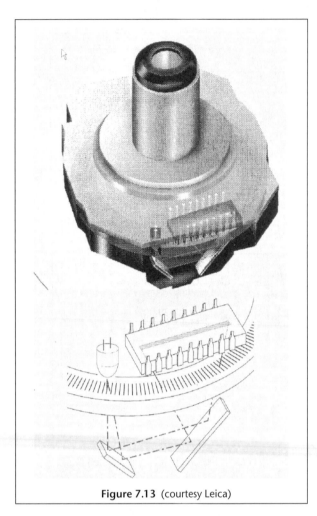

Figure 7.13 (courtesy Leica)

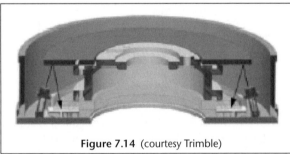

Figure 7.14 (courtesy Trimble)

(a) Plumb-bob method

1. Setting the tripod

This is probably the most important of the setup operations. If the tripod is not set properly, a great deal of time will have to be spent on subsequent operations 2 and 4.

The tripod legs are spread out and rested lightly on the ground around the survey point. If the instrument is set on a slope, two legs should be placed downhill to give a more stable set-up. In a high wind, a wide setting of the legs is essential to minimize vibration. The tripod legs are shuffled around the survey point until the plumb, which hangs from a hook on the tripod attachment screw, is within about 1 cm of the survey mark and the top of the tripod is roughly level. The

tripod legs are stamped firmly into the ground. This action will disturb the centring. The plumb-bob is returned to the centre by adjusting the telescopic tripod legs.

2. Attaching the instrument

The theodolite is removed from the box and notice is taken of how it fits, so it can be returned correctly. The instrument is then lifted by the top carry handle or the standards and attached to the tripod. It must never be lifted by the telescope. When changing stations, the instrument should be returned to its box, as damage can be caused if it is carried on the tripod.

3. Levelling

The instrument is levelled by means of either a spirit level (plate bubble) or an electronic bubble. In either case the circular bubble is levelled first of all, by using the three footscrews. The bubble always lies on the side which is highest. Care should be taken to ensure that the footscrews are not wound off the top of their threads.

Levelling using the spirit level (plate bubble)

(i)   The plate bubble is set parallel with any two footscrews, which are adjusted in equal but opposite directions to bring the bubble to the centre of its run. The bubble will follow the direction of movement of the left thumb.

(ii) The instrument is turned through 90° and the third footscrew is used to centre the bubble.

(iii) Operations (i) and (ii) are repeated until no further adjustment is required.

(iv) The instrument is turned through 180° to check if there is an error in the spirit level axis. If the bubble is off centre, an error is present.

The instrument is then levelled precisely by taking out half the bubble error, using the footscrews as in operations (i) and (ii).

The bubble error can be removed in the procedure outlined in the permanent adjustments section 7(a) in this chapter.

Levelling using an electronic bubble   More sophisticated electronic theodolites and total stations do not have a conventional plate bubble, instead they have an electronic tilt sensor called a dual axis compensator. Two possible configurations of this device are shown in Fig. 7.16. In both a vial of liquid is used to establish the horizontal plane. In the first diagram, light reflected from the liquid surface is analysed by a CCD sensor. In the other, the light projected through the liquid is displaced by refraction if tilt is present. A CCD sensor measures this displacement.

The compensator is situated in the centre of the instrument above the base unit which carries the horizontal angle encoder. The tilt is resolved into two direc-

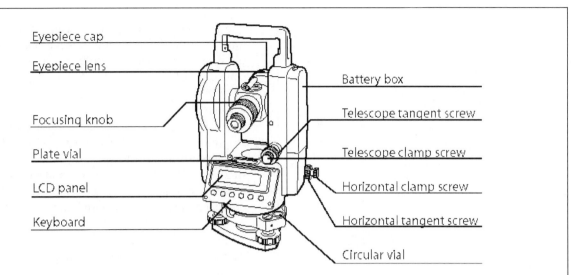

Eyepiece cap
Eyepiece lens

Focusing knob

Plate vial

LCD panel

Keyboard

Battery box

Telescope tangent screw

Telescope clamp screw

Horizontal clamp screw

Horizontal tangent screw

Circular vial

Display

Vertical angle
V=Zenith 0°, V%=Slope(VH=Horizontal 0°, Vc=Compass)

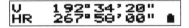

Horizontal angle   Battery indicator: 🔋
(HR=right rotation horizontal angle, HL=left rotation horizontal angle)

Keyboard

[V/%] key   (Switches the display of vertical angle)

[R/L] key   (Set the horizontal angle right
                 and left rotation alternately)

[HOLD] key   (Pressing twice holds the horizontal
                 angle shown on the display)

[0SET] key   (Pressing twice resets the horizontal angle to 0°.00;00")

[⚙] Illumination key (Illuminate LCD panel and the telescope reticle)

[ON/OFF] key (Turns the power on and off alternately)

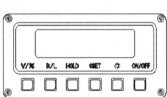

Figure 7.15 (courtesy Pentax)

tions at right angles and displayed on the instrument LCD screen.

Figure 7.17 shows a typical display. The instrument is levelled by aligning the display along any two foot screws and using all three foot screws to bring both 'bubble' images to the centre.

4 Centring

The tripod attachment screw is slightly released and the theodolite is slid in one direction until the optical or laser plummet is exactly on the survey mark. The tribrach should not be rotated as this affects the levelling.

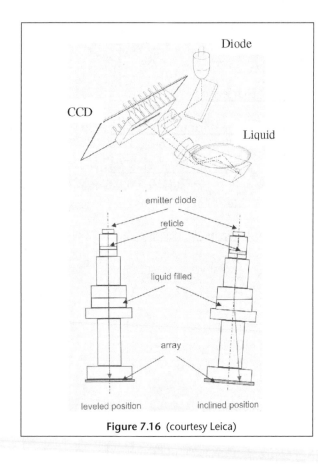

Figure 7.16 (courtesy Leica)

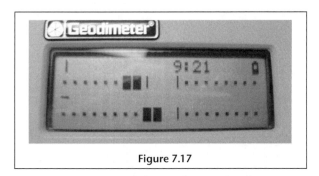

Figure 7.17

5 Parallax elimination
A piece of white paper is held in front of the telescope and the observer, sighting the paper through the telescope, turns the eyepiece until the cross-hairs are sharply focused.

(b) Optical plumb method

1 Setting the tripod
The tripod is set over the mark by eye with the top approximately level. The legs are then pressed firmly into the ground

2 Attaching the instrument
The theodolite is removed from the box and screwed to the tripod as before.

3 Centring
The instrument is tilted by adjusting the foot screws until the optical or laser plummet is exactly pointing to the survey mark. The theodolite is now centred but, of course, is not level.

4 Levelling
The circular bubble is levelled by adjusting the telescopic tripod legs. The plummet is checked and any small centring adjustment can be accomplished by sliding the theodolite on the tripod top as before. The precise levelling is carried out using the footscrews as described in part 3 of the plumb-bob method.

5 Parallax elimination
Parallax is removed as described in part 5 of the plumb-bob method.

5. Measuring horizontal angles

When exactly set over a survey mark and properly levelled, the theodolite can be used in two positions, namely:

    (a) face left or circle left,
    (b) face right or circle right.

The instrument is said to be facing left when the vertical circle is on the observer's left as an object is sighted. In order to sight the same object on face right, the observer must turn the instrument horizontally through 180° until the eyepiece is approximately pointing to the target. The telescope is then rotated about the transit axis, thus making the objective end of the telescope face the target. The vertical circle will now be found to be on the observer's right. This operation is known as transitting the telescope.

In Fig. 7.18 horizontal angle PQR is to be measured.

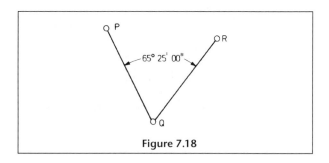

Figure 7.18

(a) Reiteration method (without setting to zero)

The possible horizontal control arrangements of a theodolite are: (a) upper and lower plate clamps, or, (b) circle setting screw (Figs 7.3, 7.4).

  1. Using the field book (Table 7.1), enter theodolite station Q in column 1 line 1.

Table 7.1

1	2	3	4	5
Observation station	Target station	Face left reading	Face right reading	Accepted mean angle
1 Q	P	25° 30′	205° 30′	
2	R	90° 55′	270° 55′	
3		65° 25′	65° 25′	65° 25′

2. Enter the left-hand target station P in column 2 line 1.
3. Enter the right-hand target station R in column 2 line 2.
4. Close the lower plate clamp, if fitted, *and do not touch it again.*
5. Set the instrument on face left.
6. Open the upper plate clamp on the alidade and the telescope clamp.
7. Turn the instrument carefully towards the left-hand target P and sight the target using the auxiliary finder sights fitted to the telescope. Lock the upper plate clamp and telescope clamp.
8. Focus the telescope on the target. The cross-wires will not be on the target but should be close. Use the slow motion screws on the upper plate clamp and telescope clamp to bisect the target accurately.
9. Read the horizontal circle and note the reading (25° 30′) in column 3 line 1.
10. *Repeat operations 6, 7, 8 and 9* for the right-hand target R, booking the horizontal circle readings (90° 55′) in column 3 line 2.
11. Subtract reading P from reading R (90° 55′ – 25° 30′ = 65° 25′) and note in column 3 line 3.

In order to measure the angle above, sixteen manipulations of the theodolite controls, two circle readings and two bookings were required. Clearly an error could easily occur with an inexperienced operator. Besides, even if these operations were conducted perfectly, the theodolite might be in poor adjustment and the angle of 65° 25′ would be incorrect.

All of these possible sources of error are eliminated by remeasuring the angle on face right.

12. Transit the telescope to set the instrument to face right and make preparations to remeasure the angle.
13. *Repeat operations 6, 7, 8 and 9,* noting the left-hand target reading P (205° 30′) in column 4 line 1. This reading should differ by 180° from that in column 3 line 1 if no errors are present.
14. *Repeat operations 6, 7, 8 and 9* for the right-hand target, noting the reading (270° 55′) in column 4 line 2.
15. Subtract reading P from reading R (270° 55′ – 205° 30′ = 65° 25′) and note in column 4 line 3.
16. Calculate the mean value of the angle and note in column 5 line 3.

EXAMPLE

2 Table 7.2 shows the field measurements of four angles of a traverse. Using the table, calculate the values of the angles.

Answer (Table 7.3)

In Example 2 the angle DAB was calculated more easily than the other three angles simply because the initial reading was 00° 00′ 00″. Many surveyors prefer this method of measuring angles and in setting out work it is standard practice.

(b) Reiteration method (setting to zero)

The actual measurement procedure is the same as for method (a) except that the initial setting of the

Table 7.2

Observation station	Target station	Face left reading	Face right reading	Accepted mean angle
B	A	89° 16′ 20″	269° 16′ 20″	
	C	185° 18′ 40″	05° 19′ 00″	
C	B	185° 39′ 40″	05° 39′ 20″	
	D	271° 38′ 20″	91° 38′ 40″	
D	C	275° 18′ 00″	95° 18′ 20″	
	A	01° 02′ 20″	181° 02′ 40″	
A	D	00° 00′ 00″	180° 00′ 00″	
	B	92° 15′ 30″	272° 15′ 30″	

Table 7.3

Observation station	Target station	Face left reading	Face right reading	Accepted mean angle
B	A	89° 16′ 20″	269° 16′ 20″	
	C	185° 18′ 40″	05° 19′ 00″	
		96° 02′ 20″	96° 02′ 40″	96° 02′ 30″
C	B	185° 39′ 40″	05° 39′ 20″	
	D	271° 38′ 20″	91° 38′ 40″	
		85° 58′ 40″	85° 59′ 20″	85° 59′ 00″
D	C	275° 18′ 00″	95° 18′ 20″	
	A	01° 02′ 20″	181° 02′ 40″	
		85° 44′ 20″	85° 44′ 20″	85° 44′ 20″
A	D	00° 00′ 00″	180° 00′ 00″	
	B	92° 15′ 30″	272° 15′ 30″	
		92° 15′ 30″	92° 15′ 30″	92° 15′ 30″

Table 7.4

	1 Observation station	2 Target station	3 Horizontal FL	4 Angle FR	5 Accepted value
1	Q	P	00° 00′ 00″	180° 00′ 00″	
2		R	65° 25′ 00″	245° 25′ 00″	
3			65° 25′ 00″	65° 25′ 00″	65° 25′ 00″

horizontal circle has to be 00° 00′ 00″. The mechanics of setting the circle varies with the type of theodolite.

(i) Theodolite with upper and lower plate clamps
Using a double centre theodolite, i.e. one fitted with upper and lower plate clamps, the procedure is:

1. Set the theodolite to face left position.
2. Set the micrometer (if fitted) to 00′ 00″.
3. Release the upper plate clamp only and set the index mark to zero degrees as closely as possible by eye. Close the clamp and, using the slow motion screw, set the circle exactly to zero. The reading is now 00° 00′ 00″.
4. Release the *lower* plate clamp and telescope clamp. Sight the left-hand station P, lock the clamps and, using the *lower plate slow motion screw*, accurately bisect the target.

(ii) Theodolite with circle-setting screw
1. Set the instrument to face left position.
2. Set the micrometer (if fitted) to 00′ 00″.
3. Release the upper plate clamp and telescope clamp. Sight the left-hand station P and, using the slow motion screws, accurately bisect the target.
4. Raise the hinged cover of the circle-setting screw and rotate the screw carefully until the horizontal circle reads exactly zero.

(iii) Electronic theodolite
1. Set the theodolite to face left position.
2. Release the upper plate clamp and telescope clamp. Sight the left-hand station P and, using the slow motion screws, accurately bisect the target.
3. Press the (zero set) key on the keypad. The horizontal circle reading will be reset to zero degrees.

In all three cases the situation has been reached where the theodolite circle reads zero and the left-hand target P is accurately bisected. The measurement of the angle is completed as follows:

1. In Table 7.4, enter the reading of 00° 00′ 00″ in column 3 line 1.
2. Open the upper plate clamp and telescope clamp.
3. Turn the instrument carefully towards the right and sight the right-hand target R.
4. Lock both clamps and, using the upper plate slow motion screw and telescope slow motion screw, accurately bisect the target.
5. Read the horizontal circle and enter the reading (65° 25′ 00″) in column 3 line 2.
6. Subtract reading P from reading R (65° 25′ 00″ − 00° 00′ 00″ = 65° 25′ 00″).
7. Transit the telescope to set the instrument on face right.
8. Resight the left-hand target and note the reading

which should be 180° 00′ 00″ if no errors have been made and if the instrument is in adjustment. Note the reading in column 4 line 1.

9. Resight the right-hand station and note the reading 245° 25′ 00″ in column 4 line 2.

10. Subtract reading P from reading R (245° 25′ 00″ − 180° 00′ 00″ = 65° 25′ 00″).

11. Calculate the mean angle and enter the value in column 5 line 3.

Note: If there is a consistent difference between face left and right readings then a collimation error is present. How this is checked and adjusted is detailed in sec. 7 of this chapter

---

EXAMPLE

3 Table 7.5 shows the field measurements of four angles of a traverse. Using the table, calculate the values of the angles.

Answer (Table 7.6)

Exercise 7.1

1 Table 7.7 shows the field measurements of two angles of a traverse survey. Calculate the values of the traverse angles.

6. Measuring angles in the vertical plane

Figure 7.19 shows the complementary relationship

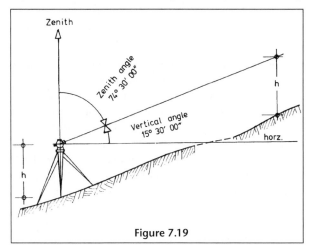

Figure 7.19

Table 7.5

Observation station	Target station	Face left reading	Face right reading	Accepted mean angle
B	A	00° 00′ 00″	180° 00′ 00″	
	C	93° 14′ 20″	273° 14′ 20″	
C	B	00° 00′ 00″	180° 00′ 20″	
	D	161° 25′ 00″	341° 25′ 00″	
D	C	00° 00′ 00″	179° 59′ 40″	
	A	175° 31′ 40″	355° 31′ 40″	
A	D	00° 00′ 00″	179° 59′ 40″	
	B	195° 32′ 20″	15° 32′ 20″	

Table 7.6

Observation station	Target station	Face left reading	Face right reading	Accepted mean angle
B	A	00° 00′ 00″	180° 00′ 00″	
	C	93° 14′ 20″	273° 14′ 20″	
		93° 14′ 20″	93° 14′ 20″	93° 14′ 20″
C	B	00° 00′ 00″	180° 00′ 20″	
	D	161° 25′ 00″	341° 25′ 00″	
		161° 25′ 00″	161° 24′ 40″	161° 24′ 50″
D	C	00° 00′ 00″	179° 59′ 40″	
	A	175° 31′ 40″	355° 31′ 40″	
		175° 31′ 40″	175° 32′ 00″	175° 31′ 50″
A	D	00° 00′ 00″	179° 59′ 40″	
	B	195° 32′ 20″	15° 32′ 20″	
		195° 32′ 20″	195° 32′ 40″	195° 32′ 30″

Table 7.7

Observation station	Target station	Face left reading	Face right reading	Mean angle
D	C	00° 00′ 00″	179° 59′ 20″	
	E	189° 39′ 50″	9° 38′ 50″	
E	D	185° 14′ 30″	5° 13′ 50″	
	F	6° 15′ 10″	186° 14′ 10″	

between zenith and vertical angles. As already explained on page 115, sec. (h), all modern theodolites measure zenith angles, i.e. the zero circle reading is in the zenith position so, in modern terminology, the terms zenith angle and vertical angle are synonymous. They are, in fact, the angles of gradient, along which linear measurements are made either by tape or EDM. They are used to calculate slope correction in traverse survey and heights in trigonometrical levelling.

(a) Observation procedure

The procedure in measuring a zenith angle is as follows:

1. Set the theodolite to face left.
2. Release the vertical and horizontal clamps.
3. Sight the target using the finder sight and lock the clamps.
4. Focus the telescope on the target. The cross-hairs will not be on the target, but should be close. Use the slow motion screws to bisect the target accurately. The horizontal hair is the important one for measuring vertical angles.
5. Take the vertical circle reading.
6. Change the instrument to face right and repeat operations 2, 3, 4 and 5.

(b) Calculation of mean vertical (zenith) angle

The face left and face right readings are on opposite sides of the vertical circle, and with perfect observations and the instrument in perfect adjustment the two readings should add up to 360°.

One method of reducing the vertical angles is to check for this as shown in the following example:

Face left reading	= 82° 10′ 04″
Face right reading	= 277° 50′ 04″
Sum	= 360° 00′ 08″
Index error	= +08″
Correcting by half the error	
Mean angle	= 82° 10′ 00″

Alternatively, the equivalent angle to the face left reading can be calculated by subtracting the face right reading from 360° and the mean is calculated in the

normal way of adding the two angles and dividing by two as follows:

Face right reading	= 277° 50′ 04″
Equivalent face right reading	= 82° 09′ 56″
Face left reading	= 82° 10′ 04″
Sum	= 164° 20′ 00″
Mean angle	= 82° 10′ 00″

A consistent large value for the index error means the instrument is out of adjustment, and the vertical collimation should be checked and adjusted as described in the next section.

EXAMPLE

4 The vertical circle readings for the zenith angle shown in Fig. 7.19 were observed as shown below.

Calculate the correct zenith angle.

Answer

Face left reading	= 74° 29′ 30″
Face right reading	= 285° 29′ 30″
Sum	= 359° 59′ 00″
Index error	= −01′ 00″
Correction	= +30″
Corrected angle	= 74° 30′ 00″

Alternatively

Equivalent face right reading	= 74° 30′ 30″
Face left reading	= 74° 29′ 30″
Mean	= 74° 30′ 00″

Exercise 7.2

1 Table 7.8 shows the field measurements of two angles of a traverse survey. Calculate the values of zenith angles.

Table 7.8

Line	Face	Vertical circle reading
XY	L	79° 30′ 50″
	R	280° 30′ 10″
YZ	L	102° 13′ 50″
	R	257° 47′ 10″

7. Errors affecting angular measurements

The errors that affect angular measurements can be considered under two headings:

(a) instrumental maladjustment,
(b) human and other errors.

(a) Instrumental maladjustments

The following procedures will measure the effect of these errors for horizontal and vertical angles. However, it is inadvisable that the inexperienced should attempt the adjustments. For complicated modern theodolites, it is best to return the instrument to a service centre.

Effect on horizontal angles
If a theodolite is to be in perfect adjustment, the relationship between the various axes should be as shown in Fig. 7.20, namely:

1. The vertical axis should be truly vertical and a right angles to the plate bubble.
2. The line of collimation of the telescope should be at right angles to the transit axis.
3. The transit axis should be at right angles to the vertical axis of the instrument.

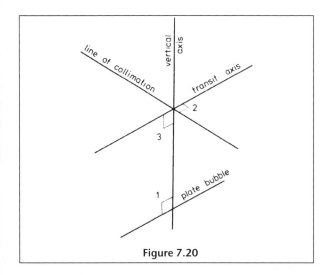

Figure 7.20

The relationships become disturbed through continuous use or misuse, and tests must be carried out before the start of any major contract and at frequent intervals thereafter to ensure that the instrument is in adjustment.

The various tests and adjustments are as follows:

1. The vertical axis must be truly vertical and at right angles to the plate bubble axis.

Test
(a) Erect the tripod firmly and screw on the theodolite. Set the plate spirit level over two screws and centralize the bubble. If the instrument is not in good adjustment the vertical axis will be inclined by the amount *e* as shown in Fig. 7.21(a).
(b) Turn the instrument through 90° and recentralize the bubble.
(c) Repeat these operations until the bubble remains central for positions (a) and (b).
(d) Turn the instrument until it is 180° from position (a). The vertical axis will still be inclined with an error *e* and the bubble of the spirit level will no longer be central. It will, in fact, be inclined to the horizontal at an angle of 2*e* (Fig. 7.21(b)). The number of divisions, *n*, by which the bubble is off-centre is noted.

Adjustment
(e) Turn the footscrews until the bubble moves back towards the centre by *n*/2 divisions, i.e. by half the error. The vertical axis is now truly vertical (Fig. 7.21(c)).
(f) Adjust the spirit level by releasing the capstan screws and raising or lowering one end of the spirit level until the bubble is exactly central. The other half *n*/2 of the error is thereby eliminated (Fig. 7.21(d)) and the spirit level is at right angles to the vertical axis.

Instruments with dual axis compensator This device eliminates this error, and will also compensate angle readings for any small alterations to instrument levelling, which may occur during the occupation of a station.

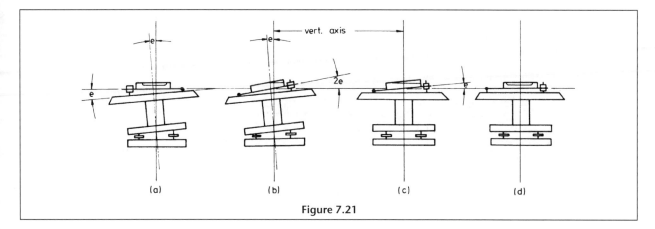

Figure 7.21

Effect of maladjustment There is no observational procedure that can be employed to eliminate this error. However, it only becomes serious for high angles of elevation or depression of the telescope.

2. The line of collimation of the telescope must be at right angles to the transit axis. The line of collimation is defined as the line joining the optical centre of the object glass to the vertical cross-hair of the diaphragm. If the position of the diaphragm has been disturbed, the line of collimation will lie at an angle to the transit axis (Fig. 7.22) with e being the error in the line of collimation.

Test
(a) After properly setting up the instrument at a point which will be designated I, sight a well-defined mark A about 100 metres away and close both clamps. The line of sight will now be pointing to the target as in Fig. 7.22(a).
(b) Transit the telescope and sight a staff B laid horizontally about 100 metres away on the other side of the instrument from A.
Note the reading, 2.100 m in Fig. 7.22(b).
Since the transit axis and line of sight make an angle of $(90 - e)$ and since the transit axis maintains its position when the telescope is trans-

mitted, the line of sight must diverge from the straight line by an amount $2e$.
(c) Change the instrument to face right and again sight mark A (Fig. 7.22(c)).
(d) Transit the telescope and sight staff B. The line of sight will again diverge from the straight line AB by an amount $2e$, but on the other side of the line. The staff reading is again noted and in Fig. 7.22(d) is 2.000 m.
(e) If the two staff readings are the same, the instrument is in adjustment and points A, I and B form a straight line.

Adjustment
(f) Since the staff readings differ, the difference represents the error $4e$, that is $4e = (2.100 - 2.000) = 0.100$ m in this case. At the outset, it was shown that the error was e, therefore $e = (0.100/4) = 0.025$ m.

The error is eliminated by bringing the vertical cross-hair of the staff reading $(2.000 + e) = 2.025$ m.

This is done by means of the antagonistic adjusting screws situated one on either side of the diaphragm.

In this case the diaphragm has to be shifted to the right (Fig. 7.23). The left screw is loosened

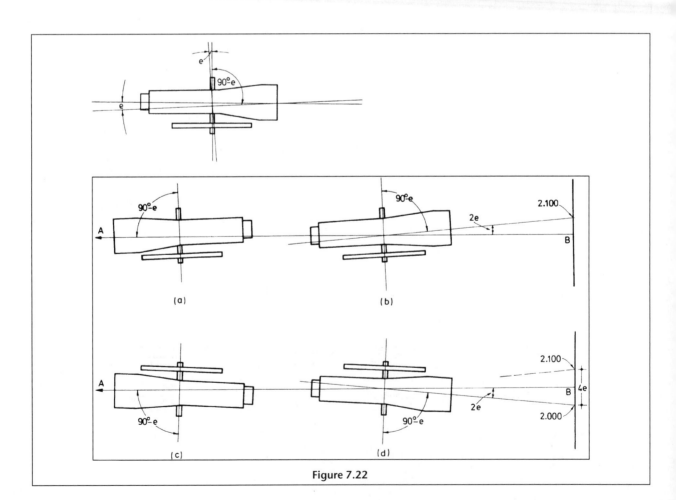

Figure 7.22

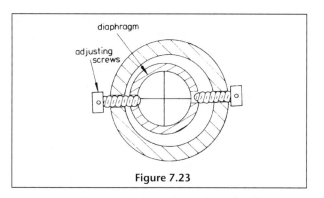

Figure 7.23

slightly and the right screw tightened until the correct staff reading is obtained.

If the cross-hair is no longer vertical, it might be necessary to loosen the diaphragm and rotate it slightly until it is perfectly vertical. The verticality is usually checked against a plumb-line.

In more sophisticated electronic theodolites and total stations, the collimation error can be checked and set in the instrument by observing a target in both faces.

Effect of maladjustment If the instrument is used in its unadjusted state, every angle will be in error. However, as will be evident from Fig. 7.22 the mean of face left and face right readings is correct. For example, the mean staff reading 1/2 (2.100 + 2.000) = 2.050 m is the correct position of B since the line of sight diverges by $2e$ on either side of the mean position for face left and face right respectively.

3. The transit axis must be truly horizontal when the vertical axis is vertical. If the instrument is not in adjustment, the transit axis will not be horizontal when the instrument is correctly set up. In Fig. 7.24, the error is e. If the telescope is inclined, the cross-hair will travel along the plane shown by the dotted line, i.e. a plane at right angles to the transit axis but *not* a vertical plane.

Test
   (a) Set up the instrument at I and level it properly.

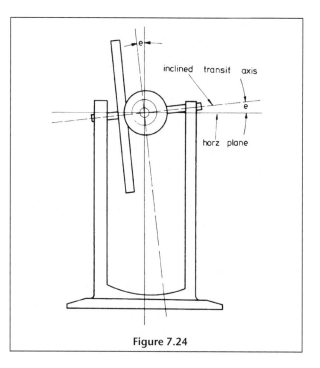

Figure 7.24

Sight a mark A at an elevation of about 60°. Close both upper and lower plate clamps (Fig. 7.25).
   (b) Lower the telescope and sight a horizontal staff or scale laid on the ground at the base of the object A. Note the reading, 'b'.
   (c) Repeat the operations on face right to obtain a second reading, 'c', on the scale. If the instrument is in adjustment both readings will be the same.

Adjustment
   (d) If the readings differ, the mean is correct since the telescope will have traversed over planes each inclined at an angle e on either side of the vertical.

   By means of either the upper or lower plate slow motion screw, set the instrument to the mean reading.

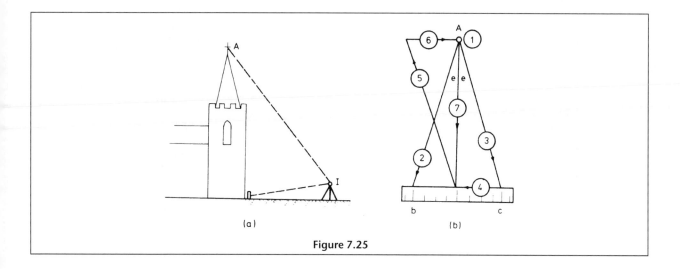

Figure 7.25

(e) Elevate the telescope until the horizontal cross-hair cuts mark A. The vertical cross-hair will lie to the side of the mark.

(f) Bring the vertical cross-hair on to mark A by means of the transit axis adjustment screw located on the standard immediately below the transit axis.

(g) Depress the telescope. If the adjustment has been carried out correctly the vertical cross-hair should read the mean staff reading.

Effect of maladjustment Angles measured between stations at considerably different elevations will be in error. However, as Fig. 7.25 clearly shows, the mean face left and face right is correct.

It must be noted at this juncture that most modern theodolites make no provision for this adjustment. The complex optical systems make the adjustment very difficult. However, since such instruments are assembled with a high degree of precision this adjustment is usually unnecessary. Besides, the mean of face left and face right observations cancels the error.

Effect on vertical angles

In a perfectly adjusted theodolite, the index reader of the vertical circle should read ninety degrees, and the bubble of the altitude spirit level should be in the centre of its run when the telescope is in the horizontal position. This is the only maladjustment that materially affects the measurement of vertical angles. On modern theodolites, employing automatic vertical circle indexing to replace the altitude spirit level, no adjustment is usually required.

Test

(a) Set up the instrument at I on face left, centralize the altitude spirit level and read a vertical levelling staff at A with the vertical circle reading zero. Note the reading, say 1.500 (Fig. 7.26(a)).

(b) Transit the telescope and repeat operation (a) on face right. Note the staff reading, say 1.400 in Fig. 7.26(b).

Adjustment

(c) The mean staff reading is correct since the face left observation is elevated and the face right reading is depressed by equal errors, *e*. Using the vertical slow motion screw, the telescope is brought to the mean reading of 1.450 m (Fig. 7.26(c)).

(d) The vertical circle will no longer read zero and must be made to do so by the clip screw which controls the altitude spirit level and vertical circle index.

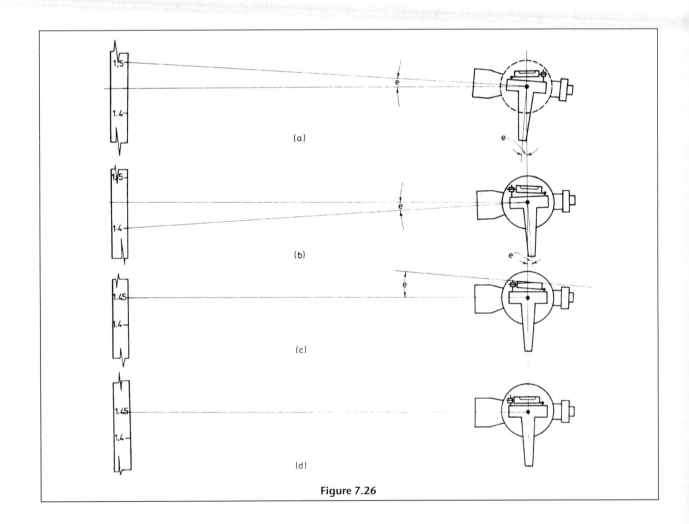

Figure 7.26

(e) The movement of the clip screw, however, will move the bubble from its central position. The bubble is recentralized by adjusting the capstan screws situated at one end of the spirit level (Fig. 7.26(d)).

In a theodolite with automatic vertical collimation the mean reading will be bisected by adjusting the cross-hairs.

In more sophisticated electronic theodolites and total stations, the collimation error can be checked and set in the instrument by observing a target in both faces.

Effect of maladjustment If the instrument is in mal-adjustment all vertical angles will be in error. However, Fig. 7.26 clearly shows that the mean of face left and face right observations is correct.

(b) Human and other errors

Human errors can be considered under two headings:

1. *Gross errors* These are mistakes on the part of the observer caused by ignorance, carelessness or fatigue. They include sighting the wrong target, measuring the anticlock-wise angle, turning the wrong screw, opening the wrong clamp, reading the circles wrongly and booking incorrectly.

These errors can only be avoided by careful observation and by observing each angle at least twice. Only then will any error show up. It is absolutely useless to measure any angle on one face only as it is open to the wildest of errors, as already pointed out.

2. *Random errors* Small errors cannot be avoided. They may be due to imperfections of human sight and touch which make it impossible to bisect the targets accurately or read the circles exactly. The errors, however, are small, and of little significance. They are minimized by taking several observations and accepting the mean.

Other errors arise from such sources as unequal expansion of the various parts of the instrument by the sun, instability of the instrument in windy weather, heat haze or mist affecting the sighting and lastly inaccurate centring of the instrument. They can only be avoided by shielding the instrument against the wind or sun and choosing times to observe which are favourable.

Lastly, if the instrument is not correctly centred, nothing can be done to eliminate or minimize the errors that must arise. Great care must be taken to position the instrument over the survey station with accuracy.

Summary
In general, gross errors cannot be eliminated or minimized by any observational system. A compensated measure will only show that there is an error. However, that should be sufficient as the observer should then take steps to trace the error. A mistake in subtraction will not necessitate any repetition of measuring but all other errors under this heading will, and in general another complete compensated measure should be made.

Systematic errors arise from instrumental maladjustments and defects. It has been shown that in almost every case a compensated measure of the angle cancels the error, the exception being that if the vertical standing axis is not truly vertical, errors cannot be eliminated. However, they are second-order errors and do not affect measurement made by a conventional theodolite, except in high elevation sights.

Small random errors are not eliminated by a compensated measurement but such action minimizes the error and the mean of a larger number of measurements would be very accurate.

It cannot be overemphasized that a compensated measure must be made of every angle regardless of its importance; otherwise the result is so uncertain as to be meaningless.

TOTAL STATIONS

If a theodolite is a car, then a total station is a motor home. It incorporates all the features of an electronic theodolite plus a distance meter which is incorporated in the telescope. The distance meter uses the principle of Electronic Distance Measurement (EDM), sending out a beam of light which is reflected back to the instrument either from a prism or directly from an object. Older instruments required a data recorder to store and process the data, but modern instruments have on-board storage and application programs. The light emitted by the instrument is at the infrared or the red end of the electro magnetic spectrum shown in Fig. 7.27. The infrared diode or laser is located in the telescope housing and either projected through the viewing telescope or by a separate optical system above the objective lens.

1. Principles of EDM

Modern total stations use two principles in analysing the light beam:

(a) phase difference measurement,
(b) pulse measurement.

Instruments can use one or the other or the two in combination.

(a) Phase difference

Electromagnetic energy is emitted in the form of a sinusoidal waveform. Figure 7.28 represents one cycle of the energy wave.

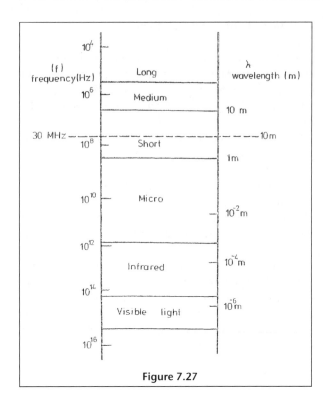

Figure 7.27

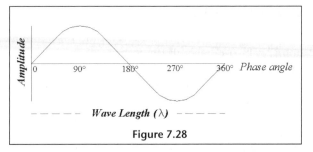

Figure 7.28

It has the following parameters:

Wavelength is the length between the start and end points of one cycle.

Frequency is the number of cycles which occur in one second.

The SI unit of frequency meaning one cycle per second is the hertz. The various multiples of one hertz are derived in the usual manner by prefixing kilo, mega, and giga to the word hertz:

$$1 \text{ hertz } = 1 \text{ Hz}$$
$$10^3 \text{ hertz } = 1 \text{ kilohertz} = 1 \text{ kHz}$$
$$10^6 \text{ hertz } = 1 \text{ megahertz} = 1 \text{ MHz}$$
$$10^9 \text{ hertz } = 1 \text{ gigahertz} = 1 \text{ GHz}$$

Phase measures the fraction of a wavelength in one cycle, being 0° and progressing to 360° at the end. e.g. 90° measures one quarter of a wavelength.

The distance meter in instruments using the phase difference measuring principle modulater the infrared or laser carrier wave to a frequency of about 1 GHz, giving a wavelength of about 30 cm. The light emitted by the diode in the instrument is transmitted as a narrow beam which is reflected from a target back along the same path as shown in Fig. 7.29.

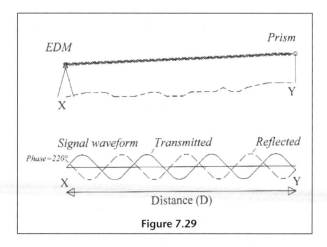

Figure 7.29

The reflecting target has traditionally been a corner cube prism as shown in Fig. 7.34, but a variety of types are now available. With the speed of light being about 300,000 km/sec the reflected light returns almost instantaneously to the distance meter where it is analysed and the distance calculated.

To explain the measuring principle it is convenient to examine the opened out waveform of the transmitted and reflected signals as shown in Fig. 7.30.

The double distance of the light path can be calculated by the formula:

$$2D = \text{number of cycles } (n\lambda) + \text{wave fraction } (\Delta\lambda)$$

The two parts of the formula are calculated separately in the distance meter.

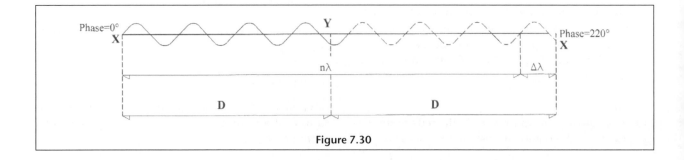

Figure 7.30

(i) *The phase difference* of the outgoing and incoming light can be measured using electrical phase detectors and the fraction of a wavelength calculated from the formula:

$$\Delta\lambda = \frac{phase\_difference}{360} \times \lambda$$

In Fig. 7.31 the outgoing light starts with a phase angle of 0° and the incoming light is measured with a phase angle of 220°. The fraction of a wavelength can be calculated as follows:

$$\Delta\lambda = (220°/360°) \times 30 \text{ cm} = 0.18 \text{ m}$$

(ii) *The wavelength count* is the more difficult part of the formula to solve. The process is commonly known as *'resolving the ambiguity'*.

Initially a method whereby the modulated wavelength of the signal was increased in multiples of 10, and the phase difference was measured for each. The wavelength fractions were of 1 m, 10 m, 100 m, 1000 m and 10 000 m and when added together gave the distance.

In early instruments the wavelength increase and the phase detection were carried out manually, and the measurement of one distance could take about five to ten minutes. The process became more automated, but still took minutes rather than seconds.

A much faster method was developed using multiple signal frequencies giving slightly different wavelengths. The formula

$$2D = n\lambda + \Delta\lambda$$

could be solved by simultaneous equation algorithms in the distance meter processing circuitry.

Today instruments use electronic circuitry to count the wave cycles automatically, the distance being resolved in seconds.

Figure 7.31 shows the configuration of the telescope in a phase shift instrument where the beam is transmitted and received through the telescope optics.

The light emitted by the diode is directed through an optical system and transmitted through the objective lens of the telescope. A small part of the emitted light passes straight down to the phase sensor. The returning reflected light is directed onto the sensor.

(b) Pulsed measurement

This system measures the time of flight (TOF) between the signal pulse being transmitted from and returning to the instrument. The formula

Distance = signal velocity × transit time

can therefore be solved. The distance meter generates more than 10 000 infrared or laser light pulses per

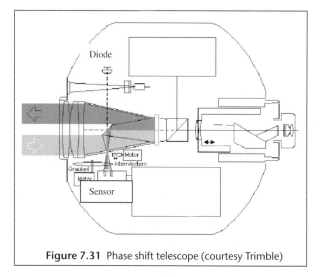

Figure 7.31 Phase shift telescope (courtesy Trimble)

second. Each pulse results in a distance measurement and these can be averaged to give an accurate final value. In principle this method is not as accurate as phase measurement but modern signal processing techniques can achieve high accuracy and a long range. The Trimble DR 300+ has a standard accuracy of 3 mm + 3 ppm.

Figure 7.32 shows the configuration of the telescope in this instrument.

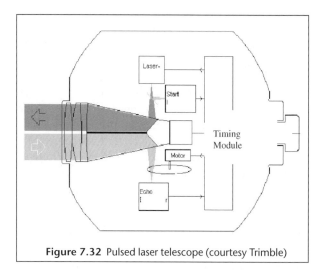

Figure 7.32 Pulsed laser telescope (courtesy Trimble)

The laser light is deflected and collimated into a narrow slightly divergent beam and transmitted through the objective lens of the telescope. A small part of the emitted light is reflected by a semi-silvered mirror to start the timing circuitry. The reflected light is collected by the objective lens of the telescope and reflected onto a sensor which completes the timing sequence.

In some instruments the telescope has to be focused before the distance is measured. The narrowness of the beam is important for accurate distance measurement, particularly when direct reflection from a surface is being used. The pulsed signal is more divergent than the phase system, but processing can minimize this error. Topcon use a dual optical system to narrow the beam.

2. Comparison of pulsed and phase systems

(a) Range

Pulsed systems are capable of much longer ranges than phase systems, up to six times greater for a similar light source. This is particularly significant in direct reflex measurement where there is no prism, and the signal is reflected from the object being measured. For the latest Trimble instruments the maximum range to a concrete surface is 400 m in pulse mode, and 100 m in phase mode.

To increase the range of phase measurement a more powerful laser can be used. Typically, pulsed lasers fall into class 1 as the pulses mean the intensity of the light does not build up so much energy when directed at an object. Phase systems have to use a class 2 or even class 3 laser to achieve the equivalent range. Maximum range using a single prism is generally about 3.5 km for a phase instrument, and 5.5 km for a pulse instrument.

(b) Laser safety

Considering the range of the distance meter in total stations leads on to laser safety standards. Most pulsed lasers fall into *Class 1* where the beam is invisible and is not harmful to the skin or to the naked eye. Phase lasers usually fall into *Class 2* where the beam is visible, and damage to the eye could be caused if the beam is directly stared at. There is no restriction on the use of such instruments at present. Some longer-range phase instruments fall into the category of *Class 3R* lasers. The safety requirements for this type of instrument are that only trained operators should use the equipment and warning signs should be erected.

(c) Accuracy

Traditionally, phase instruments are more accurate than pulse instruments, but the accuracy achievable by modern instruments has improved so greatly that the difference is insignificant in most measuring tasks. Only in the most precise instruments for industrial monitoring do phase systems have to be used.

Typical standard accuracies are:

	Phase	**Pulse**
Prism	1 mm + 1 ppm	3 mm + 3 ppm
Direct	3 mm + 2 ppm	3 mm + 3 ppm

'ppm' stands for 'parts per million' of the distance. The longer the distance the greater is the error.

As stated previously, the pulse beam has a greater divergence, which can lead to inaccuracies in direct measurement, but this can be processed out.

(d) Measuring speed

Typical measuring speeds for the best instruments are about three seconds for pulse and eight seconds for phase over a distance of 100 metres and proportionally more over longer distances. There is no difference when the distance is less than 40 metres. With such rapid measuring times, the difference might seem insignificant but the slower time causes much frustration when measurements are being made through flapping leaves or heavy traffic.

3. Factors affecting accuracy

As in all forms of surveying, care must be taken to minimize errors, particularly systematic or constant errors, which, although small, can be significant with multiple measurement as the effect is cumulative.

(a) Centring

The operation of the optical plummet should be checked occasionally while centring the instrument, by turning the instrument through the four quadrants of 90° and marking the four positions. The mean position of the four is used as a centre point and the footscrews are used to bring the optical plummet to that position. The bubble of the circular spirit level is re-centred by adjusting the capstan screws underneath the level.

(b) Atmospheric errors

The speed of light is affected by the atmospheric conditions of temperature and pressure. For long-distance measurement these can be significant. The daily atmospheric variables can be set in the instrument, which will then calculate a correction, or a ppm correction can be read from a graph or a table and set in the instrument.

(c) Prism constant

This is a correction, usually in mm, for the shape of the prism, and how it is centred over the point. For traversing, the prism sits in a tribrach plumbed over the station. For detail survey, the prism is mounted on a telescopic pole known in the trade as a 'pogo stick', which is kept vertical by means of a circular bubble.

Figs 7.33 and 7.34 show two types of prism.

(d) Sighting

Errors can be made when setting the cross-hairs in the exact centre of the prism/target. To aid this, yellow arrows are usually situated on the top and the side of the prism for an instrument where the light beam is coaxial with the telescope. For older instruments with separate optics above the main telescope, the side

Figure 7.33 Leica 360 degree prism

Figure 7.34 Geodimeter mini prism

arrows are below the prism. The more sophisticated instruments lock onto the centre of the prism automatically.

(e) Instrument calibration

Over time small maladjustments of the instrument can occur.

1. EDM errors

(i)   A *scale error* is caused if the modulation frequency varies from the design value. The error increases with the distance measured.

(ii)   A *zero (index) error* is caused if the electronic centre of the instrument does not coincide with its physical centre. In modern instruments this is designed to be negligible, but in older instruments it is included with the prism constant.

(iii) A *cyclic error* is caused by interference to the signal between the transmitter and the receiver circuits. This is manufactured to be almost negligible

in modern instruments by the design of the electrical components.

Many organizations will have a standard baseline where prism constants and atmospheric corrections can be checked, but full calibration is carried out at a laboratory-based service centre. The instrument's distance meter is bench tested. The telescope collimation and angle sensors are also checked and adjusted. If the collimation error is large the cross-hairs may also be adjusted. A large error may indicate damage to the instrument.

For major contracts, survey companies may be required to show current calibration certificates.

4. Instrument features

(a) Drive

On most instruments the traditional ball bearing rotational movement has been replaced with a friction drive, which does not require clamps and allows endless slow motion adjustment with servo motors used to assist the horizontal and vertical motion. The Trimble S6 uses 'Magdrive', a frictionless electromagnetic drive which allows very rapid turning in robotic mode.

(b) Direct reflex measurement of distance

An extra light source is added to allow measurement to objects without the need for a prism. In the latest instruments, one light source performs both prism and direct measurement.

(c) Laser pointer

If the direct light source is invisible then a red laser can be added which shows the spot to which the measurement is being made.

(d) Measuring mode

1. *Standard* for normal distance measurement.
2. *Fast* for a quick but less accurate distance.
3. *Tracking* for rapid setting out.
4. *D-Bar*, a system used by Trimble to take repeated measurement of angles and distances for high-accuracy work.

(e) Track light

The track light is a guide light unit composed of two lights, one red, the other green, which are projected from an aperture above or below the telescope. The system is used in setting out operations. The assistant who is positioning the stake-out peg sees green when he or she is to the left and red when he or she is to the right. When he or she is correctly aligned the red and green flash simultaneously.

(f) Memory

Modern instruments have internal memory capacity, varying from zero on a basic instrument to 10 000 points on the more sophisticated. Older machines required a data recorder. Storage can be supplemented with the addition of a 'flash card', which allows about 1500 points to be stored per megabyte.

(g) One-man operation

Robotic operation is possible with a keypad at the prism pole linked by radio or a laser beam to the instrument. The surveyor does not require an assistant with the instrument automatically following and locking onto the prism which the surveyor carries. When lock is lost the instrument will search for the prism using a wide-beam laser. This method facilitates coding as the surveyor follows the detail. In some localities, however, it is advisable to have an assistant guard the machine in case it is stolen.

Some manufacturers use a modular design where a machine with servo control can be upgraded to robotic with the addition of a radio control module. *Automatic locking* on to the target can be achieved with a wide-beam laser used in the search for the prism. *Automatic target recognition (ATR)* can select a specific prism, with the prism characteristics being recognized by a digital camera/sensor.

(h) Digital camera

Topcon have introduced a digital camera which shows the view through the telescope to help target identification. It can be used for edge detection when surveying, say, a window opening in a building survey.

(i) LCD display and operating system

Basic instruments have a small LCD display of about 4 lines with 24 characters. This will allow only a simple operating system. The top instruments have a large graphic display screen, and more manufacturers are standardizing their operating system on Windows CE for palmtops. The instruments can show plans and the survey plot.

(j) Programs

The application programs provided with the instrument vary. Basic machines offer only detail survey and basic setting out. Sophisticated machines offer a full range of programs, which will include full detail survey, extensive setting out including road line and level, coordinate geometry (COGO), free station/resection to fix station coordinates and even automatic scanning of a building or a quarry face.

(k) Bluetooth

This wireless communication allows the use of remote tablet or palmtop computers to control the instrument operation and display the data.

(l) Battery

A possible weakness in total station technology is the battery power source. The early NiCd batteries, which were internal in the instrument, required a large external battery to give a full day's work. They degraded rapidly, particularly with irregular charging. Modern NiMH; batteries are a great improvement, and the latest lithium–ion batteries are even better.

5. Applications of total stations

Table 7.9 shows a classification of total station instruments by usage.

(a) Construction surveying

A basic easy-to-use instrument is used for setting out foundations, vertical control and small site surveys. Prices are very competitive in this area.

(b) Engineering/land surveying

A more sophisticated instrument is used for complicated setting out such as road alignment, coordinate geometry calculations such as the height of overhead power cables, building or quarry face surveys and large detail surveys. Robotic operation results in cost savings in detail surveys.

(c) Industrial

A high-accuracy machine is required for the survey and monitoring of complex structures and installations, and in the dimensional control for the construction of large structures such as North Sea oil platforms.

Exercise 7.3

1 Make a sketch of a theodolite to show clearly the principal parts.

2 Describe briefly the tests and adjustments that should be made in order to ensure that a theodolite is in good working order.

3 Using a diagram, describe briefly any form of optical plummet built into a theodolite.

4 Describe briefly any method of measuring a horizontal angle such that errors are minimized.

5 Make a list of errors that may arise in measuring angles, using the following headings:

(a) Gross errors

Table 7.9 A classification of total stations by usage

Class	Applications	Features	Accuracy/range	Cost
Construction	Setting out	Basic setting out	Angles: 5″–10″	~£3000
	Site surveys	program	Distance: 5 mm +	∧
		Detail program	Range: 500–3000 m max	∧
		Limited storage	Reflex. 80 m	∧
		0–5000 pts		∧
		NiMH batteries		∧
				∧
				∧
Engineering	As above +	Track light	Angles: 1″–5″	∧
Land survey	More options	Motorized drive option	Distance: 2–5 mm +	~£4500
(basic)	COGO	Endless drive	Range: 3500 m max	∧
	Remote height	Large display	Reflex 250 m	∧
	Free station	Laser plummet		∧
		10000 pts		~£7000
				∧
				∧
Land survey	As above +	As above +	Angles: 1″–3″	∧
(sophisticated)	Road setting out	Large graphic display	Distance: 2–5 mm, +	∧
	Robotic	Flash card extra	Range: 3500 m max	~£20000
	Scanning	storage	Reflex 250 m	
	Grade control	Laser pointer		
		ATR. Autolock target		
		sighting.		
		Lithium–ion batteries.		
Industrial	Structural	Robotic remote control	Angles: 0.5″	~£15000
	monitoring	ATR	Distance: 2 mm	∧
		Autolock		∧
		High accuracy		~£20000

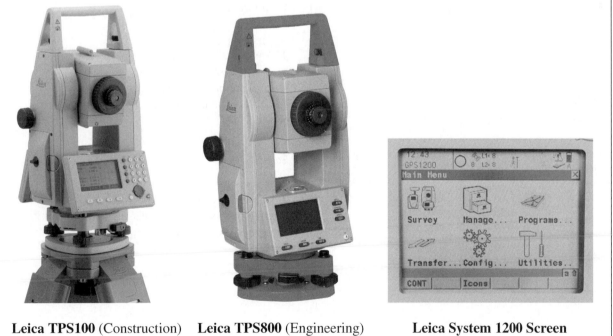

Leica TPS100 (Construction)   **Leica TPS800** (Engineering)   **Leica System 1200 Screen**
(Land Survey sophisticated)

Table 7.10

Instrument station	Target station	Face left reading	Face right reading	Mean horizontal angle	Survey line	Face	Zenith angle reading	Accepted zenith angle
B	A	00° 24′ 40″	180° 24′ 40″		AB	L	88° 20′ 20″	
	C	65° 36′ 20″	245° 36′ 20″			R	271° 39′ 00″	
C	B	00° 00′ 00″	179° 59′ 30″		BC	L	85° 50′ 50″	
	D	66° 34′ 20″	246° 33′ 50″			R	274° 08′ 30″	
D	C	332° 10′ 20″	152° 10′ 40″		CD	L	80° 33′ 00″	
	E	87° 08′ 00″	267° 08′ 20″			R	279° 26′ 20″	

(b) Systematic errors

(c) Random errors

6 Table 7.10 shows the horizontal and vertical circle readings, recorded during the measurement of angles, at survey stations B, C and D. Determine the accepted values of the angles.

Exercise 7.4

1 Outline the principle of electronic distance measurement (EDM) by:

(a) phase difference

(b) pulse.

2 List the relative advantages and disadvantages of the two methods.

3 Outline the features of a robotic total station.

6. Answers

Exercise 7.1

1 Table 7.11

Table 7.11

Observation station	Target station	Face left reading	Face right reading	Mean angle
D	C	00° 00′ 00″	179° 59′ 20″	
	E	189° 39′ 50″	9° 38′ 50″	
		189° 39′ 50″	189° 39′ 30″	189° 39′ 40″
E	D	185° 14′ 30″	5° 13′ 50″	
	F	6° 15′ 10″	186° 14′ 10″	
		181° 00′ 40″	181° 00′ 20″	181° 00′ 30″

Exercise 7.2

1 Table 7.12

Table 7.12

Line	Face	Vertical circle reading	Index error	Zenith angle
XY	L	79° 30′ 50″	−30″	79° 30′ 20″
	R	280° 30′ 10″		
		360° 01′ 00″		
YZ	L	102° 13′ 50″	−30″	102° 13′ 20″
	R	257° 47′ 10″		
		360° 01′ 00″		

Exercise 7.3

1 Answer page 114

2 Answer pages 127–131

3 Answer page 116

4 Answer page 122–125

5 (a) Gross errors: page 131

(b) Systematic errors: pages 127–131

(c) Random errors: page 131

6 See Table 7.13.

Exercise 7.4

1 Answer page 131–133

2 Answer page 134

3 Operated at prism pole using radio link; servo motion; search and lock function.

Table 7.13

Instrument station	Target station	Face left reading	Face right reading	Mean horizontal angle	Survey line	Face	Zenith angle reading	Accepted zenith angle
B	A	00° 24' 40"	180° 24' 40"		AB	L	88° 20' 20"	88° 20' 40"
	C	65° 36' 20"	245° 36' 20"			R	271° 39' 00"	
		65° 11' 40"	65° 11' 40"	65° 11' 40"		Sum	359° 59' 20"	
C	B	00° 00' 00"	179° 59' 30"		BC	L	85° 50' 50"	85° 51' 10"
	D	66° 34' 20"	246° 33' 50"			R	274° 08' 30"	
		66° 34' 20"	66° 34' 20"	66° 34' 20"		Sum	359° 59' 20"	
D	C	332° 10' 20"	152° 10' 40"		CD	L	80° 33' 00"	80° 33' 20"
	E	87° 08' 00"	267° 08' 20"			R	279° 26' 20"	
		114° 57' 40"	114° 57' 40"	114° 57' 40"		Sum	359° 59' 20"	

Chapter summary

In this chapter, the following are the most important points:

- Electronic theodolites have now largely replaced optical instruments.

- Optical instruments have a circular glass scale, the graduations of which can be viewed through a small eyepiece as the scale is illuminated by daylight reflected by a mirror.

- Electronic instruments have a circular glass scale with very fine divisions etched around its perimeter. These are illuminated by a light source and the reflected light is analysed by a sensor to count the divisions.

- The setting up of the instrument over the survey mark is known as the temporary adjustments.

- The plumb-bob is used to centre the tripod above the survey mark with the top of the tripod roughly level. The theodolite is attached and the circular bubble is levelled using the foot screws. The optical plummet is used for precise centring. Final levelling is carried out using the main spirit bubble and the foot screws.

- An alternative method is to use the optical plummet and foot screws for centring, and level the circular bubble using the tripod legs.

- Angles are measured in face left and face right positions to minimize errors.

- Consistent differences between readings on the two face positions indicate collimation or alignment errors in the instrument. These can be tested for and adjusted.

- Total station instruments measure both angles and distances using electronic angle sensors and electronic distance measurement (EDM) using light waves.

- The light source can be infrared or laser light. It is reflected from a prism held at the point to be fixed. With some instruments, it is also possible to measure to any surface by direct reflection over short ranges.

- Total station types range from simple instruments for use in building construction to sophisticated robotic types for land surveying and monitoring.

CHAPTER 8 Traverse surveys

In this chapter you will learn about

- the open and closed forms of traverse survey
- the method of conducting a traverse survey in the field
- the intricacies of making accurate linear and angular measurements and dealing with potential errors
- the methods of orientating a survey to a chosen north direction
- the various north directions to which a traverse can be orientated
- the different bearings ascribed to the lines of a traverse for calculation purposes
- the methods of converting angles into bearings
- the systems of polar and rectangular coordinates and the calculations required to change one system into the other in traverse computations
- the calculation of open and closed traverses and the method of adjusting the latter
- the plotting of a traverse using rectangular coordinates

In order to survey any parcel of ground, two distinct operations are required, namely

    (a) a framework survey and
    (b) a detail survey.

A framework survey consists of a series of straight lines, arranged in the form of triangles (linear surveys), polygons (closed traverse surveys) or vectors (open traverse surveys). A detail survey consists of a series of offsets, which are added to and supplement the framework survey.

In Chapter 3, a potential building site was surveyed by linear means (Fig. 3.13). A linear survey, however, may not be accurate enough, in which case, a theodolite traverse survey would be required.

A traverse survey consists of a series of survey lines, connected to each other, each line having length and direction. They are, therefore, vectors. The vectors, may or may not close to form a polygon.

1. Types of traverse

Theodolite traverses are classified under the following three headings.

(a) Open traverse

In Fig. 8.1, points A, B, C and D are the survey stations of an open traverse, following a stream that is to be surveyed and plotted to scale. Lines AB, BC and CD are the measured legs of the traverse and angles ABC and BCD (shown hatched) are the clockwise angles, measured using a theodolite. Together they form the survey framework.

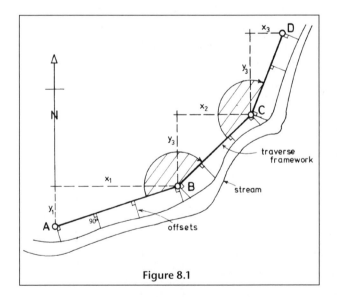

Figure 8.1

The detail survey is carried out from the legs of the traverse, by offsetting, in the same manner as for linear surveying. This type of traverse is not self-checking and errors in either angular or linear or both measurements may pass unchecked. The only check that can be provided is to repeat the complete traverse or re-survey the traverse in the opposite direction from D to A. The problem should be avoided by making a closed traverse whenever possible.

In order to plot the traverse stations, the horizontal length and direction relative to north of each line must be found. In other words, the vector data are required. These data are then transformed into rectangular coordinates x_1, y_1, etc., which are plotted to scale on plan.

(b) Closed traverse

Figure 8.2 shows a closed traverse round a building, where the vectors AB, BC, CD and DA form a closed polygon. The lengths of the lines and the values of the clockwise angles are again measured. The angles are the interior angles of the polygon and are measured as the theodolite is moved from station to station around the traverse.

The traverse is self-checking to some extent, since the sum of the interior angles should equal $[(2n - 4) \times 90°]$, where n is the number of angles. In this case, therefore, the sum should be 360°.

In a closed traverse, the exterior angles may be meas-ured in preference to the interior angles, in which case the sum should be $(2n + 4) \times 90°$.

Again the vector data, i.e. the horizontal length and direction of each line, relative to north, are required. These data are transformed into rectangular coordin-ates x_1, y_1, ..., x_4, y_4, which are plotted to scale.

EXAMPLE

1 Determine the sum of the *exterior* angles of the traverse of Fig. 8.2.

Answer

$$\text{Sum} = (2n + 4) \times 90°$$
$$= (8 + 4) \times 90°$$
$$= 1080°$$

(c) Traverse closed between previously fixed points

In this type of survey, the traverse begins on two points A and B of known bearing and ends on two different known points C and D (Fig. 8.3). The survey is once again self-checking in that the bearing of line CD deduced from the traverse angles should agree with the already known bearing of CD.

2. Basic principles of traversing

In any traverse, the survey fieldwork consists of:

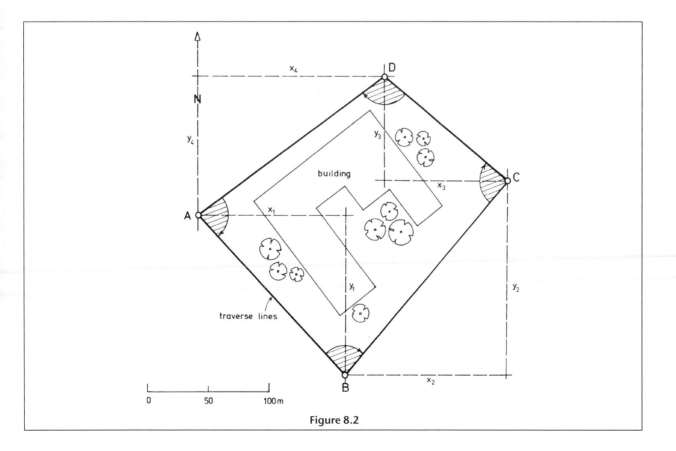

Figure 8.2

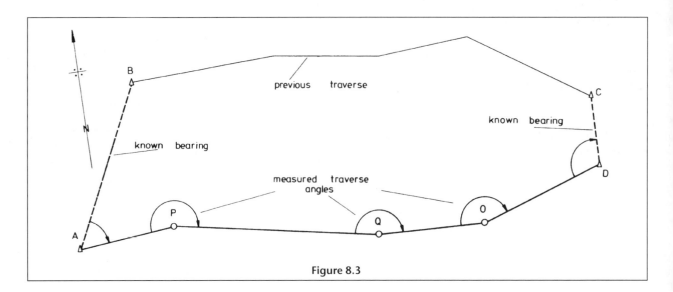

Figure 8.3

(a) measuring the slope length of every line, preferably by (i) electromagnetic means (EDM) or if unavailable by (ii) taping,
(b) measuring the angle of gradient of every line,
(c) measuring the clockwise horizontal angle between adjacent lines, using a theodolite or total station instrument,
(d) obtaining the bearing, i.e. the direction relative to north of one line of the traverse.

The processing of the fieldwork results consists of calculating:

(a) the horizontal length of every line of the traverse from the slope length and angle of gradient,
(b) the bearing of every line of the traverse from a known bearing or observed bearing and the horizontal angles,
(c) the rectangular coordinates of each station of the traverse relative to the survey origin.

3. Fieldwork

It is good practice in traversing to close the survey, but in order to facilitate an understanding of the basic principles of traversing a simple open traverse ABCD (Fig. 8.4) along the bank of a stream will be used as a basis for the fieldwork in the following description.

Usually, three or four persons are required to conduct a survey. Their duties are:

(a) to select suitable stations,
(b) to measure the distances between the stations,
(c) to erect, attend and move the sighting targets from station to station,
(d) to measure and record the angles,
(e) to reference the stations for further use.

(a) Factors influencing choice of stations

On arrival at the site, the survey team's first task is to make a reconnaissance survey of the area with a view to selecting the most suitable stations. Generally the stations have to be fairly permanent and concrete blocks are generally formed *in situ*. A bolt or wooden peg is left in the concrete to act as a centre point. Where the survey mark is to be sited on a roadway a masonry nail is hammered into the surface and a circle painted around it.

The positions of the stations are governed by the following factors:

1. Easy measuring conditions
Since the angles can be measured accurately with the theodolite or total station, the linear measurements must be made with comparable accuracy.

If EDM is the chosen method of measuring, the ground conditions are irrelevant, since the instrument can measure over any terrain.

Frequently EDM is not available in educational establishments and in small building firms, because of large numbers of students or of cost, so great care must be taken in establishing the station positions.

The measurements will be made along the ground surface, using a steel tape; therefore the surface conditions must be conducive to good measuring, for example roadways and paths which are smooth and usually regularly graded. Areas of long grass, heaps of rubble or undulating ground should be avoided.

2. Avoidance of short lines
Short lines should be avoided wherever possible. Angular errors are introduced when targets at short range are not accurately bisected. For example, taking a sight which is off target by 2 mm at 10 m range is equivalent to sighting 40 mm off target at 200 m range.

3. Sighting the actual station point
The actual station point should be visible for sighting if interchangeable tripods or EDM tripod targets are not

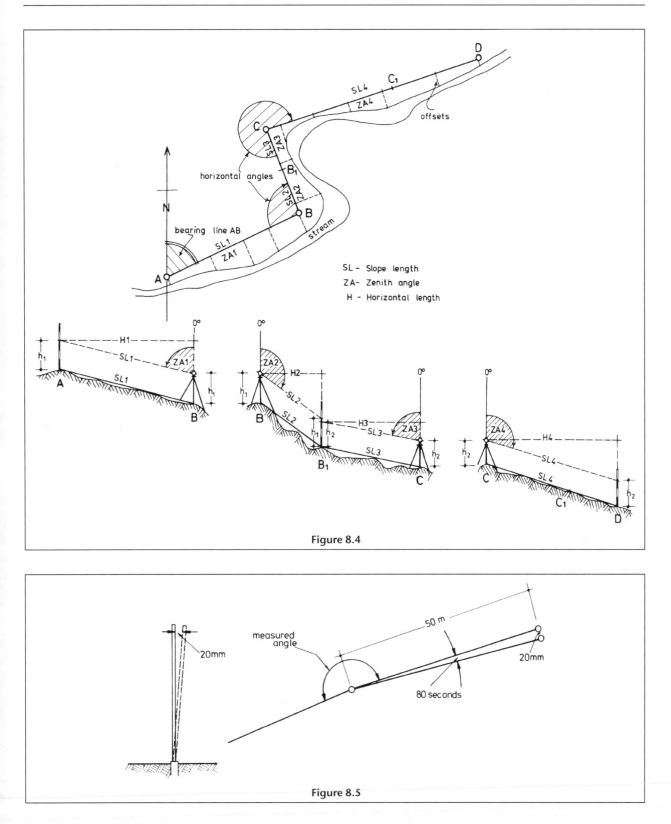

SL – Slope length
ZA – Zenith angle
H – Horizontal length

Figure 8.4

Figure 8.5

available. If the target is a ranging rod erected over a station, it must be plumbed by using a builder's spirit level and not judged by eye. EDM prism poles are fitted with spirit levels so the problem should not arise. Failure to observe this simple precaution will result in a substantial error in the measurement of the horizontal angle. For example, in Fig. 8.5 a ranging rod has been erected 20 mm off the vertical position at the top.

At 50 m range the error incurred by sighting the top of the rod instead of the ground station at the bottom is 80 seconds.

4. Referencing the stations
If possible, the stations should be chosen near to some permanent objects, e.g. trees, lampposts etc., so they can be easily found at a later date.

(b) Linear measurement

It has already been pointed out that in traverse calculations, the true plan length, i.e. the horizontal lengths of all lines of the traverse, is required. These are derived from the measured slope lengths, which are obtained by EDM or by taping.

Electromagnetic distance measurement (EDM)
EDM is undoubtedly the preferred method of measuring traverse lines. All EDM instruments can measure slope lengths. The most modern models can convert these slope lengths to plan lengths and difference in height between stations using on-board software.

At this juncture, the student is directed to study Chapter 7 (Principles of EDM), where it was convenient to explain the mechanics of measuring and the various sources of error in EDM measurement, in conjunction with the section dealing with total stations. There is therefore no need for further explanations at this or any other point in the text.

Taped measurement
Measuring by steel tape is an integral part of linear surveying, which is the subject of Chapter 3. In that chapter, an explanation of taping and errors *sufficient only for that particular form of surveying* was adequately covered. There is therefore a need to explain further the intricacies and difficulties of making accurate measurement by taping.

Measuring a line accurately by tape is the most difficult task in surveying. The surveyor must be aware, at all times, of the errors which affect linear measurement. The constant errors (misalignment, standardization and slope) and the systematic errors (temperature, tension and sag) must all be taken into account in the accurate measurement of traverse lines and in setting out construction works.

The fact that these errors must be accommodated suggests that:

1. Each tape length should be aligned using the theodolite.
2. The tape should be standardized before use, by comparing it with a new tape kept solely for that purpose and noting the difference in length. If there is a difference of more than 10 mm the tape should be discarded. However, it is common for tapes to have stretched slightly and, being expensive, they cannot easily be thrown away. They are usually used in their existing condition and the appropriate correction applied to the measured length as follows:

   Correction (c) is
   $$c = (L - l) \text{ per tape length} \qquad (1)$$
   where L = actual length of tape
   l = nominal length

3. Measurements (L) should be made on the slope and the angle of slope determined, either by finding the difference in level (h) between the top and bottom of the slope by levelling, or by observing angles of inclination (A) along the slope by theodolite.

   The plan length using the difference in level method is

   $$P = (L^2 - h^2)^{1/2}$$

   By mathematically applying a binomial expansion of this expression, it can be shown that the correction (c) to be applied to any slope length (L) is

   $$c = -h^2/2L \qquad (2a)$$

   The plan length (P) using the angle of inclination method is the familiar

   $$P = L \cos A$$

   The correction is of course the difference between slope and plan lengths, which is

   $$c = L - P$$
   $$= L - (L \cos A)$$
   $$= L(1 - \cos A) \qquad (2b)$$

   The correction is always negative.

4. Air temperature should be recorded at the time of measurement and appropriate corrections applied. A tape is the correct length when the temperature is the standard 20°C. Any rise or fall in temperature causes the tape to expand or contract. The correction involves using the coefficient of thermal expansion of steel, which is 0.000 011 2 per degree Centigrade. The correction (c) is thus

   $$c = 0.000\,011\,2\,L\,(T - t) \qquad (3)$$
   where T = actual temperature
   t = standard temperature

5. The correct tension should be applied when pulling the tape taut. The correct tension is 50 N and should be applied either by attaching a spring balance to the tape or by using a constant-tension handle. Using a spring balance is very difficult and is not recommended. A constant-tension handle is simple to use since the correct tension is always applied, thus obviating tension corrections. This is the preferred option in construction surveying since the effect of non-standard tension is negligible.

   If tension correction is absolutely necessary, the correction (c) to be applied to any length (L) is

   $$c = \frac{L\,(T - t)}{AE} \qquad (4)$$
   where T = actual tension applied
   t = standard tension
   A = cross-sectional area of tape (mm^2)
   E = Young's modulus of elasticity, which is (200 000 N per mm^2)

6. The tape should not be allowed to sag under any circumstances. There is always an alternative. If, for

example, the tape has to stretch over a hollow in the ground, the distance should be divided into two or more sections and the distances measured with their respective gradients.

If sag cannot be avoided, and sag correction becomes necessary, the correction (c) to be applied to any length (L) is

$$c = \frac{-w^2 \, L^3 \cos^2 A}{24 \, T^2} \qquad (5)$$

where w = weight of tape (N/m)

$\qquad T$ = actual tension applied.

Assuming that all of the above corrections are required, they are applied as follows:

EXAMPLE

2 The following are the results of the measurement of a survey line between two tripod heads, where the tape was allowed to sag.

Line	AB
Slope length	23.510
Vertical angle	+5° 00′
Temperature	15°C
Tension	70 N

The tape weighed 0.20 N/m; had a cross-sectional area of 2 mm^2 and was actually 30.005 m long at the time of measurement.

Calculate the corrected horizontal length of the line AB.

Answer

(i)   Standardization
   $c = (L - l)$ per tape length
   $= (30.005 - 30.000) \, 23.510/30$
   $= +0.0039$ n

(ii)   Slope
   $c = L \, (1 - \cos A)$
   $= 23.510 \, (1 - 0.996\,194\,7)$
   $= 23.510 \, (0.003\,805\,3)$
   $= -0.0895$ m

(iii)   Temperature
   $c = 0.000\,011\,2 \times 23.510 \times (15 - 20)$
   $= -0.0013$ m

(iv)   Tension
   $c = \dfrac{L \, (T - t)}{AE}$
   $= \dfrac{23.510 \, (70 - 50)}{2 \, (200\,000)}$
   $= 470.2/400\,000$
   $= +0.0012$ m

(v)   Sag
   $c = \dfrac{-w^2 \, L^3 \cos^2 A}{24 \, T^2}$
   $= \dfrac{-0.2^2 (23.510^3) \, (\cos^2 5°00')}{24 \, (70^2)}$
   $= \dfrac{-0.04 \, (12\,994.5) \, (0.992\,404)}{24 \, (4900)}$
   $= -0.0044$ m

Total correction
   $= (+0.0039 - 0.0895 - 0.0013 + 0.0012$
   $\qquad - 0.0044)$ m
   $= 0.0901$ m

Corrected horizontal length AB
   $= (23.510 - 0.086)$ m
   $= 23.420$ m

In engineering and construction surveying it is normal practice to dispense with tension and sag corrections by (i) using a tension handle and (ii) not allowing the tape to sag at any time. The description of the actual physical measurement of traverse lines follows Exercise 8.1.

Exercise 8.1

The following data provides an example of a long site baseline which has been divided into six sections of approx 50 m and measured using a nominal 50 m tape. The tape was kept on the ground and tensioned using a constant-tension handle. The difference in level of the ends of the sections and the mean temperatures have been measured.

Sect	Slope (m)	Difference In level	Temp Deg C
1	49.825	+0.316	15.0
2	49.786	+0.526	14.5
3	49.872	+0.434	14.0
4	49.846	+0.321	14.0
5	49.852	+0.100	15.0
6	35.865	-0.210	13.0

Calibrated length of tape	= 50.003 m
Standard temperature	= 20°C
Coefficient of expansion	= 11 × 10^{-6} per °C

Calculate the corrected horizontal length of the line.

Linear measuring equipment
Assuming that some or all of the corrections are to be applied, the following equipment is required to measure a traverse line:

1. A steel tape graduated throughout in metres and millimetres, checked against a standard.
2. A spring balance or preferably a constant-tension handle.
3. Measuring arrows and ranging poles.

4. Marking plates or pegs for use in soft ground. A marker plate can be made quite easily from a piece of metal, 100 mm square, if the corners are bent to form spikes. On hard surfaces, the surface is simply chalked and marked with a pencil.
5. A theodolite for lining up the pegs or marking plates.

Measuring procedure

Four persons are required to measure a line with a high degree of accuracy. Two are stationed at the rear and two at the forward end of the steel tape. In Fig. 8.4, the three lines AB, BC and CD of the traverse have to be measured. The ground conditions differ along the lines, necessitating a slightly different measuring technique for each line.

Line AB

The line is shorter than the length of the 30 m tape. The procedure for measuring is as follows:

1. The tape is unwound and laid across the two pegs A and B.
2. One person anchors the forward end by putting a measuring pin through the handle into the ground (Fig. 8.6).
3. A second person attaches the spring balance (or constant-tension handle) to the rear end of the tape, tightens it and anchors it by putting a second pin through the handle. The pin is levered back until the correct tension of 50 N is registered on the spring balance. (A constant-tension handle is simply pulled until the tape clamp clicks into position.) The command 'read' is given to the third and fourth persons, stationed at either end of the tape.
4. They read the tape against the nails in the pegs and the readings are recorded in the field book on line 1 (Table 8.1). If temperature corrections are to be made, the air temperature is noted and recorded.
5. The difference between the readings is worked out.
6. The forward anchor person moves his or her pin slightly and the whole procedure is repeated to produce a second set of readings (line 2).

Table 8.1

	Line	Rear	Forward	Length	Mean (m)
1	AB	0.419	26.150	25.731	
2		0.365	26.098	25.733	
3		0.123	25.855	25.732	25.732
4	BB_1	0.511	11.611	11.100	
5		0.483	11.583	11.100	
6		0.276	11.376	11.100	11.100
7	BC	0.375	12.576		
8		0.675	12.876		
9		0.203	12.403		
10	CC_1	0.110	29.911		
11		0.122	29.923		
12		0.131	29.933		
13	C_1D	0.567	24.077		
14		0.523	24.034		
15		0.295	23.806		

Standard tension	50 N
Nominal tape length	30.000 m
Actual tape length	30.003 m

7. A third set of readings is obtained (line 3) and the mean of the three sets is calculated.
8. The gradient of the line is measured by theodolite. (A full description of the techniques involved is given in Sec. 3(c), following.)

Line BC (Fig. 8.4)

The line BC is again shorter than the length of the tape but the ground surface undulates considerably along its length. The line must be measured in sections or bays, the lengths of the bays being governed by the occurrence of the undulations.

Line BC is divided into two bays, B to B_1 and B_1 to C. A marker plate or peg is inserted on the line to denote point B_1. Each bay is measured in exactly the same manner as line AB. The results of the measurement are noted in the field book (Table 8.1) on lines 4 to 9, where bay 1 of line BC has been reduced to produce a slope length of 11.100 metres.

The gradient of each bay is measured separately, using a theodolite (Sec. 3(c)).

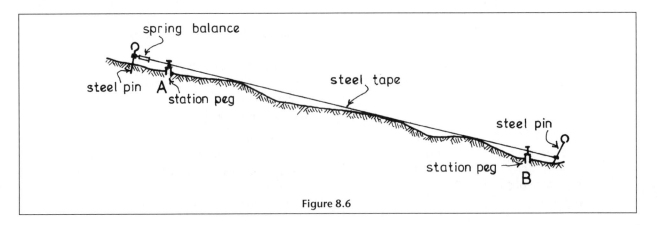

Figure 8.6

Line CD (Fig. 8.4)

Line CD is longer than the length of the tape and the ground has a regular gradient along the whole length of the line. The line must again be divided into bays C to C_1 and C_1 to D. A marker plate or peg is inserted on the line at approximately 29 metres to denote point C_1. Each bay is measured in exactly the same manner as line AB and the results noted in the field book (Table 8.1). The gradient of the whole line CD is measured using a theodolite (Sec. 3(c)). It is not necessary to measure the gradient of each bay separately in this case, although this is often done.

An alternative method of measuring is commonly used. Complete 30 m tape lengths are marked on the plates and summed at the end of the line. This method is open to more errors than the former and the line should be remeasured in the opposite direction.

EXAMPLE

3 Table 8.1 shows the results of the linear measurement of lines AB, BC and CD of the open traverse (Fig. 8.4). The table has been partially completed.
(a) Complete the reduction of the results, to produce the mean slope length of each bay of lines BC and CD.
(b) Correct the results for standardization.

Answer (Table 8.2)

Table 8.2

	Line	Rear	Forward	Length	Mean (m)
1	AB	0.419	26.150	25.731	
2		0.365	26.098	25.733	
3		0.123	25.855	25.732	25.732
4	BB$_1$	0.511	11.611	11.100	
5		0.483	11.583	11.100	
6		0.276	11.376	11.100	11.100
7	B$_1$C	0.375	12.576	12.201	
8		0.675	12.876	12.201	
9		0.203	12.403	12.200	12.201
10	CC$_1$	0.110	29.911	29.801	
11		0.122	29.923	29.801	
12		0.131	29.933	29.802	29.801
13	C$_1$D	0.567	24.077	23.510	
14		0.523	24.034	23.511	
15		0.295	23.806	23.511	23.511
	CD				53.312

Standard tension 5 kg
Nominal tape length 30.000 m
Actual tape length 30.003 m

Standardization correction

Correction = $(L - l)$ per tape length
= (30.003 − 30.000) per tape length

Line AB

Number of tape lengths = 25.732/30=0.858
Correction c = + 0.003 × 0.858
= + 0.003 m
Correct slope length = 25.732 + 0.003
= 25.735 m

Line BB$_1$

Correction c = + 0.003 × 11.100/30
= + 0.003 × 0.37
= + 0.001
Correct slope length = 11.100 + 0.001
= 11.101 m
Similarly,
correct slope length CB$_1$ = 12.201 + 0.001
= 12.202 m
and CD = 53.312 + 0.005
= 53.317 m

(c) Angular measurement

In Chapter 7, the procedure for measuring horizontal angles and zenith angles was fully described. The actual technicalities of measuring depend upon the type of theodolite.

In this case, any kind of theodolite could be used and set to zero for each face left measurement. Table 8.3 (columns 1 to 5) shows the readings obtained during the measurement of horizontal angles ABC and BCD. Lines 3 and 6 show that the difference in angular value between face left and face right is consistent and

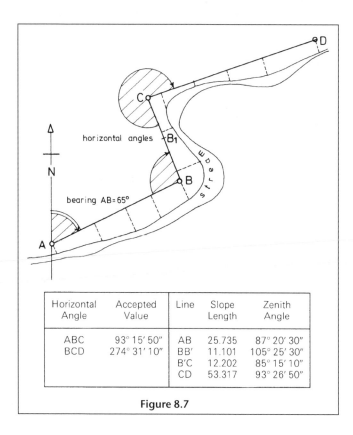

Horizontal Angle	Accepted Value	Line	Slope Length	Zenith Angle
ABC	93° 15' 50"	AB	25.735	87° 20' 30"
BCD	274° 31' 10"	BB'	11.101	105° 25' 30"
		B'C	12.202	85° 15' 10"
		CD	53.317	93° 26' 50"

Figure 8.7

Table 8.3

	1	2	3	4	5 Mean horizontal angle	6	7	8 Vertical circle reading	9
	Instrument station	Station observed	Face left (FL)	Face right (FR)		Line	Face		Zenith angle
1	B	A	00° 00′ 00″	179° 59′ 40″		BA	L	87° 20′ 40″	87° 20′ 30″
2		C	93° 15′ 40″	273° 15′ 40″			R	272° 39′ 40″	
3	AB̂C	=	93° 15′ 40″	93° 16′ 00″	93° 15′ 50″			360° 00′ 20″	
4	C	B	00° 00′ 00″	180° 00′ 00″		BB₁	L	105° 25′ 40″	105° 25′ 30″
5		D	274° 31′ 00″	94° 31′ 20″			R	254° 34′ 40″	
6	BĈD	=	274° 31′ 00″	274° 31′ 20″	274° 31′ 10″			360° 00′ 20″	
7						CB₁	L	85° 15′ 20″	85° 15′ 10″
8							R	275° 45′ 00″	
9								360° 00′ 20″	
10						CD	L	93° 27′ 00″	93° 26′ 50″
11							R	266° 33′ 20″	
12								360° 00′ 20″	

acceptable. The mean values are therefore taken to be satisfactory.

The angles of inclination along lines BA, BB₁, CB₁ and CD are shown in columns 6 to 9. It should be noted that these are zenith angles. Again the face left and face right values (column 8) show consistency and the values computed in column 9 are therefore acceptable.

Tables 8.2 and 8.3 are the records of the acceptable values of all measurements of the traverse ABCD, which are used in the calculation of the rectangular coordinates of the stations. For convenience, they are shown together as Fig. 8.7 (see page 147).

Exercise 8.2

1 Table 8.4 and Fig. 8.8 show the layout and survey data of the closed traverse ABCDEF, which was used to survey the site of Fig. 3.1.

From the table, determine

(a) the six clockwise measured angles FAB, ABC, BCD, CDE, DEF and EFA of the traverse
(b) the angular error of the traverse
(c) the slope lengths of the six lines AB, BC, CD, DE, EF and FA
(d) the vertical angle of gradient of each line

(d) Multiple tripod system

Figure 8.9 is an example of a modern traverse outfit, comprising a theodolite and several targets, each of which has its own tripod. The theodolite and targets detach from the tribrach and are interchangeable. The

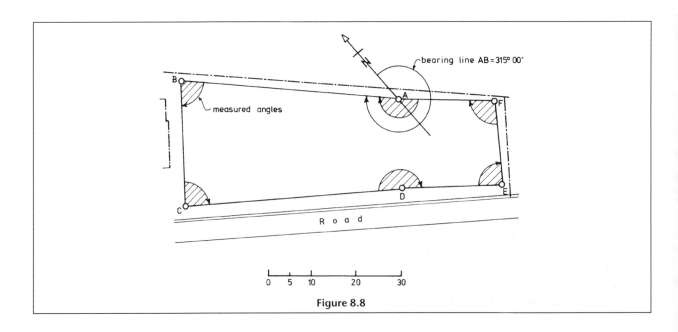

Figure 8.8

Table 8.4

Instrument station	Station observed	Face left	Face right	Mean observed angle	Line	Face	Vertical circle reading	Zenith angle	Measured length			
									Rear	Forward	Length	Mean
A	F	00° 00' 00"	180° 00' 20"		AB	L	91° 30' 40"		0.150	48.796		
	B	183° 33' 20"	3° 33' 40"			R	268° 28' 40"		0.325	48.969		
									0.200	48.842		
B	A	00° 00' 00"	179° 59' 40"		BC	L	95° 32' 00"		0.350	27.874		
	C	83° 41' 20"	263° 41' 40"			R	264° 27' 20"		0.750	28.272		
									0.523	28.049		
C	B	00° 00' 00"	180° 00' 20"		CD	L	89° 29' 10"		0.360	48.790		
	D	86° 48' 00"	266° 48' 40"			R	270° 30' 30"		0.820	49.250		
									0.600	49.030		
D	C	00° 00' 00"	180° 00' 00"		DE	L	90° 37' 20"		0.200	22.350		
	E	182° 39' 40"	2° 39' 40"			R	269° 22' 20"		0.090	22.240		
									0.325	22.475		
E	D	00° 00' 00"	179° 59' 40"		EF	L	86° 13' 30"		1.115	19.236		
	F	89° 15' 40"	269° 15' 40"			R	273° 45' 50"		1.300	19.416		
									0.800	18.911		
F	E	00° 00' 00"	180° 00' 00"		FA	L	90° 02' 50"		0.326	22.121		
	A	94° 00' 40"	274° 00' 40"			R	269° 56' 30"		0.250	22.045		
									0.200	21.995		

Standard tape tension   5 kg
Nominal tape length   50.000 m
Actual tape length   50.000 m

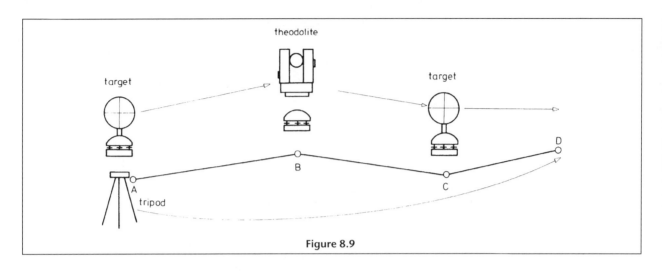

Figure 8.9

tribrach, of course, is fitted with an optical plummet, three levelling screws and plate bubble.

For the measurement of any angle, three plumbing operations are necessary regardless of the form of target being used, resulting in the expenditure of much time and effort and, of course, with such a system small plumbing errors must inevitably occur. When three tripods are available the errors are eliminated. In the measurement of angle ABC in Fig. 8.9, the tripods are plumbed and levelled over the stations.

Targets are placed at stations A and C and the theodolite at station B.

The angle ABC is measured in the normal way. Angle BCD is measured by leap-frogging tripod A to station D; target A to B; theodolite B to C; and target C to D. There is no need to centre the theodolite at C since the tripod

is already over the mark. Similarly, the target at back station B is correctly centred.

(e) Detail survey

The details that have to be added in the case of the traverse ABCD (Fig. 8.4) are simply the banks of the stream. These are surveyed by offsetting at frequent intervals along the various traverse lines. All of the offsets are shown on three pages of booking in Fig. 8.10. The method of offsetting is explained in Chapter 1, Sec. 5, and Chapter 3, Sec. 3.

(f) Summary of fieldwork in traversing

The first three sections of this chapter have dealt fully with the fieldwork involved in making a traverse survey.

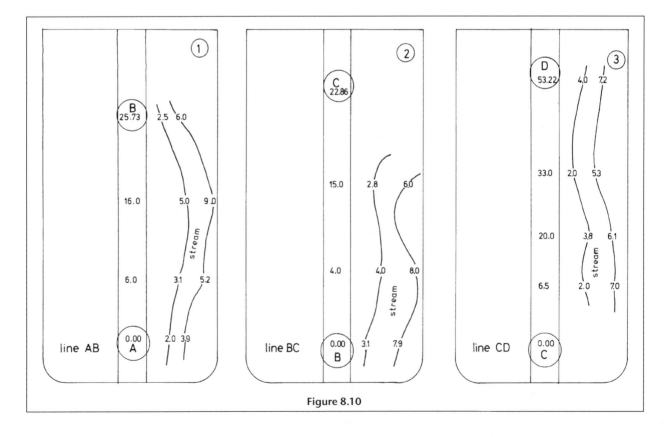

Figure 8.10

Section 1 described the three forms which a traverse may take, namely:

(i) *Open*. This form of traverse is not recommended as there is no check on the fieldwork. It was used in this chapter as a teaching tool in order to present the basic principles of traversing without distraction.

(ii) *Closed*. In this form of traverse the lines form a polygon; therefore there is a field check on the angular measurements and a coordinate closure check when complete calculation has taken place (following).

(iii) *Closed on previously fixed points*. The traverse begins on one line which has been previously coordinated and finishes on a second coordinated line. There should therefore be agreement between the bearings of the two lines and a coordinate check on completion of the calculations (following).

Section 2 listed the fieldwork quantities which have to be obtained, namely:

(i) the slope distance and gradient along each line,
(ii) the horizontal clockwise angles between lines,
(iii) the bearing of one line of the traverse.

It also listed the steps required to convert the fieldwork data to rectangular coordinates in later calculations.
   Section 3 dealt in detail with:

(i) the factors influencing station positioning,
(ii) the methods of measuring the lines of a traverse, namely EDM and ground taping,
(iii) the various errors associated with taping and the corrections applied to minimize them,
(iv) the measurement of horizontal and zenith angles,
(v) the booking of fieldwork data,

(vi) the multiple tripod system,
(vii) detail survey.

Section 4 of this chapter will concentrate on how the fieldwork is translated into rectangular coordinates.

4. Traverse calculations

The principal objective of a traverse is the production of a scaled drawing showing the traverse stations. Two methods are available for plotting the survey, namely:

(i) Graphical, in which the stations are plotted using a scale rule and protractor. This method is not practised commercially and will not be considered further.

(ii) Mathematical, in which the rectangular coordinates of each station are computed either manually, by using the polar/rectangular function of a calculator, or by computer, usually in the form of a spreadsheet.

In a coordinate calculation, the horizontal (plan) length of every line must be calculated from the field data as a first step.

(a) Plan lengths from zenith angles

Figure 8.11 shows the general situation, where a theodolite or total station, set up at a station X, is sighting a staff or target on point Y at the correct instrument height h. The zenith angle of θ degrees is recorded.

In the right-angled triangle formed through the theodolite telescope, length S equals the slope length XY and length H is the plan length of the line XY.

In the right-angled triangle

$$H/S = \text{sine } \theta$$
$$\text{Therefore } H = S \times \sin \theta$$

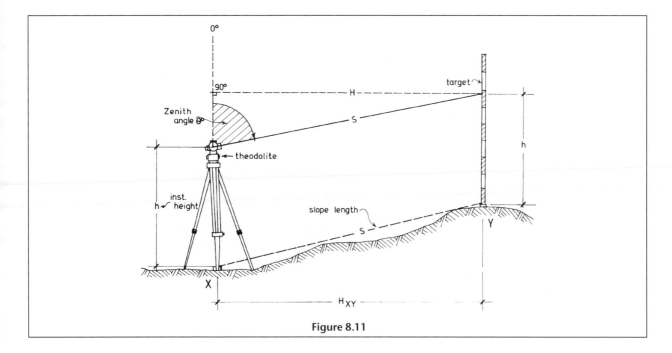

Figure 8.11

i.e. Horz. length = slope length × sin (zenith angle)
This formula holds good for all zenith angles.

EXAMPLE

4 Refer to Fig. 8.4 showing the zenith angles and slope lengths of traverse ABCD. The field data are as follows:

Line	Slope length	Zenith angle
AB	25.735	87° 20′ 30″
BB′	11.101	105° 25′ 30″
B′C	12.202	85° 15′ 10″
CD	53.317	93° 26′ 50″

Calculate the plan lengths of lines AB, BC and CD.

Answer

Plan length = slope length × sin (zenith angle)
Plan AB = 25.735 × sin (87° 20′ 30″)
     = 25.707 m
Plan BC = 11.101 × sin (105° 25′ 30″)
     + 12.202 × sin (85° 15′ 10″)
     = 10.701 + 12.160
     = 22.861 m
Plan CD = 53.317 × sin (93° 26′ 50″)
     = 53.221 m

(b) Orientation of traverse surveys

The second step in the calculation of a traverse is the calculation of the bearing of each line. The bearing of a line is its direction relative to some north point, of which there are several.

The theodolite in its conventional form does not measure bearings directly, that is, it cannot measure angles from north, so some method must be used to find a north direction. The following are four widely accepted methods of doing so.

Use of an Ordnance Survey plan

An OS plan is gridded at 100 metre intervals with the National Grid, the northward lines of which point to Grid North.

On a traverse, one of the lines is plotted as accurately as possible on an OS plan, from measurements taken to points of detail, such as road and fence junctions, corners of buildings etc. (Fig. 8.12). The angle between the traverse line and a Grid line can be measured on the plan by protractor with a small margin of error, less than one degree. This accuracy is acceptable for a small construction site. It simply means that the complete traverse will be wrongly orientated by the error incurred. Strictly, therefore, the survey is in error absolutely but not relatively. If the survey must be absolutely correct, say in a roadway engineering project, the following second method would be used.

Survey based on two known points

Two coordinated points (hopefully in the neighbourhood) are chosen on which to begin the survey. These may be OS stations or points of some other previous traverse. The bearing and distance between the points are required. Figure 8.12 illustrates the situation where point Q has coordinates of 1432.25 m East and 1010.32 m North while point P has coordinates of 1056.30 m East and 753.19 m North.

The difference in eastings is ΔE, while the difference in northings is ΔN.

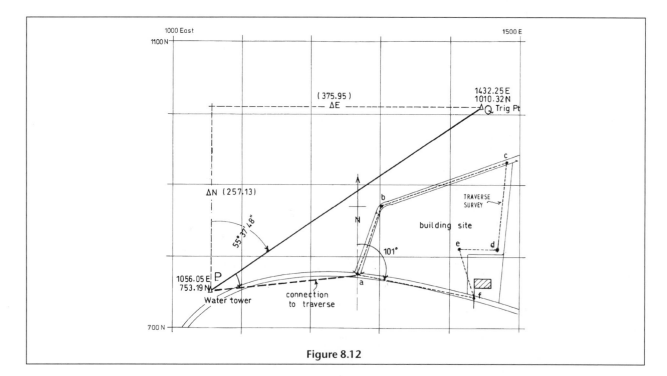

Figure 8.12

$$\Delta E_{PQ} = 1432.25 - 1056.30$$
$$= 375.95 \text{ m}$$

and $\Delta N_{PQ} = 1010.32 - 753.19$
$$= 257.13 \text{ m}$$

By the theorem of Pythagoras,

$$\text{Length PQ} = \sqrt{375.95^2 + 257.13^2}$$
$$= \underline{455.47 \text{ m}}$$

Bearing of line PQ = Θ
$$\text{Tan } \Theta \ = \Delta E/\Delta N = (375.95/257.13)$$
$$= 1.462\ 100\ 9$$
$$\Theta = 55.629\ 928\ 0$$
$$= \underline{55° \ 37' \ 48''}$$

Compass bearings
The bearing of any line of a traverse can be found within a few degrees by using a compass. The bearing will be relative to Magnetic North. This method is not recommended since it is prone to error but it is useful in remote areas, woodlands etc.

Assumed bearings
On local surveys of small construction sites, it may not be important or necessary to relate the survey to any form of north. Some arbitrary point is chosen as reference object (RO) and treated as being the equivalent of the North Pole. Common reference objects are tall chimneys, church steeples or simply pegs hammered into the ground in locations where they can be found easily. A clockwise angle is then measured at a traverse station from the RO to a second traverse station, thus establishing the arbitrary bearing between the stations.

(c) Bearings

1. Whole circle bearings
No matter which of the above three north points is chosen as the reference direction for the survey, the angle measured in a clockwise direction between that north direction and any survey line is the *whole circle bearing (WCB)* of the line. [From this point forward in the text, all bearings will be called whole circle bearings, regardless of their north point reference.]

Figure 8.13 shows a north point of a survey drawn from a survey station X.

The line XA makes an angle of 40° with the north direction XN while lines XB and XC make angles of 160° and 290° respectively. The maximum value of any angle is 360°, i.e. the angle can have any value on the whole circle of graduations provided it is measured in a clockwise direction. The angle of 160° formed by the north point XN and the line XB is the whole circle bearing of line XB.

EXAMPLE

5 Refer to Fig. 8.13 and estimate the whole circle bearings of lines XD, XE, XF and XG.

Answer

XD	10°
XE	100°
XF	200°
XG	350°

---

2. Back and forward bearings
Figure 8.14 shows a line joining two stations A and B where the whole circle bearing from A to B is 63°. If an observer were to stand at station A and look towards station B he or she would be looking in a direction 63° relative to the chosen north point. In surveying terms, the observer is looking *forward* towards station B from station A, and the forward whole circle bearing of the line AB is 63°. If the observer now stood at station B and looked towards station A he or she would be looking *back* along the line, i.e. in exactly the reverse direction.

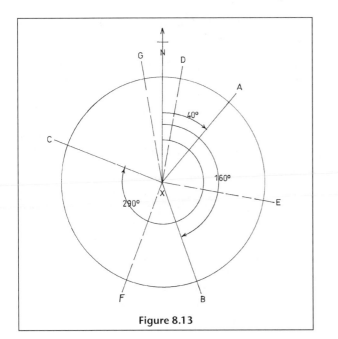

Figure 8.13

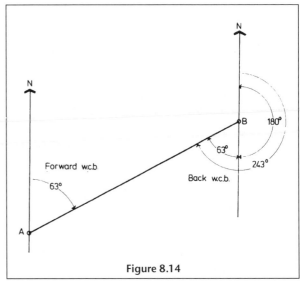

Figure 8.14

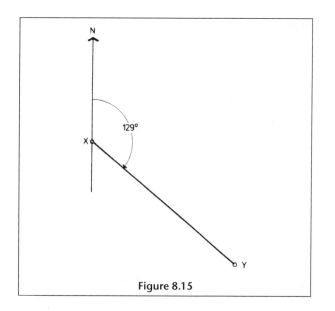

Figure 8.15

This direction BA is known as the *back* whole circle bearing. The angular value is the clockwise angle from the north point to the line BA, namely (180° + 63°) = 243°.

From this simple example an equally simple rule can be deduced.

Rule To obtain back bearings from forward bearings, or vice versa, add or subtract 180°.

In Fig. 8.15 the forward whole circle bearing of line XY is 129°. Using the rule above the back whole circle bearing YX is (180° + 129°) = 309°. The reader should now complete the diagram by drawing the north point through station Y and proving that 309° is the correct answer.

EXAMPLE

6 The forward bearings of four lines are as shown below:

AB. 31°   BC. 157° 30′   CD. 200°   DE 347° 15′

Calculate the back bearings.

Answer

Line	Forward		Back
AB	31° 00′	+180°	211° 00′
BC	157° 30′	+180°	337° 30′
CD	200° 00′	−180°	20° 00′
DE	347° 15′	−180°	167° 15′

3. Quadrant bearings

For good reasons the term 'quadrant bearing' seems to have lost favour with certain sections of the surveying fraternity but a knowledge of quadrant bearings is still required in certain traverse calculations. As an example, when calculating bearings from known coordinates, many calculators do not return the value

of the angle as a whole circle bearing but simply as a positive or negative value between 0° and 180°.

The surveyor has to recognize the quadrant in which the bearing lies before he or she can deduce the whole circle bearing.

In Fig. 8.16, when the cardinal points of the 360° circle are drawn and labelled North, South, East and West, the circle will have been divided into four 90° quadrants, known as the north-east, south-east, south-west and north-west quadrants. The quadrant bearing of any line is the angle that it makes with the north–south axis. This angle is given the name of the quadrant into which it falls. In Fig. 8.17, the whole circle bearings of four lines AB, AC, AD and AE are 40°, 121°, 242° and 303° respectively.

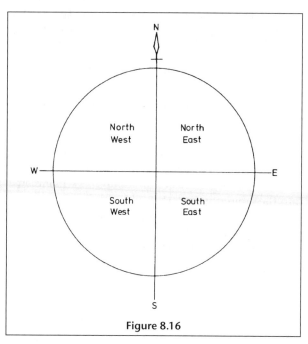

Figure 8.16

The angles that the lines make with the north–south axis are respectively:

AB	40°	quadrant bearing	N 40° E
AC	59°	quadrant bearing	S 59° E
AD	62°	quadrant bearing	S 62° W
AE	57°	quadrant bearing	N 57° W

The conversion of whole circle to quadrant bearings, sometimes called reduced bearings, can be readily carried out by the following rules:

1. When the whole circle bearing lies between 0° and 90°, its quadrant bearing has the same numerical value and lies in the NE quadrant.
2. When the whole circle bearing lies between 90° and 180°, the quadrant bearing is (180° − WCB) and lies in the SE quadrant.
3. When the whole circle bearing lies between 180° and 270°, the quadrant bearing is (WCB − 180°) and lies in the SW quadrant.

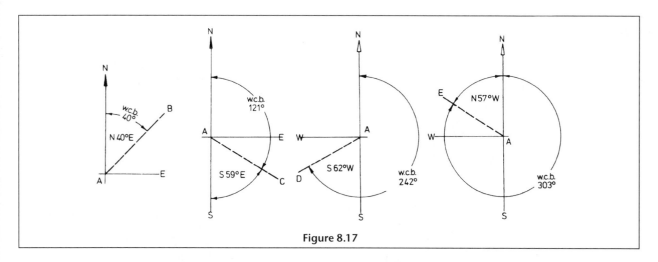

Figure 8.17

4. When the whole circle bearing lies between 270° and 360°, the quadrant bearing is (360° − WCB) and lies in the NW quadrant.

EXAMPLES

7 Convert the following whole circle bearings into quadrant bearings.

60° 30'   240° 10'   352° 10'   131° 00'

Answer

Whole circle
bearing   *Quadrant bearing*
 60° 30'   N 60° 30' E
240° 10'   (240° 10' − 180°)   S 60° 10' W
352° 10'   (360° − 352° 10')   N 07° 50' W
131° 00'   (180° − 131° 00')   S 49° 00' E

8 Convert the following quadrant bearings into whole circle bearings

N 40° 30' 20" E   S 75° 15' 45" E
S 39° 18' 17" W   N 64° 59' 58" W

Answer

Quad bearing   *WCB*
N 40° 30' 20" E   40° 30' 20"
S 75° 15' 45" E   (180° − 75° 15' 45")   104° 44' 15"
S 39° 18' 17" W   (180° + 39° 18' 17")   219° 18' 17"
N 64° 59' 58" W   (360° − 64° 59' 58")   295° 00' 02"

Exercise 8.3

1 The whole circle forward bearings of the lines of a traverse are as follows:

Line	
MN	35° 00'
NO	175° 36'
OP	214° 14'
PQ	267° 32'
QR	356° 20'

Calculate the back bearings of the lines.

2 Convert the following whole circle bearings into quadrant bearings:

311° 10'   247° 30'   060° 10'   093° 00'   270° 00'
167° 50'   111° 10'   264° 50'   359° 10'   179° 00'

3 Convert the following quadrant bearings into whole circle bearings.

N 30° 10' W   S 60° 30' W   S 07° 45' E   N 10° 00' E
East   S 79° 10' W   S 89° 50' E   S 64° 30' E

(d) Conversion of angles into bearings

In order to plot the positions of traverse stations, using the method of rectangular coordinates, the forward bearing of every line of any traverse is required.

The bearing of one line only (usually the first) need be known. As already explained, it may be a Grid bearing, a magnetic bearing or an assumed bearing. The bearing of every other line is calculated.

Open Traverse
Figure 8.18 shows three lines of an open traverse. The forward bearing of line AB is 65° 00'. The traverse angles are

Angle ABC = 100° 00' and
Angle BCD = 135° 00'

The forward bearings of line BC and line CD are required.

Line AB The required forward bearing is angle NAB. It has been measured directly as 65° 00'.

Line BC The required forward bearing is angle NBC. In Fig. 8.18, line AB has been projected forward, through station B.

Angle NBC = 65° + 180° (hatched) + 100°
          = (65° + 180°) + 100°
          = 245° + 100°
          = back bearing BA
            + traverse angle ABC   (1)
          = 345° 00'

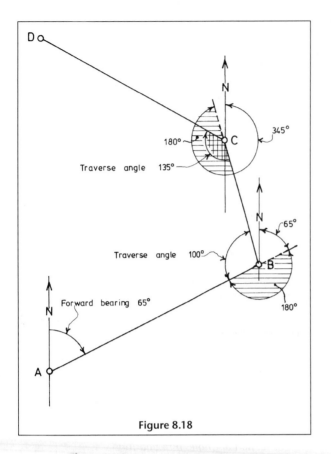

Figure 8.18

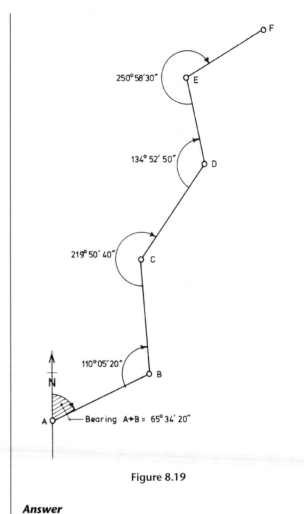

Figure 8.19

Line CD   The required forward bearing is angle NCD. In Fig. 8.18, line BC has been projected forward, through station C.

Angle NCD = 345° − 180° (hatched)
+ 135° (cross-hatched)
= (345° − 180°) + 135°
= 165° + 135°
= back bearing CB
  + traverse angle BCD   (2)
= 300° 00′

From these calculations, formulae (1) and (2) provide a general formula for use in bearings calculation.

> *Forward* bearing of *any* line
>   = *back* bearing of *previous* line
>   + *clockwise* angle between the lines

Frequently, it will be found that forward bearings, obtained from the above formula, will exceed 360°. In those cases, 360° is subtracted from the values to produce the correct bearings.

EXAMPLE

9 Figure 8.19 shows an open traverse along the pegs of a site boundary. The forward bearing of line AB of the traverse is 65° 34′ 20″. Calculate the forward bearing of each line of the survey.

Answer

		Whole circle bearing
AB		65° 34′ 20″
Back bearing BA	245° 34′ 20″	
+ ABC	110° 05′ 20″	
BC	355° 39′ 40″	355° 39′ 40″
Back bearing CB	175° 39′ 40″	
+ BCD	219° 50′ 40″	
	395° 30′ 20″	
	−360°	
CD	35° 30′ 20″	35° 30′ 20″
Back bearing DC	215° 30′ 20″	
+ CDE	134° 52′ 50″	
DE	350° 23′ 10″	350° 23′ 10″
Back bearing ED	170° 23′ 10″	
+ DEF	250° 58′ 30″	
	421° 21′ 40″	
	−360°	
EF	61° 21′ 40″	61° 21′ 40″

Many surveyors prefer to use an amended version of the rule in which the calculation of the back bearing is omitted. Compensation is made at the end of the calculation for any line by adding or subtracting 180°.

The rule then becomes: 'To the *forward* bearing of the previous line add the next clockwise angle. If the sum is greater than 180°, subtract 180°. If the sum is less than 180°, add 180°. If the sum is greater than 540°, subtract 540°. The result is the forward bearing of the next line.'

In Fig. 8.19, the forward bearing of line BC is calculated as follows:

Forward bearing AB =	65° 34' 20"
+ ABC	110° 05' 20"
	175° 39' 40"

Since the sum is less
than 180° add 180°   +180° 00' 00"
Forward bearing BC =   355° 39' 40"

which agrees with the previous calculation.

Using this method the computerized solution of bearings is simplified.

Exercise 8.4

1 Figure 8.20 is a reproduction of Fig 8.7 showing a short traverse along the north bank of a stream. The previously determined bearing of line AB is 65° 00'. Calculate the forward bearings of lines BC and CD.

Closed traverse
The fieldwork of a closed traverse is self-checking to some extent. In any closed traverse the sum of the internal angles should equal $(2n - 4) \times 90°$, where n is the number of angles. When exterior angles are measured the sum should be $(2n + 4) \times 90°$.

In a closed traverse of six stations, therefore, the sum of the exterior angles should be

$$(2n + 4) \times 90° = (12 + 4) \times 90°$$
$$= 1440° 00' 00"$$

It is very unlikely that the angles will sum to this amount because of the small errors inherent in angular measurements. However, the sum should be very close to this amount and should certainly be within the following limits:

$\pm 40 \sqrt{n}$ seconds, in the case of a single-second reading theodolite

and

$\pm \sqrt{n}$ minutes, in the case of a twenty-second reading theodolite

In each case, n is the number of instrument stations. In all cases, where the sum of the observed angles is not $(2n \pm 4) \times 90°$ the angles must be adjusted so that they do sum to this figure.

EXAMPLE

10 In Fig. 8.21, the values of the exterior angles, measured by a one second theodolite, are:

Angle	Mean observed value
ABC	272° 03' 10"
BCD	272° 05' 51"
CDE	104° 50' 31"
DEF	261° 11' 06"
EFA	266° 10' 15"
FAB	263° 38' 25"

(a) Determine the sum of the measured angles.
(b) Determine the angular error of the traverse.
(c) Adjust the angles of the traverse to eliminate the error.
(d) Given that the whole circle bearing of line AB is 43° 40' 45", calculate the bearings of all other lines of the traverse.

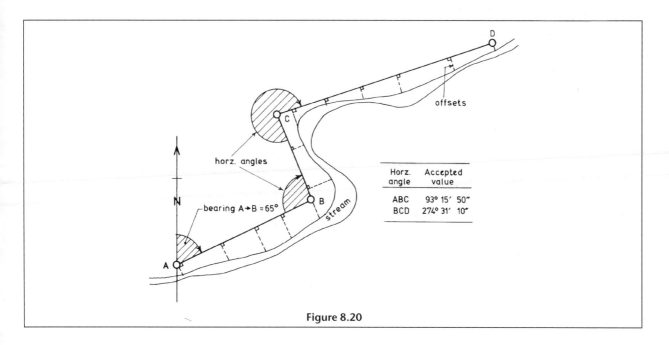

Figure 8.20

Horz. angle	Accepted value
ABC	93° 15' 50"
BCD	274° 31' 10"

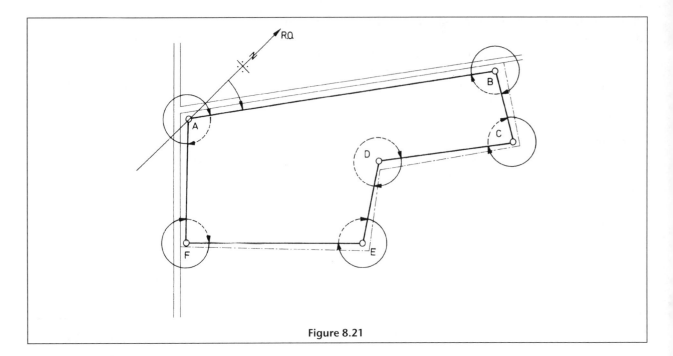

Figure 8.21

Answer

(a) Sum of exterior angles $= (2n + 4) \times 90°$

$$= 16 \times 90°$$

$$= 1440°\ 00'\ 00''$$

Sum of observed angles $= 1439°\ 59'\ 18''$

(b) Therefore angular error $= -42''$

Correction per angle $= +\frac{1}{6}$ of $42''$

$$= +07''$$

(c) Corrected angular values are:

Angle	Mean observed value	Correction	Corrected angle
ABC	272° 03′ 10″	+07″	272° 03′ 17″
BCD	272° 05′ 51″	+07″	272° 05′ 58″
CDE	104° 50′ 31″	+07″	104° 50′ 38″
DEF	261° 11′ 06″	+07″	261° 11′ 13″
EFA	266° 10′ 15″	+07″	266° 10′ 22″
FAB	263° 38′ 25″	+07″	263° 38′ 32″
	1439° 59′ 18″	+42″	1440° 00′ 00″

(d)

		Whole circle bearing
AB		43° 40′ 45″
Back bearing BA	223° 40′ 45″	
+ ABC	+272° 03′ 17″	
	495° 44′ 02″	
	−360°	
BC	135° 44′ 02″	135° 44′ 02″
Back bearing CB	315° 44′ 02″	
+ BCD	+272° 05′ 58″	
	587° 50′ 00″	
	−360°	
CD	227° 50′ 00″	227° 50′ 00″
Back bearing DC	47° 50′ 00″	
+ CDE	+104° 50′ 38″	
DE	152° 40′ 38″	152° 40′ 38″

Back bearing ED	332° 40′ 38″	
+ DEF	+261° 11′ 13″	
	593° 51′ 51″	
	−360°	
EF	233° 51′ 51″	233° 51′ 51″
Back bearing FE	53° 51′ 51″	
+ EFA	+266° 10′ 22″	
FA	320° 02′ 13″	320° 02′ 13″
Back bearing AF	140° 02′ 13″	
+ FAB	+263° 38′ 32″	
	403° 40′ 45″	
	−360°	
AB	43° 40′ 45″	

Agrees with initial bearing

Exercise 8.5

1 Figure 8.22 is a reproduction of Fig. 8.8, which shows the layout and measured angles of the closed traverse of a small building site.

(a) Determine the sum of the measured angles.

(b) Determine the angular error of the traverse.

(c) Adjust the angles of the traverse to eliminate the closing error.

(d) Given that the whole circle bearing of line AB is 315° 00′ 00″, calculate the forward bearings of all lines of the traverse.

2 Figure 8.23 shows the mean observed angles of traverse MNOP. Calculate the whole circle bearing of each line of the traverse.

Traverse closed on previously fixed points

In Fig. 8.24 (page 160), the bearings of lines BA and DC are known from a previous traverse to be 204° 11′ 05″

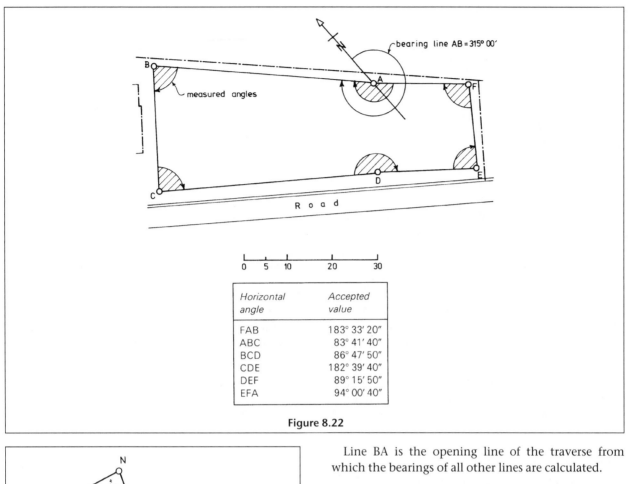

Figure 8.22

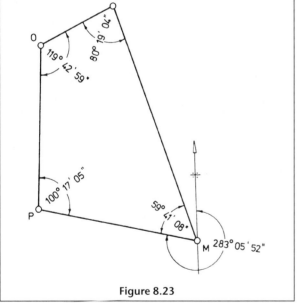

Figure 8.23

Line BA is the opening line of the traverse from which the bearings of all other lines are calculated.

		Whole circle bearing
BA		204° 11′ 05″
Back bearing AB	24° 11′ 05″	
+ BAP	+72° 39′ 42″	
AP	96° 50′ 47″	96° 50′ 47″
Back bearing PA	276° 50′ 47″	
+ APQ	+187° 40′ 12″	
	464° 30′ 59″	
	−360°	
PQ	104° 30′ 59″	104° 30′ 59″
Back bearing QP	284° 30′ 59″	
+ PQO	+169° 23′ 47″	
	453° 54′ 46″	
	−360°	
QO	93° 54′ 46″	93° 54′ 46″
Back bearing OQ	273° 54′ 46″	
+ QOD	+161° 58′ 20″	
	435° 53′ 06″	
	−360°	
OD	75° 53′ 06″	75° 53′ 06″
Back bearing DO	255° 53′ 06″	
+ ODC	+106° 17′ 21″	
	362° 10′ 27″	
	−360°	
DC	02° 10′ 27″	02° 10′ 27″

and 02° 10′ 47″ respectively. The observed angles of traverse BAPQODC are:

Angle	Observed value
BAP	72° 39′ 42″
APQ	187° 40′ 12″
PQO	169° 23′ 47″
QOD	161° 58′ 20″
ODC	106° 17′ 21″

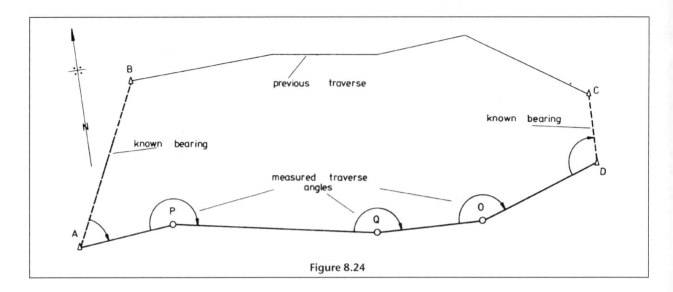

Figure 8.24

Known whole circle bearing line DC = 02° 10′ 47″
Traverse error   = −20″
Correction per angle   = + 20/5″
                                                              = + 04″

The observed angles could be recalculated by adding 04″ to each and the bearings could be found from the corrected angles as before. The corrected bearing of line AP will therefore be 04″ greater than the original while the bearing of the next line PQ will be 08″ greater. In fact the corrected bearings of each line will increase successively by 04″, until finally bearing DC will be (5 × 04)″ greater than the original uncorrected bearing.

The correction is therefore much more quickly done by simply finding the correction per angle and adding it successively to the uncorrected bearings.

Line	Uncorrected bearing	Correction	Corrected bearing
BA			204° 11′ 05″
AP	96° 50′ 47″	+ 04″	96° 50′ 51″
PQ	104° 30′ 59″	+ 08″	104° 31′ 07″
QO	93° 54′ 46″	+ 12″	93° 54′ 58″
OD	75° 53′ 06″	+ 16″	75° 53′ 22″
DC	02° 10′ 27″	+ 20″	02° 10′ 47″

EXAMPLE

11 AB and FG are two survey lines with bearings fixed from a primary traverse as 39° 40′ 20″ and 36° 18′ 30″ respectively. The bearings are to be unaltered.

A secondary traverse run between the above stations produced the following results:

Angle	Mean observed value
ABC	179° 59′ 40″
BCD	210° 05′ 40″
CDE	149° 44′ 40″
DEF	177° 46′ 40″
EFG	179° 02′ 20″

Calculate the corrected bearings of each line.

Answer

		Whole circle bearing	Correction	Corrected bearing
AB		39° 40′ 20″	–	39° 40′ 20″
Back bearing BA	219° 40′ 20″			
+ ABC	+179° 59′ 40″			
	399° 40′ 00″			
	−360°			
BC	39° 40′ 00″	39° 40′ 00″	−10″	39° 39′ 50″

Back bearing CB	219° 40′ 00″			
+ BCD	+ 210° 05′ 40″			
	429° 45′ 40″			
	−360°			
CD	69° 45′ 40″	69° 45′ 40″	−20″	69° 45′ 20″
Back bearing DC	249° 45′ 40″			
+ CDE	+ 149° 44′ 40″			
	399° 30′ 20″			
	−360°			
DE	39° 30′ 20″	39° 30′ 20″	−30″	39° 29′ 50″
Back bearing ED	219° 30′ 20″			
+ DEF	+ 177° 46′ 40″			
	397° 17′ 00″			
	−360°			
EF	37° 17′ 00″	37° 17′ 00″	−40″	37° 16′ 20″
Back bearing FE	217° 17′ 00″			
+ EFG	+ 179° 02′ 20″			
	396° 19′ 20″			
	−360°			
FG	36° 19′ 20″	36° 19′ 20″	−50″	36° 18′ 30″

Correct bearing FG = 36° 18′ 30″
Traverse error   = +50″
Angular correction = −(50/5)″
        = −10″

5. Rectangular coordinates (terminology)

[It is important that the student fully understands the rules of trigonometry (Ratios of angles of any magnitude) on page 9 of Chapter 1 before studying the remainder of the text of this chapter.]

If it can be imagined, for the moment, that a large sheet of graph paper could be unrolled over a site, with the y axis coinciding with the north direction, then it would be possible to define any point on the site by conventional X and Y coordinates.

In Fig. 8.25, station A is the origin of a survey and has the coordinates (0.00E, 0.00N). [It is usual practice to make the starting coordinates 1000E, 1000N, but in order to make clear the basic steps in calculation, the zero coordinates are used to deliberately introduce negative values into the computation.] The coordinates of station B are (x_B, y_B) while those of station C are (x_C, y_C) etc. These are the *total coordinates* of the stations. The x coordinate always precedes the y coordinate.

In construction and engineering surveying the x coordinates are called *eastings* while the y coordinates are called *northings*.

Following logically from those are the terms *difference in eastings* and *difference in northings*, denoted in Fig. 8.25 as Δx and Δy respectively.

6. Traverse coordinate calculations

The following data relate to the open traverse ABCDE of Fig. 8.25. The rectangular coordinates of each station are required.

Line	AB	BC	CD	DE
WC Bearing	30°	110°	225°	295°
Length (m)	50.0	70.0	82.0	31.2

(a) Basic calculation – open traverse

1. Line AB is the first line of the traverse. It has direction (30°) and magnitude (50.0 m). These quantities are the *polar coordinates* of line AB. They are converted into partial *rectangular coordinates* (Δx, Δy) as follows:

In triangle IAB,

$$\Delta x_1 = \text{IB} \qquad\qquad \Delta y_1 = \text{IA}$$
$$= \text{AB} \sin 30° \qquad\quad = \text{AB} \cos 30°$$
$$= 50.0 \times \sin 30.0 \qquad = 50.00 \times \cos 30.0$$
$$= +25.0 \text{ m} \qquad\qquad = +43.3 \text{ m}$$

Note that both signs are positive since the line travels eastwards (to the right) along the x axis and northwards (upwards) along the y axis.

The formulae for changing polar coordinates to partial rectangular coordinates are therefore:

$$\Delta x = (Length\ of\ line) \times (sin\ WC\ Bearing)$$
$$\Delta y = (Length\ of\ line) \times (cos\ WC\ Bearing)$$

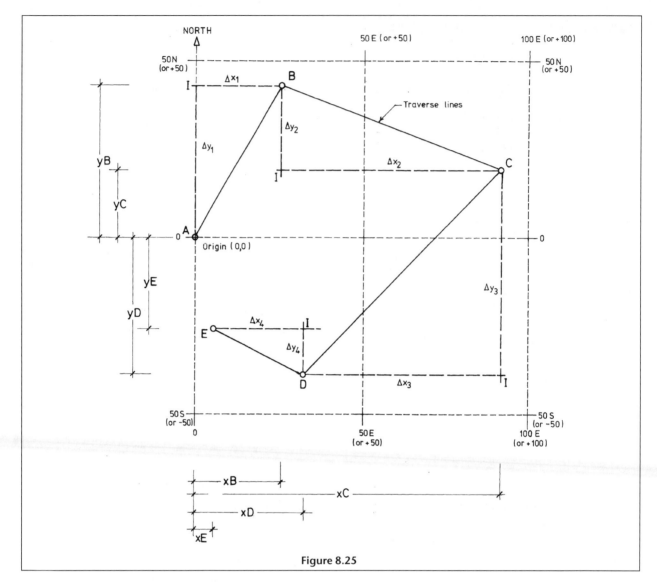

Figure 8.25

These formulae hold for every line of any traverse. The trigonometric values of sine and cosine of the whole circle bearing will automatically produce the correct positive or negative partial coordinates of the line.

2. The whole circle bearing of traverse line BC is 110° and since this is a second quadrant angle in mathematical terms, sin 110° = +sin 70° and cos 110° = –cos 70°.

The partial coordinates of line BC are therefore

ΔE = length BC × sin WCB
   = 70.0 × sin 110°
   = 70.0 × 0.939 69
   = +65.8 m
ΔN = length × cos WCB
   = 70.0 × cos 110°
   = 70.0 × –0.342 02
   = –23.9 m

3. Similarly the partial coordinates of line CD are

ΔE = 82.0 × sin 225°
   = 82.0 × –0.707 11
   = –58.0 m

ΔN = 82.0 × cos 225°
   = 82.0 × –0.707 11
   = –58.0 m

and the partial coordinates of line DE are

ΔE = 31.2 × sin 295°
   = 31.2 × –0.906 31
   = –28.3 m
ΔN = 31.2 × cos 295°
   = 31.2 × +0.422 62
   = +13.2 m

(b) Partial coordinates

As already stated, the differences in eastings (Δx_1, Δx_2, Δx_3, Δx_4) and the differences in northings (Δy_1, Δy_2, Δy_3, Δy_4) are collectively the partial coordinates. These coordinates could, at this stage, be plotted on a rectangular grid, each station being plotted from the previous station. However, the disadvantage would be that if one station were plotted wrongly, the succeeding stations would also be wrong.

(c) Total coordinates

To overcome this disadvantage, all the stations are plotted from the first station or origin of the survey by converting the partial coordinates to 'total co-ordinates'. The total coordinates of any station are obtained by algebraically adding the previous partial coordinates.

Treating station A as the origin of the survey, i.e. its coordinates are 0.0E and 0.0N, the eastings of successive stations B, C, D and E are found as in Table 8.5, with the numerical values as given in Table 8.6. The northings of the stations are found in an identical manner (Table 8.7) with the numerical values as given in Table 8.8. The total coordinates (eastings and northings) of each station are therefore as shown in Table 8.9.

Table 8.5

Line	Δx	Easting	Station
		0.0	A
AB	Δx_1	$0.0 + \Delta x_1$	B
BC	Δx_2	$0.0 + \Delta x_1 + \Delta x_2$	C
CD	Δx_3	$0.0 + \Delta x_1 + \Delta x_2 + \Delta x_3$	D
DE	Δx_4	$0.0 + \Delta x_1 + \Delta x_2 + \Delta x_3 + \Delta x_4$	E

Table 8.6

Line	Δx	Easting			Station
		0.0	=	0.0	A
AB	+ 25.0	0.0 + 25.0	=	+ 25.0	B
BC	+ 65.8	0.0 + 25.0 + 65.8	=	+ 90.8	C
CD	− 58.0	0.0 + 25.0 + 65.8 + (− 58.0)	=	+ 32.8	D
DE	− 28.3	0.0 + 25.0 + 65.8 + (− 58.0) + (− 28.3)	=	+ 4.5	E

Table 8.7

Line	Δy	Northing	Station
		0.0	A
AB	Δy_1	$0.0 + \Delta y_1$	B
BC	Δy_2	$0.0 + \Delta y_1 + \Delta y_2$	C
CD	Δy_3	$0.0 + \Delta y_1 + \Delta y_2 + \Delta y_3$	D
DE	Δy_4	$0.0 + \Delta y_1 + \Delta y_2 + \Delta y_3 + \Delta y_4$	E

Table 8.8

Line	Δy	Northing			Station
		0.0	=	0.0	A
AB	+ 43.3	0.0 + 43.3	=	+ 43.3	B
BC	− 23.9	0.0 + 43.3 + (−23.9)	=	+ 19.4	C
CD	− 58.0	0.0 + 43.3 + (−23.9) + (−58.0)	=	− 38.6	D
DE	+ 13.2	0.0 + 43.3 + (−23.9) + (−58.0) +13.2	=	+ 25.4	E

(d) Traverse table

All of the calculations are set out in a traverse table shown in Table 8.10(a). It is completed as follows:

1. Enter the station designations in column 8.
2. Enter the survey lines in column 1 beginning on line 2.

Table 8.9

Station	Easting	Northing
A	0.0	0.0
B	+ 25.0	+ 43.3
C	+ 90.8	+ 19.4
D	+ 32.8	− 38.6
E	+ 4.5	− 25.4

Table 8.10(a)

	1	2 Whole circle bearing	3 Distance (m)	4 Partial coordinates ΔE	5 Partial coordinates ΔN	6 Total coordinates Eastings	7 Total coordinates Northings	8 Station
1						00.0	00.0	A
2	AB	30°	50.0	+ 25.0	+ 43.3	+ 25.5	+ 43.3	B
3	BC	110°	70.0	+ 65.8	– 23.9	+ 90.8	+ 19.4	C
4	CD	225	82.0	– 58.0	– 58.0	+ 32.8	– 38.6	D
5	DE	295°	31.2	– 28.3	+ 13.2	+ 4.5	– 25.4	E
				+ 90.8	+ 56.5	+ 4.5	– 25.4	
				– 86.3	– 81.9			
				+ 4.5	– 25.4			

3. Enter the whole circle bearing of each line in column 2.
4. Enter the plan length of each line in column 3.
5. Enter the difference in eastings in column 4 and the difference in northings in column 5.
6. Calculate the total coordinates by successively adding the partial coordinates. Enter the eastings in column 6 and the northings in column 7.

(e) Use of P/R function of a calculator

On all scientific calculators, the rectangular coordinates can be computed directly from the polar coordinates and vice versa, by using the P→R and the R→P function buttons.

Unfortunately, manufacturers of calculators have their own preferences in their methods of operation and the reader is directed to the instruction manual of his or her particular type of calculator where full instructions are given on the use of the functions.

The following instructions for the Casio type of calculator demonstrate the general principle, where the polar coordinates (length 50.0 m and bearing 30°) are changed into rectangular coordinates ΔE and ΔN thus:

Instruction	Calculator operation
1. Enter length	Input 50.0
2. Operate buttons	INV P→R
3. Enter WCB (Decimalized)	Input 30.0
4. Operate button	=
5. Read answer (ΔN)	43.3
6. Operate button	X→Y
7. Read answer (ΔE)	25.0
8. Repeat for remaining lines of traverse	

(f) Accuracy of computations

Many surveying calculations are nowadays carried out by computers but on site and in class and lecture rooms, most calculations are done using eight- or ten-digit hand-held calculators, mentioned in Sec. (e). When using these calculators, it is important to remember that, while they give answers to many decimal places, the calculated quantity cannot be more precise than the given data. There is therefore little point in accepting answers calculated to millimetres when the given data is only measured with metre accuracy. It is worth emphasizing that such answers are not wrong but neither are they strictly correct.

(g) Computer spreadsheet solution

A traverse survey can have many stations for which coordinates are required. The calculation is therefore repetitive and is easily carried out using a computer spreadsheet of which Excel and Microsoft Works are probably the most commonly used. Table 8.10(a) is the manual solution to the open traverse of Fig. 8.25. The spreadsheet solution using Microsoft Works is shown in Table 8.10(b). The spreadsheet has been used in a very basic manner to illustrate the possibilities of programming. It is not intended that this textbook become or replace a good computing textbook. However, a little explanation is obviously desirable for readers who may not be familiar with spreadsheets.

(i) Table 8.10(b) is Table 8.10(a) with the addition of a grid of letters and numbers. In the table, the headings are entered on lines 1 and 2 and columns A to L. All other *known* data are typed in their respective cells, e.g. the line DE, bearing (295° 00′ 00″) and distance 31.2 m are entered on line 7, columns A, B, C, D and E respectively. The known coordinates of station A (0.0East and 0.0North) are entered on line 3, columns H and I.

(ii) In computing, all bearings must be decimalized firstly then converted to radian measure to allow the program to run correctly. This can be done in either one column or two columns (K and L) as in Table 8.10(b).

As an example, a bearing of 10° 20′ 30″ is converted as follows:

Bearing = [10 + 20/60 + 30/3600] × [PI() / 180]
= 0.180 496 1 radians

These columns are 'hidden' and will not be printed if the *File, page setup* menu is set up correctly.

Table 8.10(b)

Traverse Table (Microsoft WORKS)   Workings (Hidden)

	A	B	C	D	E	F	G	H	I	J	K	L
1	Line	Whole Circle Bearing			Distance	Δ East	Δ North	**East**	**North**	Stn.	Wcb	Wcb
2	**	Deg	Min	Sec	**	**	**			**	Decimal	Radians
3								0.0	0.0	A	0.00	0.000000
4	AB	30	0	0	50.0	25.0	43.3	25.0	43.3	B	30.00	0.523599
5	BC	110	0	0	70.0	65.8	−23.9	90.8	19.4	C	110.00	1.919862
6	CD	225	0	0	82.0	−58.0	−58.0	32.8	−38.6	D	225.00	3.926991
7	DE	295	0	0	31.2	−28.3	13.2	4.5	−25.4	E	295.00	5.148721

(iii) The width of columns, number of decimal points and centring (or not) of entries follow standard spreadsheet practice.

(iv) The coordinate formulae are entered in their respective cells to complete the calculation as follows:

Cell	Formula entry
K4	$= B4 + C4/60 + d4/3600$
K5, K6 and K7	Copy formula of K4
L4	$= K4 \times PI() \div 180$
L5, L6 and L7	Copy formula of L4
F4	$= E4 \times \sin(L4)$
F5, F6 and F7	Copy formula of F4
G4	$= E4 \times \cos(L4)$
G5, G6 and G7	Copy formula of G4

H4	$= H3 + F4$
H5, H6 and H7	Copy formula of H4
I4	$= I3 + G4$
I5, I6 and I7	Copy formula of I4

7. Plotting rectangular coordinates

Figure 8.26 is a portion of a rectangular grid drawn to scale 1:1000.

Point A is the origin of the survey and is therefore the intersection of the zero easting and zero northing lines.

Point C (90.8E, 19.4N) is to be plotted.

1. The square in which the point is to be plotted is determined by inspection.
2. The easting, 90.8 m, is scaled along the zero north line and the point is marked c_1.

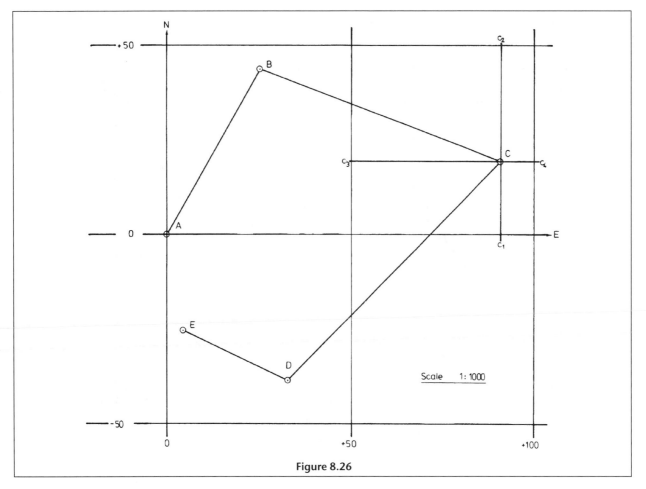

Figure 8.26

3. Similarly, the easting is scaled and marked c_2 along the 50 m north line.
4. A pencil line is drawn to join the points c_1 and c_2, and the line is checked to ensure that it is parallel to the north–south grid lines.
5. The northing, 19.4 m, is marked along the 50 m and 100 m easting lines (points c_3 and c_4) and the line joining them is checked for parallelism along the east–west grid lines.
6. The intersection of lines c_1c_2 and c_3c_4 forms the point C.

All other stations of the survey are plotted in the same manner and marked with a small circle, cross or triangle.

EXAMPLE

12 The following data refer to an open traverse ABCDE. Calculate and plot the coordinates of the traverse. [*Note* – This example is intended as a plotting exercise. Bearings are therefore given in degrees only and distances to one place of decimals.]

Line	AB	BC	CD	DE
WCB	35°	94°	58°	17°
Length (m)	79.0	59.0	52.0	44.0

Answer Table 8.11 and Figure 8.27

Table 8.11

Line	Whole circle bearing	Distance (m)	ΔE	ΔN	Eastings	Northings	Station
					0.0	0.0	A
AB	35°	79.0	+ 45.3	+ 64.7	+ 45.3	+ 64.7	B
BC	94°	59.0	+ 58.9	− 4.1	+ 104.2	+ 60.6	C
CD	58°	52.0	+ 44.1	+ 27.6	+ 148.3	+ 88.2	D
DE	17°	44.0	+ 12.9	+ 42.1	+ 161.2	+ 130.3	E
			+ 161.2	+ 130.3	+ 161.2	+ 130.3	

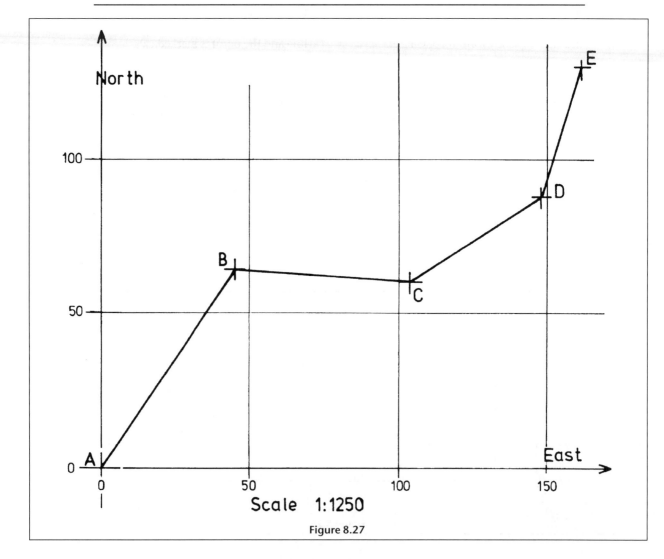

Scale 1:1250

Figure 8.27

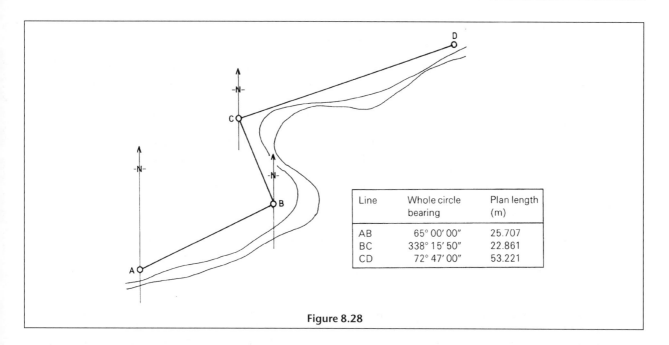

Line	Whole circle bearing	Plan length (m)
AB	65° 00′ 00″	25.707
BC	338° 15′ 50″	22.861
CD	72° 47′ 00″	53.221

Figure 8.28

Exercise 8.6

1 Figure 8.28 shows the traverse of Fig. 8.7, which was the subject of 'traverse fieldwork' covered earlier in this chapter. The traverse details are as follows:

Line	Whole circle bearing	Plan length (m)
AB	65° 00′ 00″	25.707
BC	338° 15′ 50″	22.861
CD	72° 47′ 00″	53.221

Given that the coordinates of station A are 100.000E and 100.000N, calculate the coordinates of stations B, C and D and plot the survey to scale 1:500.

8. Calculation of a closed traverse

Basically there is no difference between the calculation of a closed traverse and an open traverse. In a closed traverse, the initial and final coordinates of the first station should be identical. This will very seldom occur and the resultant closing error must be eliminated.

E X A M P L E

13 Calculate the total coordinates of all stations in the following traverse given that station A is the origin:

Line	AB	BC	CD	DE	EA
WC bearing	29° 30′	100° 45′	146° 30′	242° 00′	278° 45′
Length (m)	83.500	59.400	62.000	50.300	90.400

This is an example of a closed traverse which has a closing error. In this example the sexagesimal

---

measure of bearings occurs for the first time in trigonometrical calculations. All of the bearings must be converted to decimal form. Using the scientific calculator a bearing of, say, 30° 53′ 15″ is decimalized using the converter button which appears as DMS or ° ′ ″. The answer appears on screen as 30.887500.

Note. The method of converting sexagesimal measure to centesimal measure varies with the type of calculator.

Answer Table 8.12

(a) Closed traverse – closing error

The difference between the initial and final coordinates of station A determines the errors in eastings and northings. Figure 8.29 shows the initial and final coordinates plotted on a very much exaggerated scale.

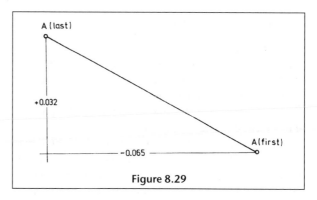

Figure 8.29

The error of closure AA is found by the theorem of Pythagoras to be

$$AA = \sqrt{0.065^2 + 0.032^2} = 0.072 \text{ m}$$

Table 8.12

Line	Whole circle bearing	Length (m)	ΔE	ΔN	Eastings	Northings	Station
					0.000	0.000	A
AB	29° 30′ 0″	83.500	41.117	72.675	41.117	72.675	B
BC	100° 45′ 0″	59.400	58.358	−11.080	99.475	61.595	C
CD	146° 30′ 0″	62.000	34.220	−51.701	133.695	9.894	D
DE	242° 00′ 0″	50.300	−44.412	−23.614	89.283	−13.720	E
EA	278° 45′ 0″	90.400	−89.348	13.752	−0.065	0.032	A

Table 8.13

Line	Whole circle bearing	Length	ΔE	ΔN	ΔE correction	ΔN correction	Corrected ΔE	Corrected ΔN	Eastings	Northings	Station
									0.000	0.000	A
AB	29° 30′ 0″	83.500	41.117	72.675	0.016	−0.008	41.133	72.667	41.133	72.667	B
BC	100° 45′ 0″	59.400	58.358	−11.080	0.011	−0.006	58.369	−11.086	99.502	61.581	C
CD	146° 30′ 0″	62.000	34.220	−51.701	0.012	−0.006	34.232	−51.707	133.734	9.874	D
DE	242° 00′ 0″	50.300	−44.412	−23.614	0.009	−0.004	−44.403	−23.618	89.331	−13.744	E
EA	278° 45′ 0″	90.400	−89.348	13.752	0.017	−0.008	−89.331	13.744	0.000	0.000	A
		345.600									
Errors			−0.065	0.032	0.065	−0.032	0.000	0.000			
Corrections			0.065	−0.032							

Correction $(k_1) = + 1.885 \times 10^{-4}$
Correction $(k_2) = - 9.197 \times 10^{-5}$
Accuracy 1 in 4800

The total length of the traverse is 345.60 metres; therefore the fractional accuracy of the survey is

0.072 m in 345.6 m

which equals

1 in 4800 approximately

(b) Adjustment by Bowditch's rule

The closing error of the survey is due to the accumulation of errors along each draft of the traverse. It should be remembered that the angular and linear errors are assumed to be equally inaccurate. As a result the closing error is distributed throughout the traverse in proportion to the lengths of the various drafts.

In the rectangular coordinate method, it is the separate ΔE and ΔN errors that have to be distributed and not the closing error of distance. Each error is treated separately and is distributed throughout the traverse in proportion to the lengths of the drafts, as given in Table 8.13.

In Example 13, the error in eastings = −0.065 m; therefore the correction in eastings = + 0.065 m.

Easting correction at station B

$$= \frac{\text{length of line AB}}{\text{total length of survey}} \times \text{total correction}$$

and in fact at any station the easting correction (c_E) is

$$c_E = \left(\frac{\text{length of draft}}{\text{total length of survey}} \right.$$
$$\left. \times \text{total correction in easting} \right)$$

$$= \frac{l}{L} \times C_E$$

$$= \frac{C_E}{L} \times l$$

Therefore $c_E = k_1 \times l$ (since C_E and L are constants for every line).

The northing correction (c_N) for every station is derived exactly as above:

$$c_N = \frac{l}{L} \times C_N$$

$$= \frac{C_N}{L} \times l$$

$$= k_2 \times l$$

In Example 13, the correction procedure is as follows:

1. Total easting correction (C_E) = +0.065 m

Total length of survey (L) = 345.60 m

$$\text{Therefore } k_1 = \frac{+0.065}{345.60}$$
$$= 1.885 \times 10^{-4}$$

2. Easting correction at

B = $1.885 \times 10^{-4} \times 83.50 = 0.016$ m
C = $1.885 \times 10^{-4} \times 59.40 = 0.011$ m
D = $1.885 \times 10^{-4} \times 62.00 = 0.012$ m
E = $1.885 \times 10^{-4} \times 50.30 = 0.009$ m
A = $1.885 \times 10^{-4} \times 90.40 = \underline{0.017}$ m
Sum = $\overline{0.065}$ m

3. Total northing correction ($C_N = -0.032$ m

Total length of survey (L) = 345.60 m

$$\text{Therefore } k_2 = \frac{-0.032}{345.60}$$
$$= -9.197 \times 10^{-5}$$

4. Northing correction at

B = $-9.197 \times 10^{-5} \times 83.50 = -0.008$ m
C = $-9.197 \times 10^{-5} \times 59.40 = -0.006$ m
D = $-9.197 \times 10^{-5} \times 62.00 = -0.006$ m
E = $-9.197 \times 10^{-5} \times 50.30 = -0.004$ m
A = $-9.197 \times 10^{-5} \times 90.40 = \underline{-0.008}$ m
Sum = $\overline{-0.032}$ m

5. The corrected partial coordinates are obtained from the algebraic addition of the calculated partial coordinates (Table 8.13) and the corrections. For example, corrected partial coordinates of B are

$$\Delta E$$
$$+41.117 + 0.016 = 41.133$$
$$\Delta N$$
$$+72.675 - 0.008 = 72.667$$

Table 8.13 is simply an extension of a normal traverse table.

Exercise 8.7

1 Figure 8.30 shows the reduced data (horizontal lengths and whole circle bearings) of the building site already used in Fig 8.23.

(a) Calculate the coordinates of the stations and the closing error of the traverse, given that the coordinates of station A are 100.000E, 100.000N.
(b) Adjust the traverse using Bowditch's rule.

9. Miscellaneous coordinate problems

(a) Closing bearing and distance

In survey work, it is often necessary to calculate the bearing and distance between two known coordinated points. For example, in Fig. 8.31, the coordinates of all

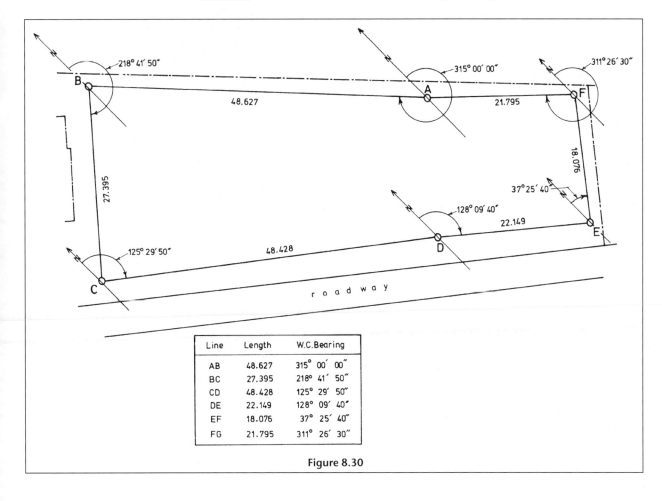

Line	Length	W.C. Bearing
AB	48.627	315° 00′ 00″
BC	27.395	218° 41′ 50″
CD	48.428	125° 29′ 50″
DE	22.149	128° 09′ 40″
EF	18.076	37° 25′ 40″
FG	21.795	311° 26′ 30″

Figure 8.30

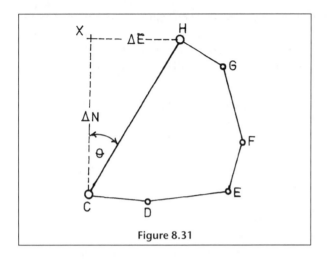

Figure 8.31

stations of the traverse C to H have been calculated and for some reason, e.g. setting out a building, the bearing and distance between two of the points is required. The following text shows how the closing bearing and distance CH is computed.

[*Note*: When the bearing of any line (from C to H in this instance) is to be calculated, the coordinates of C, the starting point, are subtracted algebraically from the coordinates of H, the final point.]

Coords of Station C	Coords of station H
300.20 E	463.30 E
320.00 N	595.30 N

In triangle CHX, the difference in easting between the points C and H is ΔE

$$\Delta E = 463.30 - 300.20$$
$$= + 163.10$$

The difference in northing between the points C and H is ΔN.

$$\Delta N = 595.30 - 320.00$$
$$= 275.30$$

The bearing of line CH is angle θ:

$$\tan \theta = \frac{\Delta E}{\Delta N}$$
$$= \frac{163.10}{275.30}$$

Therefore $\theta = 30° 38' 40''$

On most calculators, the angle θ will appear as a positive or negative value. The whole circle bearing has to be deduced either from a rough plotting of the two points C and H or from a knowledge of quadrant bearings.

From Fig. 8.31, the plot shows that the bearing lies in the north-east quadrant, hence the whole circle bearing (WCB) of line CH is 30° 38' 48".

The distance between points C and H is, by Pythagoras,

$$= \sqrt{163.10^2 + 275.30^2}$$

$$= \sqrt{102\,391.70}$$

$$= 319.99 \text{ m}$$

Alternatively, once the bearing has been calculated, the distance between the points C and H is found in either of the following ways:

$$\frac{\Delta N}{CH} = \cos \theta \qquad\qquad \frac{\Delta E}{CH} = \sin \theta$$

Therefore $CH = \dfrac{\Delta N}{\cos \theta}$ and $CH = \dfrac{\Delta E}{\sin \theta}$

$$= \frac{275.30}{0.860\,347} \qquad\qquad = \frac{163.10}{0.509\,709}$$

$$= 319.99 \qquad\qquad\qquad = 319.99 \text{ m}$$

Closing bearing and distance using $\boxed{R \to P}$ *function on calculator*

The calculation is easily carried out on a scientific calculator using the rectangular to polar conversion facility, (R→P). The calculation varies with the type of calculator and users should consult their calculator manual for full instructions.

(b) Solution of triangles

It should be noted that the majority of problems connected with coordinates finish with an unsolved triangle in which the coordinates of two points and the bearing of one or more sides is known. Generally the closing bearing and distance between the coordinated points must be found before the triangle can be solved.

EXAMPLE

14 In Fig. 8.32 two tunnels AB and EF are being driven forward until they meet in order to accommodate telephone cables.

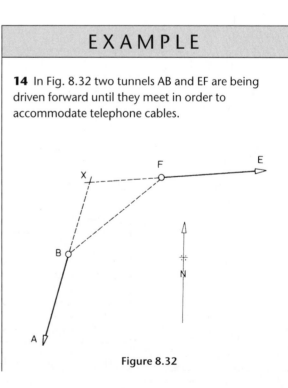

Figure 8.32

Calculate the distance still to be driven in each tunnel given the following information:

Bearing AB = 15° 00'

| | Coordinates B | 624.30 E |
| | | 1300.50 N |

Bearing EF = 265° 00'

| | Coordinates F | 845.90 E |
| | | 1482.30 N |

Solution

(a) Produce EF and AB until they meet at X. Join BF to form triangle FBX.

(b) Bearing XB = bearing BA

$\qquad$ = 195° 00'

$\quad$ Bearing XF = bearing FE

$\qquad$ = 85° 00'

Therefore angle FXB = 195° 00' – 85° 00'

$\qquad$ = 110° 00'

(c) Easting F = 845.90 $\quad$ Northing F = 1482.30

$\qquad$ B = 624.30 $\qquad\qquad$ B = 1300.50

$\qquad$ $\Delta E = \underline{221.60}$ $\qquad\qquad$ $\Delta N = \underline{181.80}$

$\tan$ bearing BF $= \dfrac{221.60}{181.80}$

$\quad$ bearing BF = 50° 38' 05"

$\quad$ Distance BF $= \dfrac{221.60}{\sin 50° 38' 05"}$

$\qquad$ = 286.63 m

(d) Angle XBF = bearing BF – bearing BX

$\qquad$ 50° 38' 05"

$\qquad$ –15° 00' 00"

$\qquad$ = $\underline{35° 38' 05"}$

(e) Angle BFX = bearing FX – bearing FB

$\qquad$ = 265° 00' 00"

$\qquad$ –230° 38' 05"

$\qquad$ = $\underline{34° 21' 55"}$

Check angles of triangles FXB

$\quad$ Sum = 110° 00' 00"

$\qquad$ 35° 38' 05"

$\qquad$ $\underline{34° 21' 55"}$

$\qquad$ $\underline{180° 00' 00"}$

(f) In triangle FXB, by sine rule,

$$\frac{XB}{\sin F} = \frac{BF}{\sin X}$$

Therefore $XB = \dfrac{BF \sin F}{\sin X}$

$\qquad = \dfrac{286.63 \times \sin 34° 21' 55"}{\sin 110° 00' 00"}$

$\qquad = \dfrac{286.63 \times 0.564\ 466\ 9}{0.939\ 692\ 6}$

$\qquad$ = 172.17 m

$$\frac{FX}{\sin B} = \frac{BF}{\sin X}$$

Therefore $FX = \dfrac{BF \sin B}{\sin X}$

$\qquad = \dfrac{286.63 \times \sin 35° 38' 05"}{\sin 110° 00' 00"}$

$$= \frac{286.63 \times 0.582\ 615\ 6}{0.939\ 692\ 6}$$

$$= 177.71 \text{ m}$$

Exercise 8.8

1 The following angles were measured on a closed theodolite traverse:

Instrument station	Target station	Face left
A	D	00° 10' 20"
	B	89° 26' 40"
B	A	89° 26' 40"
	C	189° 05' 20"
C	B	189° 05' 20"
	D	269° 29' 40"
D	C	269° 29' 40"
	A	00° 09' 00"

Calculate:

(a) the measured angles,

(b) the angular correction,

(c) the corrected angles,

(d) the whole circle bearing of each line, given that the bearing of AD is 45° 36' 00".

2 Calculate the closing error and accuracy of the following traverse:

Line	Length	Bearing
AB	110.20	156° 40' 00"
BC	145.31	75° 18' 00"
CD	98.75	351° 08' 00"
DE	163.20	276° 29' 00"
EA	52.34	187° 27' 00"

3 The partial coordinates of a closed traverse are:

Line	Length	Difference in Easting	Difference in Northing
PQ	252.41	0.00	252.41
QR	158.75	–110.76	–113.82
RS	153.50	–25.24	–151.41
SP	136.74	136.15	12.67

Distribute the closing error by Bowditch's rule and calculate the corrected partial coordinates.

4 A closed traverse ABCDEA produced results shown below:

Back station	Instrument station	Fore station	Clockwise horizontal angle
A	B	C	283° 31' 40"
B	C	D	329° 06' 50"
C	D	E	90° 47' 20"
D	E	A	299° 43' 00"
E	A	B	256° 50' 20"

(a) Calculate the error of closure of the angles.

(b) Adjust the angles for complete closure.

(c) Calculate the corrected bearings of lines BC, CD, DE and EA, given that the bearing of AB is 152° 24' 40".

10. Answers

Exercise 8.1

Bay	*Slope length* *L (m)*	*Slope corr* $(-h^2/2L)$ *m*	*Temperature corr* $(T_m - T_s) \times L \times 11 \times 10^{-6}$	*Total corr* *(m)*	*Horizontal length (m)*
1	49.825	–0.001	–0.003	–0.004	49.821
2	49.786	–0.003	–0.003	–0.006	49.780
3	49.872	–0.002	–0.003	–0.005	49.867
4	49.846	–0.001	–0.003	–0.004	49.842
5	49.852	0.000	–0.003	–0.003	49.849
6	35.865	–0.001	–0.003	–0.004	35.861
	285.046	–0.008	–0.018	–0.026	285.020

Total horz. length D = 285.020 m
Standardization corr. = $(L - l)$ per 50 m
= (50.003 – 50) × 285.02/50
= 0.003 × 5.700
= +0.017 m
Horz. length of baseline = + 285.020 + 0.017
= 285.037 m

Exercise 8.2

1 See Table 8.14

Exercise 8.3

1

Line	*Back bearings*
MN	215° 00′
NO	355° 36′
OP	34° 14′
PQ	87° 32′
QR	176° 20′

2 N 48° 50′ W; S 67° 30′ W; N 60° 10′ E;
S 87° 00′ E; West;
S 12° 10′ E; S 68° 50′ E; S 84° 50′ W;
N 00° 50′ W; S 01° 00′ E

3 329° 50′; 240° 30′; 172° 15′; 10° 00′;
90° 00′; 259° 10′; 270° 10′; 115° 30′

Exercise 8.4

Forward AB = 65° 00′ 00″
Back BA = 245° 00′ 00″
+ Angle B 93° 15′ 50″
Forward BC = 338° 15′ 50″
Back CB = 158° 15′ 50″
+ Angle C 274° 31′ 10″
432° 47′ 00″
Forward CD = 72° 47′ 00″

Exercise 8.5

1

Angle	*Measured angle*	*Adjustment*	*Adjusted angle*
FAB	183° 33′ 20″	+ 10″	183° 33′ 30″
ABC	83° 41′ 40″	+ 10″	83° 41′ 50″
BCD	86° 47′ 50″	+ 10″	86° 48′ 00″
CDE	182° 39′ 40″	+ 10″	182° 39′ 50″
DEF	89° 15′ 50″	+ 10″	89° 16′ 00″
EFA	94° 00′ 40″	+ 10″	94° 00′ 50″
Sum =	719° 59′ 00″	+ 01′ 00″	720° 00′ 00″

Forward AB = 315° 00′ 00″
Back BA = 135° 00′ 00″
+ ABC 83° 41′ 50″

Forward BC = 218° 41′ 50″
Back CB = 38° 41′ 50″
+ BCD 86° 48′ 00″

Forward CD = 125° 29′ 50″
Back DC = 305° 29′ 50″
+ CDE 182° 39′ 50″
= 488° 09′ 40″

Forward DE = 128° 09′ 40″
Back ED = 308° 09′ 40″
+ DEF = 89° 16′ 00″
397° 25′ 40″

Forward EF = 37° 25′ 40″
Back FE = 217° 25′ 40″
+ EFA = 94° 00′ 50″

Forward FA = 311° 26′ 30″
Back AF = 131° 26′ 30″
+ FAB 183° 33′ 30″
Forward AB = 315° 00′ 00″

Table 8.14

Instrument station	Station observed	Face left	Face Face right	Mean observed angle	Line	Face	Vertical circle reading	Zenith angle	Measured length			
									Rear	Forward	Length	Mean
A	F	00° 00' 00"	180° 00' 20"		AB	L	91° 30' 40"	91° 31' 00"	0.150	48.796	48.646	
	B	183° 33' 20"	3° 33' 40"			R	268° 28' 40"		0.325	48.969	48.644	48.644
	FÂB	183° 33' 20"	183° 33' 20"	183° 33' 20"			359° 59' 20"		0.200	48.842	48.642	
B	A	00° 00' 00"	179° 59' 40"		BC	L	95° 32' 00"	95° 32' 20"	0.350	27.874	27.524	
	C	83° 41' 20"	263° 41' 40"			R	264° 27' 20"		0.750	28.272	27.522	27.524
	AB̂C	83° 41' 20"	83° 42' 00"	83° 41' 40"			359° 59' 20"		0.523	28.049	27.526	
C	B	00° 00' 00"	180° 00' 20"		CD	L	89° 29' 10"	89° 29' 20"	0.360	48.790	48.430	
	D	86° 48' 00"	266° 48' 00"			R	270° 30' 30"		0.820	49.250	48.430	48.430
	BĈD	86° 48' 00"	86° 47' 40"	86° 47' 50"			359° 59' 40"		0.600	49.030	48.430	
D	C	00° 00' 00"	180° 00' 00"		DE	L	90° 37' 20"	90° 37' 30"	0.200	22.350	22.150	
	E	182° 39' 40"	2° 39' 40"			R	269° 22' 20"		0.090	22.240	22.150	22.150
	CD̂E	182° 39' 40"	182° 39' 40"	182° 39' 40"			359° 59' 40"		0.325	22.475	22.150	
E	D	00° 00' 00"	179° 59' 40"		EF	L	86° 13' 30"	86° 13' 50"	1.115	19.236	18.121	
	F	89° 15' 40"	269° 15' 40"			R	273° 45' 50"		1.300	19.416	18.116	18.116
	DÊF	89° 15' 40"	89° 16' 00"	89° 15' 50"			359° 59' 20"		0.800	18.911	18.111	
F	E	00° 00' 00"	180° 00' 00"		FA	L	90° 02' 50"	90° 03' 10"	0.326	22.121	21.795	
	A	94° 00' 40"	274° 00' 40"			R	269° 56' 30"		0.250	22.045	21.795	21.795
	EF̂A	94° 00' 40"	94° 00' 40"	94° 00' 40"			359° 59' 20"		0.200	21.995	21.795	

Sum = 719° 59' 00"
Traverse error = −01' 00"

2 Sum of interior angles $= (2n - 4) \times 90°$

$$= \underline{360° \ 00' \ 00''}$$

Sum of observed angles $= 360° \ 00' \ 16''$

Angular error $= +16''$

Correction per angle $= 16/4$

$$= -04''$$

Angle	Observed value	Correction	Correction value
PMN	59° 41′ 08″	−04″	59° 41′ 04″
MNO	80° 19′ 04″	−04″	80° 19′ 00″
NOP	119° 42′ 59″	−04″	119° 42′ 55″
OPM	100° 17′ 05″	−04″	100° 17′ 01″
Sum =	360° 00′ 16″	− 16″	360° 00′ 00″

Line	Whole circle bearing
MN	342° 46′ 56″
NO	243° 05′ 56″
OP	182° 48′ 51″
PM	103° 05′ 52″

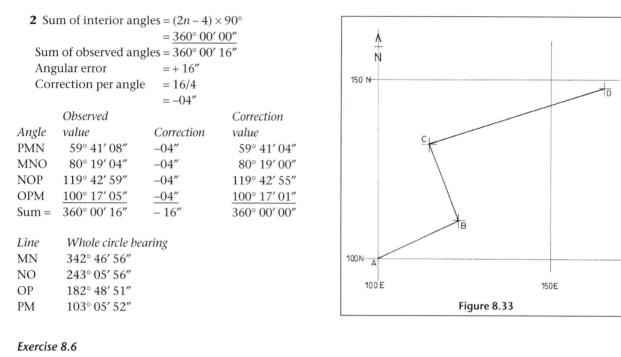

Figure 8.33

Exercise 8.6

1. Table 8.15 and Figure 8.33.

Table 8.15

Line	Whole circle bearing	Length	Difference in eastings	Difference in northings	Eastings	Northings	Station
					100.000	100.000	A
AB	65° 0′ 0″	25.707	23.298	10.864	123.298	110.864	B
BC	338° 15′ 50″	22.861	−8.466	21.236	114.832	132.100	C
CD	72° 47′ 0″	53.221	50.836	15.753	165.668	147.853	D

Exercise 8.7

1. Table 8.16.

Table 8.16

Line	Length	Whole circle bearing	Difference in eastings	Difference in northings	ΔE correction	ΔN correction	Corrected ΔE	Corrected ΔN	Eastings	Northings	Station
									100.000	100.000	A
AB	48.627	315° 0′ 0″	−34.384	34.384	0.006	0.006	−34.379	34.390	65.621	134.379	B
BC	27.395	218° 41′ 50″	−17.127	−21.381	0.003	0.003	−17.124	−21.377	48.497	113.013	C
CD	48.428	125° 29′ 50″	39.427	−28.120	0.006	0.006	39.433	−28.115	87.930	84.898	D
DE	22.149	128° 9′ 40″	17.415	−13.685	0.003	0.003	17.418	−13.683	105.348	71.216	E
EF	18.076	37° 25′ 40″	10.986	14.355	0.002	0.002	10.988	14.357	116.336	85.572	F
FA	21.795	311° 26′ 30″	−16.338	14.425	0.003	0.003	−16.336	14.428	100.000	100.000	A
	186.470	Errors	−0.022	−0.022	0.022	0.022	0.000	0.000			
		Corrections	0.022	0.022							

Correction $(k_1) = 0.000 \ 116 \ 26$

Correction $(k_2) = 0.000 \ 119 \ 21$

Accuracy of survey is 1 in 5993

Exercise 8.8

1

	(a)	(b)	(c)
	Observed angle	*Correction*	*Corrected angle*
DAB	89° 16′ 20″	+20″	89° 16′ 40″
ABC	99° 38′ 40″	+20″	99° 39′ 00″
BCD	80° 24′ 20″	+20″	80° 24′ 40″
CDA	90° 39′ 20″	+20″	90° 39′ 40″
	359° 58′ 40″	+1′ 20″	360° 00′ 00″

(d)

AD	45° 36′ 00″
AB	134° 52′ 40″
BC	54° 31′ 40″
CD	314° 56′ 20″

2 Closing error $\Delta E = +0.036$ m

$$\Delta N = -0.212 \text{ m}$$

$$\text{Distance} = \sqrt{0.036^2 + 0.212^2}$$

$$= 0.215 \text{ m}$$

$$\text{Accuracy} = 1 \text{ in } 2650$$

3

ΣE	ΣW	ΣN	ΣS
136.15	136.00	265.08	265.23
$\Delta E = +0.15$		$\Delta N = -0.15$	

Corrected partial coordinates

Line	*Easting*	*Northing*
PQ	− 0.054	252.464
QR	− 110.794	− 113.686
RS	− 25.273	− 151.477
SP	136.121	12.699

4

Angle	*Measured value*	*Adjustment*	*Adjusted angle*
ABC	283° 31′ 40″	+ 10″	283° 31′ 50″
BCD	329° 06′ 50″	+ 10″	329° 07′ 00″
CDE	90° 47′ 20″	+ 10″	90° 47′ 30″
DEA	299° 43′ 00″	+ 10″	299° 43′ 10″
EAB	256° 50′ 20″	+ 10″	256° 50′ 30″
	1259° 59′ 10″	+ 50″	1260° 00′ 00″

Line	*Whole circle bearing*
AB	152° 24′ 40″
BC	255° 56′ 30″
CD	45° 03′ 30″
DE	315° 51′ 00″
EA	75° 34′ 10″

Chapter summary

In this chapter, the following are the most important points:

- A traverse survey is a form of control survey, consisting of a string of points, called stations. If the string is made to close, the traverse is a *closed traverse*, otherwise it forms an *open traverse*. It is common that a particular traverse may begin on a line of a previous traverse and close on some other known line. This is a *polygonally closed traverse*.

- The *slope* distance of the lines joining successive stations of any traverse is measured either manually by tape or by EDM. The direction of those lines, relative to some form of north, is also measured, but not directly from north, as is possible but not practicable, using a compass. The direction of a line is called the *bearing* and since its value obviously lies between 00° and 360°, it is called a whole circle bearing (WCB).

- The bearing, length and gradient (zenith angle) of each line must be obtained very accurately, so great care is taken when measuring with a tape and theodolite to ensure that all sources of error (gross, constant, systematic and random) are eliminated as far as possible. Numeric examples are shown on pages 144 to 147.

- The horizontal (plan) length of every line is obtained by converting the slope length to plan length by the formula:

    plan length = slope length × sin (zenith angle)

- The bearing of *one* line of the traverse is obtained from an OS plan, either by protractor or from two sets of coordinates. A compass bearing (relatively inaccurate) may be used as a last resort or the bearing may simply be assumed.

    Every line has two bearings, namely, a *forward bearing* in one direction and a *back bearing* in the opposite direction, which differ, of course, by 180°.

- The clockwise horizontal angle between each pair of stations is measured and the bearing of any line is calculated from a knowledge of the back bearing of the previous line and the clockwise angle between the two lines:

    Fwd WCB of any line = Back bearing of previous line + clockwise angle between the lines

- The WCB bearing and horizontal length of every line are together called the *polar coordinates* of the line. In order to plot the stations the *rectangular coordinates* of the stations are calculated from the polar coordinates.

- The origin of the survey has, or is given, a set of rectangular coordinates. The x coordinate is the *easting* (E) and the y coordinate is the *northing* (N). The differences between the coordinates of the two stations of any line of a traverse are known as ΔE and ΔN.
- The coordinates of all stations of the traverse are computed relative to the origin in two stages.

  1. For any line, ΔE and ΔN are calculated from the formulae:

      $\Delta E = horz\ dist \times sin\ WCB$
      $\Delta N = horz\ dist \times cos\ WCB$

  2. The E and N of each station is found by successively adding the previous ΔE and ΔN to the coordinates of the origin.

- Open and closed traverses are computed in the same manner. The coordinates of the final station of a closed traverse should agree with the coordinates of the first station. That situation is unlikely and there is therefore a *closing error* in the coordinates. The error is distributed amongst the stations in proportion to the lengths of the traverse lines.

  Numerous examples of coordinate calculations are given on pages 167 to 171.

- Rectangular coordinates are the 'bread and butter' of site surveying and the reader is advised to revise the relevant section of Chapter 1 (basic trigonometry) which will help to give an insight into the intricacies of traverse calculations.

CHAPTER 9 Global Positioning System (GPS)

In this chapter you will learn about:

- the principles of GPS and its use in land surveying
- the satellite systems used in GPS
- the principle of measuring distance to fix position by GPS
- the sources of error and how they are minimized
- the instruments and field procedures used
- the spheroids, projection systems and coordinates used in terrestrial reference frameworks (TRF)
- likely future developments

In the late 18th century triangulation was introduced as a method of control survey for national surveys. A primary framework was broken down with secondary frameworks. In GPS, satellites in precise fixed orbits round the Earth are the primary framework. Electronic ranges to a receiver fix its position by trilateration. Coordinates can be fixed anywhere in the world independent of weather with no restrictions of intervisibility. The first survey application was for control survey. Now, with real-time processing, GPS can be used for accurate detail survey and setting out.

Initial development was carried out by the US Department of Defense, who had the resources to cope with the enormous expense involved. The Navigational System for Timing and Ranging (NAVSTAR) has, at present, 24 operational satellites. To date, the US policy has been that access to the system is freely open to civilian users. The Russian system, Global Navigation Satellite System (GLONASS), is also now available with 14 satellites, but in some opinions it is somewhat less reliable. It gives better coverage north of 50°N latitude.

Satellites transmit two binary timing codes on two frequencies from which the distance from the satellite to the receiver can be calculated. One code is called the *coarse acquisition code* and gives a position to better than 100 m accuracy. The other code is the precise code. In the NAVSTAR system it was periodically encrypted and became unavailable to civilian users. This restriction has now been removed. There have been no restrictions on GLONASS signals. A single receiver can give a navigation position to within about 10 m; if it is left static for an hour or more, this can improve to 1 m. Height accuracy is about half that of plan.

The early specialized applications of GPS have now expanded to be part of everyday life. The accuracy and compactness of receivers have increased dramatically. With survey equipment costs having decreased considerably, GPS has now become standard kit with many organizations; however, blocking of the signal by high buildings and trees still limits its use for land surveying.

1. Principles of GPS

(a) The satellites

Each system consists of a constellation of satellites orbiting the Earth at a high altitude to give global coverage (Fig. 9.1). The systems are as follows:

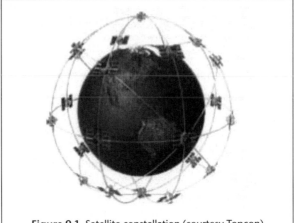

Figure 9.1 Satellite constellation (courtesy Topcon)

Figure 9.2 NAVSTAR satellite (courtesy Topcon)

Navstar (Fig. 9.2)
US Dept of Defense.
24 operational satellites (29 total).
20 180 km. altitude.
6 orbital planes, 55 degree inclination.
12 hr orbital period.
7.5 year lifespan.
21 satellite replacement system in year 2012.

Glonass
Russian Space Federation.
14 satellites. 18 by 2005.
19 100 km altitude.
3 orbital planes.
65 degree inclination.

Galileo
The new European system will have its first launch in 2005, and hopefully be operational by 2010. More details are given at the end of the chapter in Sec. 8(b).

The satellites are monitored by ground stations, checking that they are functioning correctly and tracking their orbit. Corrections to their orbital constants (ephemeris) and clock offset are transmitted back to the satellite.

NAVSTAR monitoring stations, known as the control segment, are located at Hawaii (Pacific Ocean), Ascension Island (Atlantic Ocean), Diego Garcia (Indian Ocean), Kwajalein (SE Asia) and Colorado Springs (central USA), which is the master station. The satellite can store ephemeris data for up to 14 days before requiring update.

(b) Ranging

The fixation of a satellite receiver on the surface of the Earth is by trilateration. The distance or range to the satellite is measured by timing the radio signal from the satellite to the receiver. Because of errors inherent in the system they are known as *pseudo ranges*.

One satellite distance gives position on a sphere of that radius, as in Fig. 9.3.

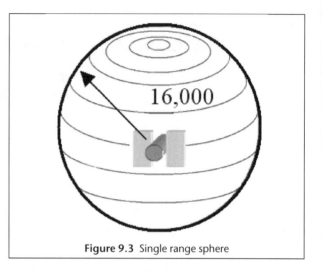

Figure 9.3 Single range sphere

Two satellite distances give position on the circle of intersection of the two satellite range spheres, as in Fig. 9.4.

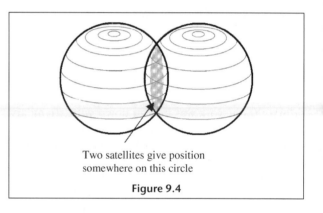

Two satellites give position
somewhere on this circle

Figure 9.4

Three satellite distances gives position as two possibilities on either side of the circle of intersection of the two range spheres, as in Fig. 9.5. One of these will not be close to the Earth, hence position is fixed.

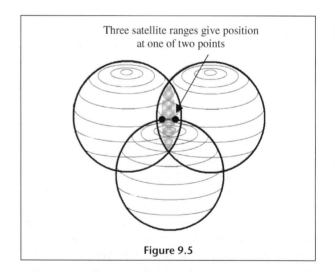

Three satellite ranges give position
at one of two points

Figure 9.5

In practice, a fourth and preferably a fifth satellite is required to eliminate errors.

(c) Distance measurement

The principle of EDM discussed in Chapter 7 is again used. *Distance = velocity × time.*

In GPS, a radio signal, which travels at the speed of light (~300 000 km/sec), is used. If the satellite was overhead, then the transit time would be about 0.06 sec. Therefore GPS clocks have to be very precise.

A satellite emits a unique pseudo random code on a carrier frequency. When the receiver locks onto the signal, it synchronizes a matching code, measuring the delay in the signal as shown in Fig. 9.6.

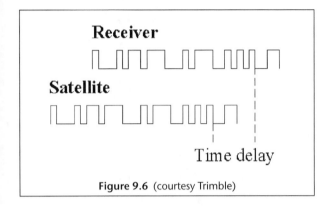

Figure 9.6 (courtesy Trimble)

The principle of satellite distance measurement is like someone listening to music on a radio at a football match, and someone else also listening to the same station at the opposite end of the stadium. There would be a slight delay in the sound from the radio at the far end reaching the first listener. Matters are made worse by the fact that the stadium is filled with a cheering crowd, which is analogous to the other radio noise which a receiver has to filter out. Fortunately the pseudo random code is a simple pulsed signal, more like morse code than music. The receiver can use amplification to pick out the peaks of the pulses over the other radio noise. The signal is divided into time segments and analysed to find a match to the pseudo random code. This is known as chipping the signal. Because of this amplification and the simplicity of the signal, GPS receivers need only a small satellite dish, unlike SKY TV.

NAVSTAR satellites transmit on two carrier frequencies. The L1 carrier frequency is 1575.42 MHz (19 cm wavelength) and the L2 carrier frequency is 1227.60 MHz (24 cm wavelength). Each satellite has its own pseudo random code. The GLONASS satellites have the same code, but transmit at slightly different frequencies from each other. L1 is 1602 + 0.5625N and L2 is 1246 + 0.4375N.

The L1 frequency carries two codes. The first is the Coarse Acquisition (C/A) code. It modulates at 1 MHz and repeats every 1023 bits. The second pseudo random code is called the P (precise) code. It modulates at 10 MHz and repeats on a seven-day cycle. The P code requires a more sophisticated receiver.

The L2 frequency carries only a P code.

Code phase measurement of distance by synchronizing the pseudo random codes to measure time delay as in Fig. 9.6 cannot achieve accuracies of better than 1 m with even the most sophisticated receivers. Survey receivers therefore use *carrier phase measurement* to give centimetre accuracy. This is analogous to phase measurement EDM discussed in Chapter 7, but the problem of ambiguity in the wavelength count is reduced since the distance is already known to about a metre from the code phase measurement. Therefore only a small portion of the carrier wave has to be analysed, as shown in Fig. 9.7.

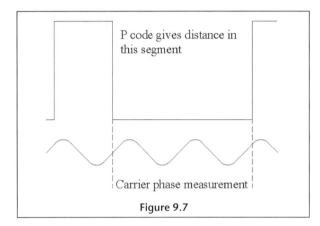

Figure 9.7

A low-frequency status signal is also transmitted which gives information on satellite orbit parameters, clock corrections and other system information. An almanac which allows the receiver to predict where a satellite will be in the sky is included in the data signal.

2. Errors in GPS

(a) Timing

An error of a thousandth of a second would result in a positional error of 300 km. In Fig. 9.8 the triangle of error produced by clock errors is shown.

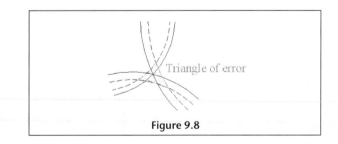

Figure 9.8

In the satellite an atomic clock is used, but if this were also used in the receiver the cost would be prohibitive. Instead the receiver clock can be continually calibrated by taking a fourth satellite range. Three perfect measurements give an accurate fix, but so do four imperfect measurements. In practice at least five are required for centimetre accuracy.

(b) Ephemeris

The high altitude of the satellites means that there is very little disturbance to their orbits but errors can be produced by the gravitational pull of the sun and the moon. Periodic solar flares can also cause wobbles in the satellite orbit. The ground monitoring stations use very precise radar to continuously measure satellite position and speed. The resulting *ephemeris corrections* are relayed up to the satellite from the master control station. This is stored and transmitted to the receiver in the status signal.

(c) Atmosphere

Atmospheric conditions cause small variations in the speed of light, which result in an error similar to the timing error. The GPS signal is slowed by the charged particles in the ionosphere and water vapour in the troposphere. These errors are more variable than clock errors and also vary with the angle of the signal from the satellite. It is more difficult to compute an error than in the clock correction. Two techniques can be applied.

(i) Modelling   Atmospheric data can be input into the system and a mathematical model of the likely timing errors for a particular angle of signal to a receiver calculated. The receiver can then apply this correction.

Measurements at known base stations can produce atmospheric parameters in the locality. If there is a network of base stations, then a local model of atmospheric corrections can be produced.

(ii) Dual frequency measurement   Survey receivers are dual frequency and an atmospheric correction for the signal speed can be calculated by comparing the signal delay of the L1 and L2 frequencies.

(d) Multipath

This is an error caused by the satellite signal taking an indirect path to the receiver antenna.

At the ground, any surface which reflects light will reflect the satellite signal. In the transmission of television signals this effect can be seen in the TV picture as 'ghosting'. The result in GPS is a distance error. In urban areas, buildings, particularly those with large glass areas, will cause this error. A choke ring antenna can be used to reduce this or in the latest receivers the reflections are filtered out with proprietary circuitry (e.g. Pulse Aperture Correlation and Pinwheel in Sokkia receivers).

(e) Geometric dilution of precision (GDOP)

The accuracy of the trilateration fix from the satellites depends on a wide distance between the satellites in the sky as in Fig. 9.9. In the top diagram the number of satellites available is limited by tree and building

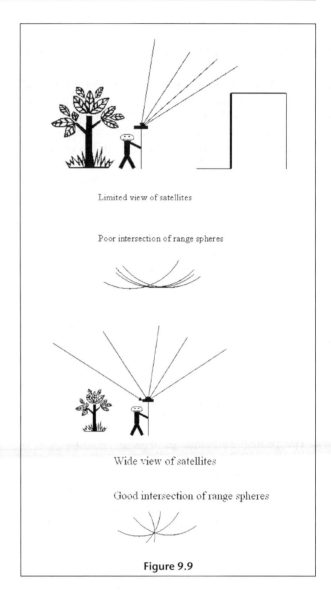

Limited view of satellites

Poor intersection of range spheres

Wide view of satellites

Good intersection of range spheres

Figure 9.9

obstructions producing a poor intersection of range spheres. In the lower diagram there are no obstructions, allowing a wide spread of satellites to be observed, producing a good intersection of the range spheres. This gives a low GDOP.

In European latitudes in the northern hemisphere, if the view to the south is obstructed, there may be a limited view to the satellites. This will compromise accuracy. Good receivers will choose the satellites which give the lowest GDOP. Additional satellite availability from GLONASS and eventually GALILEO will improve the GDOP.

3. Differential GPS (DGPS)

Although elaborate precautions are taken to minimize errors, small errors do occur, which are not accounted for. DGPS is a technique used to minimize errors for accurate position fixes.

It is somewhat reminiscent of the old technique of barometric heighting used in dense jungle, where the conventional technique of trigonometrical levelling was not practicable. A barometer/altimeter was kept at a bench mark. Roving instruments gave height readings, which were then corrected by the readings taken at the bench mark at the same time.

(a) DGPS Procedures

In DGPS a receiver is set up on a known base station. Rover receivers are moved in the locality and their positions are corrected by the base station data. With the global scale of GPS it is assumed that for a small local survey the corrections will be the same over the whole area at a particular moment in time. The reference receiver works in reverse. Knowing its exact position, it can work out what the exact distance to each satellite should be. It compares this with the measured range and calculates the timing correction for the available satellites. If a radio or phone link is fitted on both receivers, the corrections can be transmitted to the roving receiver and positions can be corrected immediately ('on the fly'). Otherwise final coordinates are calculated by post processing back in the office, using the reference station's correction files. The reference receiver does not know which satellites the rover is using; therefore it saves timing corrections for all the available satellites at time intervals during the operation of the survey. The link between the reference station and the rover forms a *baseline* from which the vector data of bearing and distance can be used for subsequent processing and adjustment.

(b) Reference stations

The static receiver can be either a receiver set up by the survey team or one of a national or international network of receivers, which are now available.

(i) Single reference station   In this system, two receivers are required. A reference receiver is set up on the known station and a compact roving receiver moves to fix positions. It is assumed that the corrections calculated at the base station apply also to the rover. The link between the receivers can be achieved with a Positioning Data Link Radio (PDL), which is limited to a maximum range of 10 km with line of sight conditions or by mobile phone for a longer range. OS trig pillars can be used as reference stations, but these have largely been replaced by the National GPS Network passive stations, which are more easily accessible. They are generally located at the side of roads. There are 900, with an average interval of 30 km.

(ii) Reference networks   The use of an existing Continuously Recording Reference Station Network (CORS) of GPS receivers has obvious advantages, not the least being that only one receiver is required.

One of the primary uses of GPS is in navigation, and a number of networks have been set up to service this type of user.

Hydrographic: The General Lighthouse Authority (GLA) has a network with receivers located at lighthouses throughout the UK. Accuracy is low (1–2 m) and a special radio is required.

Aeronautical: These are satellite systems, which are independent of GPS and provide augmented positional information. The Wide Area Augmentation System (WAAS) is a worldwide system. The European Geographical Navigation Overlay System (EGNOS) is free and gives accuracies of 1 m in plan and 2 m in height. A spare channel in the receiver can be used to receive corrections from the system's satellites.

Commercial systems available on subscription include *Landstar* from Thales and *Omnistar* from Fugro. These have a network of GPS receivers, and a footprint of corrections for a particular region is broadcast by geostationary satellites. Accuracies can be at the sub-metre level.

OSNET: This is the ongoing development of the *OS active GPS network* for centimetre positioning. Initially for use by OS surveyors, it is designed to have a network of 150 fixed receiving stations, giving an average spacing of about 100 km. Figure 9.10 shows the distribution of the 47 stations which are presently available.

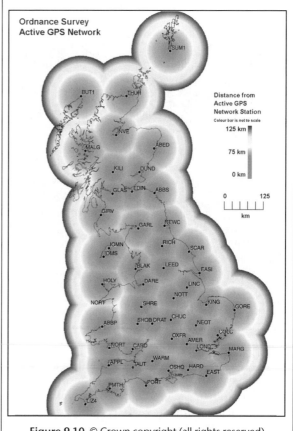

Figure 9.10 © Crown copyright (all rights reserved)

Atmospheric modeling will allow corrections and the rate of change of corrections to be passed to roving receivers. The OS server will be connected to by GSM or

GRSM phone, using a service provider. The correction data can then be transmitted to the receiver by a 'bluetooth' connection.

The integration of different manufacturers' equipment and software with the reference network can cause problems. A Receiver Independent Exchange (RINEX) data format has been developed, but there still may be differences in antenna heights and offsets which have to be corrected for in the processing. L1 and L2 signals have different offsets. The OS active stations use a mixture of Leica, Ashtec and Trimble receivers. At the moment corrections are available retrospectively as a RINEX file for each station at one-hourly intervals. This can be downloaded free from the OS GPS website at www.gps.gov.uk.. Some stations may not be available due to maintenance

Figure 9.11 shows an OS active station antenna installed on the roof of a building.

Figure 9.11 © Crown copyright (all rights reserved)

(c) Post processing of DGPS data

Post processing involves the calculation of corrected station coordinates after the field observations have been completed. With older receivers this was the only method possible.

Modern GPS receivers and software allow calculations of positions to be carried out almost instantaneously, but for precise control survey it is desirable to gather as much precise correction information as possible. This is not necessarily available online.

Processing may include some or all of the following:

(i)   Decode RINEX data for OS Active Stations.
(ii)   Apply antenna offset corrections.
(iii)   Obtain and apply precise ephemeris corrections.
(iv)   Apply ionospheric and troposheric model corrections.
(v)   Apply least squares control network adjustment to produce 'best fit' coordinates.
(vi)   Convert GPS coordinates to local and/or national coordinate system.
(vii)   Convert GPS heights to local and/or national datum.

4. GPS procedures

Although the technology differs greatly from conventional observations in surveying, the basic principles should not be forgotten. Checks should be built into the survey by taking extra measurements. On a small site is used more than one base station. For larger areas, extra measurements will allow checks and adjustments of a network of points and a value of the accuracy obtained.

(a) Static

The receiver is set up on a tripod over the station and measurements are taken. The receiver will take observations about every 15 seconds for an hour or more to reduce ambiguities and systematic errors in the carrier phase measurement. A large number of satellites can be observed to obtain the best GDOP and a precise average position is calculated.

This method is used for the most accurate work, such as control survey, using DGPS. It has largely replaced conventional traversing methods. A similar procedure of working from the known station to the unknown station is carried out. The network is built up with primary, secondary and tertiary control points as in conventional survey methods.

Two receivers are used. The reference receiver takes observations at the known station simultaneously with the roving receiver at the station to be fixed. The reference receiver can then 'leap frog' to the next station to be fixed and the role of the receivers is reversed. This is commonly referred to as a baseline survey.

Checks should be built into the framework observations. With three receivers, work can be carried out most efficiently with one receiver moving as the other two are recording. It is possible to use two as reference receivers and the rover is fixed from this baseline. Similar to a conventional traverse, observations should close to a known station or baseline, and other known stations incorporated along the way. To eliminate a constant scale error, EDM check lines can be observed and incorporated into the least squares adjustment in the post processing.

Check heights can be measured by conventional methods as the height accuracy of GPS is typically half that of the planimetric accuracy.

In the post processing, ETRS89 (European system) coordinates are often converted to OSGB36 (OS National Grid) via the OSTN02 transformation and heights converted to OS Datum Newlyn via the OSGM02 geoid model. Both are available on the OS GPS website.

The precise control network so formed can be used for a reference GPS network in subsequent detail survey and setting out. The spacing of the framework points should be related to intervisibility and the communication method to be used in the subsequent work (PDF radio, GSM phone etc.).

Software systems such as *Leica Spider* and *Topcon TopNET* integrate the data from a network of reference receivers, calculating parameters for algorithms to model GPS corrections, which are transmitted to the roving receivers. This results in a *'Virtual Reference Station'* (VRS), which simulates having a reference station next to the rover. Accuracy is significantly improved.

The processing steps are shown below:

Collate reference station data at central server

Resolve ambiguities between reference stations

Calculate correction parameters for model

Derive/transmit corrections for rover's position

Variations on the *static method* include *rapid static* where the receiver will record for less than an hour, and *stop and go* where the receiver will record for a few minutes. These procedures can be used for less accurate applications such as tertiary control, photo control for aerial surveys, fixing boundary markers and locating telecom towers.

(b) Kinematic

Here the receiver is moving with the aerial carried on a pole by the surveyor or on a vehicle or boat. After initialization, the receiver is kept continuously locked onto the satellites and a point can be fixed in seconds. A warning will be given if lock is lost or if the available satellites do not give the required accuracy. Corrections can be carried out 'on the fly' (*real-time kinematic*) or by post processing. In the latest systems, the antenna is divided into quadrants, and some of the processing is carried out in the antenna, which speeds processing time. With modern receivers, initialization of the rov-

ing receiver to the reference receiver takes a matter of seconds, and the receiver continually checks its position with the reference station (e.g. Leica Smartcheck).

This method is used in detail survey, setting out and earth moving machine control with centimetre accuracy.

5. GPS equipment

The equipment required to survey ground positions using GPS consists of an antenna, a receiver and optionally a controller. The various types can be conveniently categorized in terms of the accuracy that they are designed to achieve.

(a) Centimetre accuracy

These are dual frequency, receiving both L1 and L2 signals, and use carrier phase measurement to achieve centimetre precision. The latest types will also receive the more powerful L2c signal. They typically have 24 channels. The Topcon GPS+ receiver has 40 channels, allowing the use of the GLONASS satellites. A reference and a rover receiver are available. Geodetic or RTK antenna types are available, depending on accuracy requirements. The receiver and antenna can now be an integrated unit carried on top of a pole, to which is attached the controller. In the Sokkia IG1000, the controller is a tablet computer with touch screen graphics. Top-end geodetic receivers can be used for a CORS network.

Figure 9.12 shows the components of the Leica GPS 1200 system.

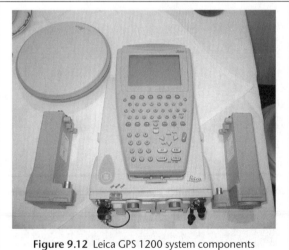

Figure 9.12 Leica GPS 1200 system components

The antenna can be tripod or pole mounted. The receiver can be carried in a back pack with the controller attached to the pole as shown in Fig. 9.13. The option of the radio link or phone link to the reference station is shown on either side.

The Leica 1200 Smartstation is a total station with an integrated GPS on top.

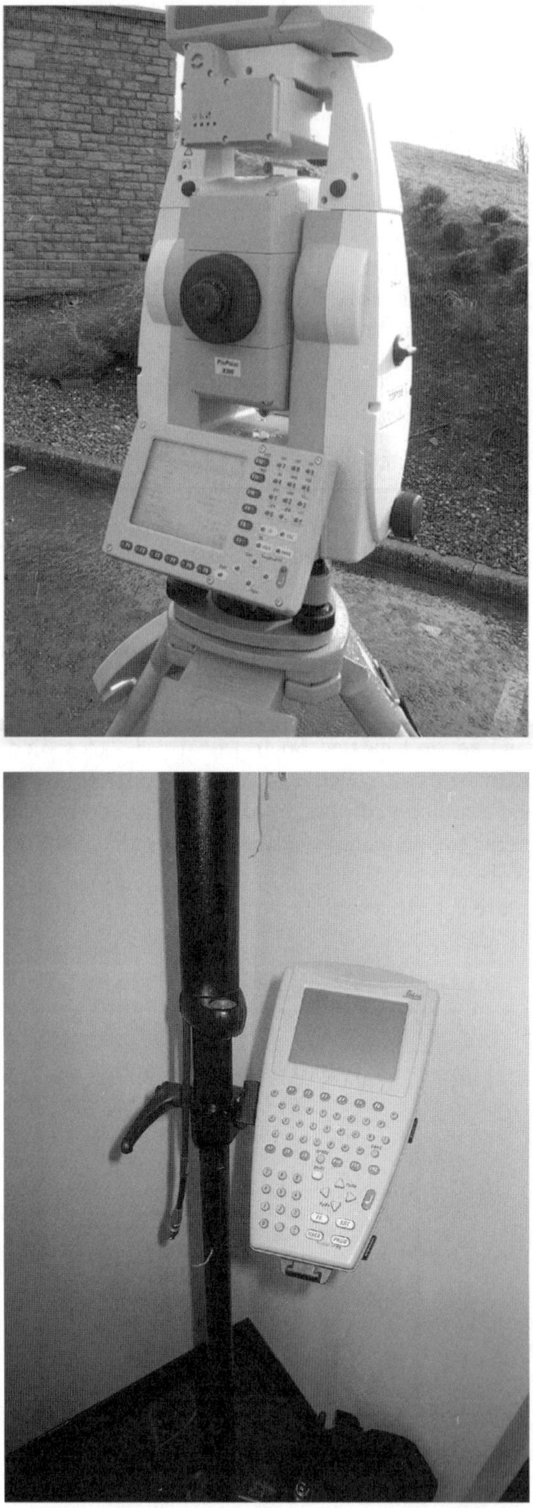

Figure 9.13 Leica System 1200

(b) Sub-metre accuracy

These are single channel receivers with typically 12 channels, one of which is enabled to receive the WAAS

or EGNOS differential correction. They are cheaper and more compact, often installed in a back pack with the antenna on a pole. Trimble systems are upgradable to dual frequency. These systems are capable of accuracies up to 0.3 m with differential correction. They are used for map updating and GIS data collection.

(c) Hand-held metre accuracy

These are compact versions of sub-metre systems. The antenna is integrated into the top of the unit, which has a large colour touch screen. With differential corrections they are now approaching the performance of sub-metre systems for real time kinematic (RTK) work. They are widely used for GIS data collection such as utilities mapping.

Figure 9.14 shows the Trimble GeoExplorer CE.

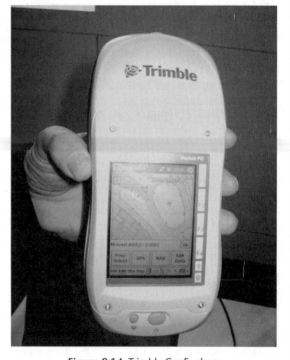

Figure 9.14 Trimble GeoExplorer

6. Terrestrial reference frame (TRF)

A terrestrial reference system is a another word for a coordinate system or datum. Chapter 2 contains details of the Ordnance Survey system of coordinates used in maps. This now goes by the initials *OSGB36*. It uses a Transverse Mercator Projection and the shape of the Earth is approximated to an ellipse with a major and minor axis specified by Airy in 1830. This is a local system which best suits Great Britain. GPS is a global system and must use a worldwide system, which can be realized with a framework of precisely fixed GPS stations.

The *World Geodetic System 1984 (WGS84)* is used for GPS positions. To use GPS, the WGS84 latitude, longitude and ellipsoid height will usually be transformed into a local system of Cartesian coordinates. An in-depth description of this is beyond the scope of this textbook, but a glossary of terms is given below at (a).

(a) Reference surface

Ellipsoid: The Earth is not a perfect sphere, having a slight bulge at the equator. An ellipse can be specified which best fits the shape of the Earth's crust. WGS84 uses a global best fit.

Geoid: This is roughly the shape of the earth at sea level and is described as a line of constant gravitational potential. In other words, a line at right angles to the direction of gravity. It is an undulating line due to the gravity anomalies caused by the thickening of the crust with the less dense rocks of the continental land masses.

Figure 9.15 shows the relationship between the two surfaces.

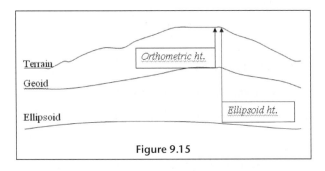

Figure 9.15

WGS84: World Geodetic System. This is a global reference system with the 'xyz' origin at the centre of the earth and the 'x' axis originally passing through the Greenwich meridian. The orientation of the axis is, however, set so that the mean drift of the continental plates is zero. The prime meridian, therefore, is slowly drifting away from Greenwich. Points in the system do not have fixed coordinates. In Northern Europe continental drift is about 2 cm per year. Heights in the system are above the ellipsoid.

ETRF89: European Terrestrial Reference Frame 1989. The drifting coordinates of WGS84 are impractical for surveying. In Europe, a number of points fixed in WGS84 1989 were taken as the basis of this coordinate 'frame'.

ETRS89: European Terrestrial Reference System is the practical realization of ETRF89.

ODN: Ordnance Datum Newlyn. This is the traditional vertical TRF for Britain realized in a system of fundamental benchmarks (FB).

OSGM91: OS Geoid Model 1991. This is based on the OS FBs.

OSGM02: OS Geoid Model 2002. This has replaced OSGM91 for precise GPS heighting. It is a numerical model of the differences of the orthometric heights of the FBs from their ETRS89 heights, which are taken from the ellipsoid of WGS84. OSGM02 heights differ

by as much as 45 m in the south-east of the UK, to 56 m in the north-west. An interpolator is available to calculate the geoid–ellipsoid separation at any location (www.gps.gov.uk). Ellipsoid heights using Airy 1840 differ by only metres from OSGM02.

(b) Change of coordinate system

For surveying work it is normal to use a local coordinate system. In the case of the UK this is likely to OSGB36. GPS processing software will contain a number of options for transforming WGS84/ETRS89 into local, national or international coordinate systems. A transformation will accomplish all or some of the following:

(i)   geographical latitude and longitude to eastings and northings, and vice versa,
(ii) change of map projection,
(iii) coordinate axis rotation,
(iv) origin shift,
(v)   scale change,
(vi) change of height datum.

In the UK scale increases away from the 400 000 mE grid line. For small sites, a common transformation is to convert ETRS89 into OSGB36 but leave the scale factor at unity.

For larger sites a full classical or Helmert transformation will be performed, involving coordinate axis rotation, origin shift and scale change.

This software function was formerly achieved by post processing, but in the latest systems it can be completed in seconds.

7. GPS survey project example

The following is an outline of the establishing of a control network by GPS along the route of a 15 km section of new motorway to the south west of Glasgow. Control stations were established to facilitate the production of site plans, to set out road details and for machine control in the construction of the road. The information is supplied courtesy of Survey Solutions (Scotland).

(a) Control survey

Primary, secondary and tertiary stations were established. Coordinates were required in OSGB36 and a local grid. The layout of the stations was by necessity linear, following the route of the road. Stations were sited to avoid multi-path errors and to minimize the GDOP.

(i)   *A primary control* baseline was established using five OS active stations as reference. These were Mallaig, Glasgow, Edinburgh, Carlisle and Isle of Man north as shown in Fig. 9.16

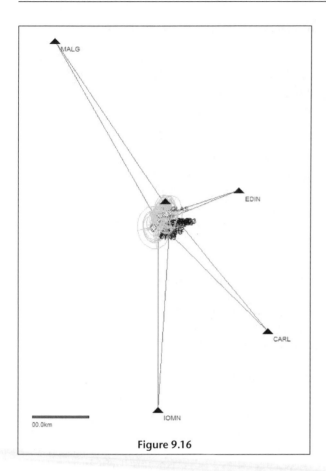

Figure 9.16

from ETRS89 to OSGB36, which tends to be somewhat inhomogeneous.

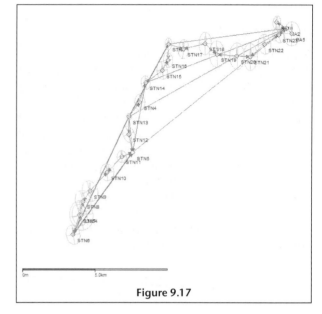

Figure 9.17

Dual frequency static measurements were recorded over a six-hour period. Post processing corrections were carried out using the OS RINEX data for the five stations and precise ephemeris data. Coordinates were converted from ETRS89 into OSGB36 and the local grid.

(ii) *Secondary control* stations were established from the primary baseline using static or fast static GPS at approximately 1 km intervals along the line of the route. These stations would be used for day-to-day survey operations. Redundant observations were included in the network to keep a check on accuracy. This network and the link to the baseline are shown in Fig. 9.17.

(iii) *Tertiary control* stations were established to supplement the secondary network. The main consideration in choosing the location of these points was intervisibility to facilitate radio communication in RTK GPS setting out and machine control of graders and bulldozers. The stations were also to be used for optical observation and levelling.

The stations were established by a variety of methods. Fast static GPS, RTK GPS, total station traversing and levelling were used. Coordinates were established in OSGB36 and the local system with heights on the OS Newlyn datum.

The project was completed with centimetre accuracy, although some problems were encountered with the OSTN02 transformation

8. Future developments

(a) Improvements to existing systems

In 1998 the US announced an upgrade to its GPS system with 16 new satellites by 2012. These will broadcast an additional code L2c for civilian users from 2005, and an additional frequency L5 from 2006. The L5 signal will be more powerful and give increased accuracy and reliability.

The GLONASS satellites are undergoing a similar upgrade.

(b) Galileo

This is a new system sponsored by the European Union under a public–private partnership. The first launch was due in 2005 and it is planned to be fully operational by 2010.

It will consist of:

30 satellites, altitude 24 000 km,
3 orbits at 56-degree inclination to the equator,
10 satellites per orbit,
14-hour period,
2 European control centres (UK and Italy),
3 frequencies: 1164–1215 MHz
           1260–1300 MHz
           1559–1591 MHz
           + integrity check.

The planned benefits of the system are likely to be improved accuracy with reduced GDOP and improved performance in built-up areas.

A disadvantage will be that surveyors will have to pay to use the system.

(c) Miniaturization

As with all forms of electronic equipment, the size of GPS instruments is likely to diminish. This is witnessed by the latest integrated antenna and receiver units with palmtop controllers, and the integrated total station and GPS of the Leica 1200 system. Surveyors will eventually be using mobile phones for GPS surveying.

Exercise 9.1

1 Describe the principle of distance measurement and ranging to give a position fix using GPS.

2 Outline the sources of error in GPS measurement, and describe procedures by which they can be minimized.

3 Describe the field procedure for *real-time kinematic* differential GPS detail survey.

9. Answers

1 Distance is measured by the transit time of the radio signal from the satellite (Fig. 9.4). Knowing the distance from one satellite fixes the position on a sphere (Fig. 9.1). For centimetre accuracy the phase difference of the carrier wave is also measured.

Two range spheres give the position on the circle of contact of the spheres (Fig. 9.2). A third satellite range gives two possible positions on opposite sides of the circle, one of which is on the surface of the Earth.

2 *Clock error*: produces lack of synchronization of the satellite signal and the receiver's signal. A correction for this can be calculated by observing four or more satellites.

Atmospheric error: this produces variations in the speed of the signal passing through the layers of varying density in the Earth's atmosphere. Two techniques of correcting the error can be applied.

(i) Atmospheric modelling from a single base station or a network of base stations can give parameters for the conditions at a particular time.

(ii) Dual frequency receivers measure on two different frequency signals; the difference in the distance calculated allows the atmospheric parameters to be estimated.

Ephemeris error: the satellite orbit may wobble slightly. This is continuously monitored and relayed to the satellite, which includes it in its signal.

Multipath error: reflections cause distorted signal transit time, particularly with a weak signal. These reflections can be filtered out using a choke ring antenna or proprietary circuitry in the receiver.

GDOP: see Fig. 9.6

3 The equipment used consists of a reference receiver at the base station, which calculates corrections. The roving receiver is linked to the base station by radio or mobile phone and the coordinates calculated by the rover are instantly corrected by the base station data. At the start of the survey, the rover is first initialized to establish the reference with the base station, and this is continually checked during the survey even if lock is lost.

Chapter summary

In this chapter, the following are the most important points:

- Global Position System (GPS) satellites transmit radio signals to receiver's on the ground, which allows positions to be calculated.

- At present, two satellite constellations are available, the US NAVSTAR system and the Russian GLONASS system. The European Galileo system will be introduced hopefully in 2010.

- The distance of a receiver from a satellite is measured by timing the delay in the radio signal from the satellite. The delay is measured by synchronizing matching pseudo random codes from the satellite and the receiver.

- A minimum of 5 distances (ranges) are required for a precise fix.

- Survey receivers use phase measurement of the carrier frequency to give centimetre accuracy.

- The distances measured are known as pseudo ranges as they are inaccurate due to the presence of many errors.

- The main errors are: timing; ephemeris, atmospheric; multipath; Geometric Dilution of Precision (GDOP).

- Errors can be minimized by using dual frequency receivers and differential GPS (DGPS) measuring procedures.

- GPS uses a global coordinate system, WGS84 or ETRS89 in Europe. This is then transformed into a national or local coordinate system, e.g. OSGB36 for the UK.

CHAPTER 10 Detail Survey

In this chapter you will learn about:

- three methods of surveying detail for a plan: radial survey; GPS and photogrammetry
- radial survey using a total station involving:
  - setting up and orientating the instrument
  - fieldwork
  - manual calculations
  - computer processing
  - plotting
  - controlling accuracy
- the fieldwork and processing for GPS detail survey
- the methods of producing detail plans from aerial photography using photogrammetric measurement
- other software associated with detail survey:
  - CAD to produce a neat finished drawing
  - digital terrain models to allow a 3D model to be displayed, and associated sections and volume calculations to be carried out
  - geographic information systems to add data to your plan

Detail survey involves the fixing of ground detail such as roads, fences and buildings for inclusion on a map or a plan. It may also involve a contour survey.

Methods of detail survey follow the basic survey principle of '*working from the whole to the part*'. First a precise control survey is carried out, which is then followed by the less accurate methods of the detail survey. The less accurate methods of the detail survey are, however, sufficient to meet the requirements for the map or plan. This follows the survey principle of '*economy of accuracy*'.

Chapter 3, Linear survey, illustrates this with the tri-lateration control framework and the short offset measurements to pick up the detail. In this case control and detail survey are carried out simultaneously, but in other methods a control traverse or triangulation survey would come first.

Plane tabling was a traditional method used in the 19th and early 20th centuries. A control plot was fixed on a table which was set up on a tripod set up over a control point. The table was rotated to orientate the plot with a neighbouring station and detail was fixed by direction and distance (radiation) or by direction (intersection). The method had the advantage that the

map could be seen in the field, but the wet climate of the UK was a distinct disadvantage.

Stadiatacheometry was a method widely used in the 20th century. The difference between the top and bottom stadia hair readings in the theodolite telescope onto a staff when multiplied by 100 gave the distance, and horizontal and vertical angles allowed positions to be fixed by radiation. A theodolite with special stadia hairs called a self-reducing tacheometer was developed to minimize the calculations. Detail was recorded from the control stations. To conserve accuracy, the length of sights was limited. This method was the forerunner of the modern method of radiation survey using total stations.

Today there are three methods commonly used for detail survey:

1. radiation with total station,
2. real-time kinematic GPS,
3. photogrammetry.

In this chapter, the first two of these methods are described, but because photogrammetry is beyond the scope of this book, only an outline of the principles is included.

1. Principles of radial positioning

(a) Orientation of survey

Figure 10.1 illustrates the principles. Point A is a survey station with known (or assumed) coordinates. The reference object (RO) is the point chosen as the starting point, from which the bearings of all lines of the survey will be measured. The RO may be a second survey station or some prominent object such as the spire of a church, a pylon or a road sign. The direction (bearing) from station A to the RO must be known or assumed. The direction may be relative to magnetic, true or Grid north, or may simply be assumed to point to north, in which case the bearing from station A to the RO is zero degrees.

Points 1 to 24 are the points of detail of the survey, e.g. spot heights, manholes, etc. The three-dimensional coordinates of the points are required.

(b) Objectives

A minimum of two surveyors is required to conduct a radial survey. Their duties are:

1. To select suitable survey stations. The only criterion for the selection is that as many detail points as possible should be visible from any one survey station. If all of the points of detail are not visible from a single station, a traverse is made to establish more survey stations, e.g. in Fig. 10.1 the required traverse stations are A, B, C.
2. To measure the horizontal angle, the vertical or zenith angle and slope distance, using EDM means, to every point (numbered 1 to 24), of the survey.
3. To record the survey data, either manually or automatically, on an electronic field book and to process these data. The data may be calculated manually or post-processed on computer.

(c) Fieldwork

The RO is assumed to be the north point, bearing zero degrees (00° 00′ 00″).

1. The theodolite/EDM is set over the point A and accurately centred and levelled.
2. The height of the instrument, i, from point A to the transit axis is measured, using a tape or rule, and noted in the field book (Table 10.1).
3. The coloured target, or the actual reflector if the EDM telescope is coaxial, is set to this height and taken to point B (Fig. 10.2) by an assistant, where it is held vertically by utilizing the attached spirit level. The target height is noted in the field book on the same line as station B (line 2).
4. The RO is sighted by the observer and the instrument is set to zero by means of the zero-set key (electronic theodolite or total station) or by means of the upper and lower plate clamps (optical theodolite). This reading of 00° 00′ 00″ is recorded in the field book on line 1. The horizontal circle reading to any other point will therefore be the whole circle bearing from station A to the point.
5. Point B is then sighted and the horizontal circle reading and vertical circle reading are noted in the

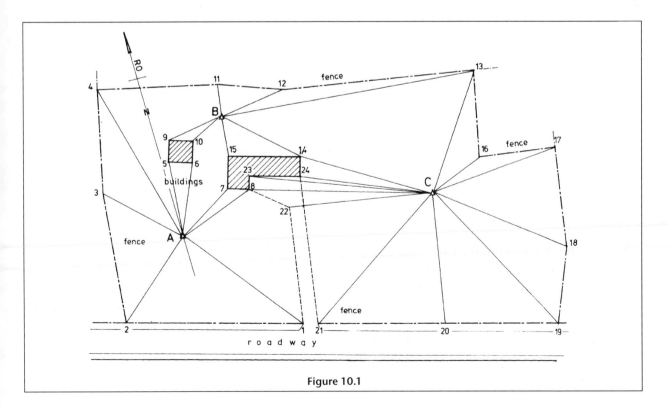

Figure 10.1

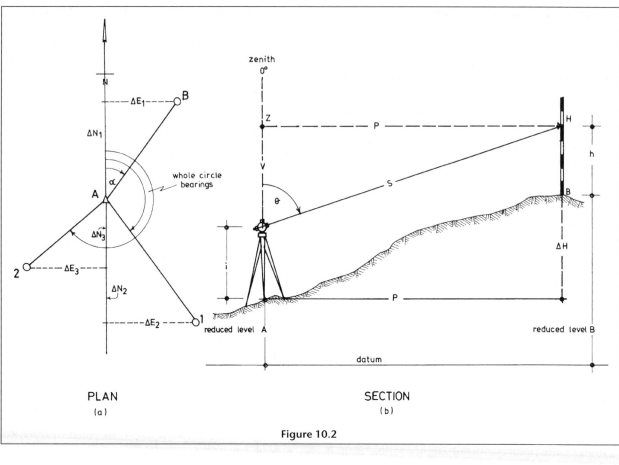

Figure 10.2

Table 10.1

Instrument station	A
Easting	100.000 m
Northing	200.000 m
Reduced level	35.210 m AD
Instrument height	1.350 m

	Target point	Target height	Horizontal circle reading	Zenith angle	Slope distance	Remarks
1	RO		00° 00′ 00″			Reference object
2	B	1.350	36° 30′ 00″	71° 30′ 10″	18.325	Survey station
3	1	1.350	147° 29′ 10″	92° 04′ 00″	21.110	Fence
4	2	0.850	231° 15′ 30″	87° 13′ 40″	14.676	Fence

field book on line 2. The slope distance is then measured to the reflector, using the EDM, and also noted on line 2. This completes the field work for point B.

6. The assistant is directed to the next point and the procedures 3 and 4 are repeated for that point. The field data for points 1 and 2 are shown in Table 10.1 on lines 3 and 4.

2. Calculation and plotting of survey (manually)

The field data may be calculated and plotted manually or may be post processed on computer.

(a) Derivation of formulae

Calculations are carried out in the following order since each part-calculation depends upon the result from the previous part.

1. *Plan length* In the right-angled triangle (Fig. 10.2(b)) formed by the slope length (S), the plan length (P) and the vertical (V),

$$P/S = \sin \theta$$
$$\text{Therefore } P = S \sin \theta \qquad (1)$$

i.e. plan length = slope length × sin zenith angle

2. *Vertical height V* In the triangle, the vertical dimension (V) is the height from the transit axis of the theodolite/EDM to the target:

$$V/S = \cos \theta$$
$$\text{Therefore } V = S \cos \theta \qquad (2)$$

i.e. vertical height = slope length × cos zenith angle

3. *Difference in height (ΔH)* The difference in height between the ground at instrument station A and the ground at target station B is ΔH. In Fig. 10.2(b),

$$\Delta H = i + V - h \qquad (3)$$

i.e. difference in height between stations
= instrument height + vertical height − target height

For angles of depression (i.e. downhill sights) the vertical dimension (V) is negative. Since the zenith angle (θ) in such a case would be greater than 90°, the calculation $V = P \cos \theta$ would naturally result in a negative answer.

In order to reduce the calculation of $\Delta H = i + V - h$, the target height (h) is made equal to the instrument height (i), as in this case.

$$\text{Therefore } \Delta H = V \qquad (3a)$$

4. *Reduced level of target station* In Fig. 10.2(b) station B lies at a higher level than station A.

$$\text{Therefore, } RL(B) = RL(A) + \Delta H \qquad (4)$$

i.e. reduced level target station = reduced level of instrument station + rise (ΔH)

Station B could, of course, be at a lower level than station A, in which case

Reduced level (B) = reduced level (A) − fall (ΔH)

5. *Coordinates of target station* In Fig. 10.2(a) the whole circle bearing of the line from station A to station B is α. From previous theory on traversing (Sec. 5 in Chapter 8) the difference in eastings between the stations is ΔE and the difference is northings is ΔN:

$$\Delta E_1 = AB \times \sin \alpha \qquad (5)$$
$$= plan \text{ length } AB \times \sin \text{ bearing } AB$$

and

$$AN = AB \times \cos \alpha \qquad (6)$$
$$plan \text{ length} \times \cos \text{ bearing } AB$$

These formulae hold good for any line between the instrument station and target station.

Again from previous theory on traversing (Sec. 5 in Chapter 8), the total coordinates (E and N) of any target station are found by adding the difference in easting (ΔE) between the instrument station and target station to the easting of the instrument station, and likewise for northings, i.e.

Easting (target station) = easting (instrument station) + difference in eastings

$$E_T = E_1 + \Delta E \qquad (7)$$

and

Northing (target station)
= northing (instrument station) + difference in northings

$$N_T = N_1 + \Delta N \qquad (8)$$

(b) Summary of formulae

Plan length = slope length × sin zenith angle
$$P = S \sin \theta \qquad (1)$$

Vertical V = slope length × cos zenith angle
$$V = S \cos \theta \qquad (2)$$

Difference in height ΔH
= instrument height + vertical V − target height
$$\Delta H = i + V - h \qquad (3)$$

RL target station = RL instrument station + difference in height
$$RL(T) = RL(I) + \Delta H \qquad (4)$$

Difference in eastings
= plan length × sin bearing
$$\Delta E = P \sin \alpha \qquad (5)$$

Difference in nothings
= plan length × cos bearing
$$\Delta N = P \cos \alpha \qquad (6)$$

Eastings of target = easting instrument + difference in eastings
$$E_T = E_1 + \Delta E \qquad (7)$$

Northings of target = northing instrument + difference in northings
$$N_T = N_1 + \Delta N \qquad (8)$$

(c) Calculations

The calculations of point B are shown in Table 10.2.

(d) Field data/calculation table

Usually, the calculations are processed on the field data table as in Table 10.3, which should be self-explanatory.

E X A M P L E

1 Table 10.1 shows the data observed to points B, 1 and 2. (The coordinates of point B are calculated in Tables 10.2 and 10.3.) Calculate the coordinates (east, north, reduced level) of points 1 and 2.

Answer (Table 10.4)

These answers are summarized on the field data/calculation table (Table 10.3).

Table 10.3

Instrument station   A
Easting   100.000 m
Northing   200.000 m
Reduced level   35.210 m
Instrument height   1.350 m

Target point	Target height (m)	Horizontal circle (α) reading	Zenith angle (θ)	Slope distance S (m)	P ($S\sin\theta$)	V ($S\cos\theta$)	ΔH ($i+V-h$)	RL target ($RL_i+\Delta H$)	ΔE ($P\sin\alpha$)	ΔN ($P\cos\alpha$)	E ($E_i+\Delta E$)	N ($N_i+\Delta N$)	Remarks
RO		00° 00′ 00″		–	–	–							Reference object
B	1.350	36° 30′ 00″	71° 30′ 10″	18.325	17.378	5.814	5.814	41.024	10.337	13.969	110.337	213.969	Survey Station
1	1.350	147° 29′ 10″	92° 04′ 00″	21.110	21.096	−0.761	−0.761	34.449	11.339	−17.789	111.339	182.211	Fence
2	0.850	231° 15′ 30″	87° 13′ 40″	14.676	14.659	0.710	1.210	36.420	−11.434	−9.174	88.563	190.826	Fence

Table 10.2

Formula number	Formula	Point B calculations Calculation	Result
1	$P = S \sin \theta$	$P = 18.325 \sin 71° 30' 10''$	17.378
2	$V = S \cos \theta$	$V = 18.325 \cos 71° 30' 10''$	5.814
3	$\Delta H = i + V - h$	$\Delta H = 1.350 + 5.814 - 1.350$	5.814
4	$RL(B) = RL(A) + \Delta H$	$RL(B) = 35.210 + 5.814$	41.024
5	$\Delta E_{AB} = P \sin a$	$\Delta E_{AB} = 17.378 \sin 36° 30' 00''$	10.337
6	$\Delta N_{AB} = P \cos a$	$\Delta N_{AB} = 17.378 \cos 36° 30' 00''$	13.969
7	$E_B = E_A + \Delta E_{AB}$	$E_B = 100.000 + 10.337$	110.337
8	$N_B = N_A + \Delta N_{AB}$	$N_B = 200.000 + 13.969$	213.969

Table 10.4

	Point 1	Point 2
$P = S \sin \theta$	$21.110 \sin 92° 04' 00'' = 21.096$	$14.676 \sin 87° 13' 40'' = 14.659$
$V = S \cos \theta$	$21.110 \cos 92° 04' 00'' = -0.761$	$14.676 \cos 87° 13' 40'' = 0.710$
$\Delta H = i + V - h$	$1.350 + (-0.761) - 1.350 = -0.761$	$1.350 + 0.710 - 0.850 = 1.210$
$RL_T = RL_A + \Delta H$	$35.210 - 0.761 = 34.449$	$35.210 + 1.210 = 36.420$
$\Delta E_{AT} = P \sin a$	$21.096 \sin 147° 29' 10'' = 11.339$	$14.659 \times \sin 231° 15' 30'' = -11.434$
$\Delta N_{AT} = P \cos a$	$21.096 \cos 147° 29' 10'' = -17.789$	$14.659 \times \cos 231° 15' 30'' = -9.174$
$E_T = E_A + \Delta E_{AT}$	$100.00 + 11.339 = 111.339$	$100.000 - 11.437 = 88.563$
$N_T = N_A + \Delta N_{AT}$	$200.00 - 17.789 = 182.211$	$200.000 - 9.174 = 190.826$

Exercise 10.1

1 Figure 10.3 shows the layout of part of a radial positioning survey and Table 10.5 shows the partially completed field data of the survey.

(a) Calculate the three-dimensional coordinates (east, north, height) of each point, given that station A is the origin of the survey and the bearing from A to the RO is zero degrees.
(b) Plot the survey to scale 1:250.

3. Calculation and plotting using microcomputer-based mapping systems

(a) Introduction

The previous calculations are lengthy, laborious and repetitious – exactly the disadvantages that computers were designed to overcome. Every manufacturer of surveying instruments provides a microcomputer-based mapping system which solves these problems speedily, accurately and efficiently. Typical of these computer packages are Liscad from Leica, Terramodel from Trimble, Mapsuite+ from Sokkia, SSS from Topcon and LSS from McCarthy Taylor. Most can be purchased on a modular basis with common elements being: field data processing; contouring; CAD; sections; areas and volumes; terrain modelling and visualization. They will run on most computers, although a good graphics card and large screen are advisable. For plotting, a large format inkjet plotter is desirable.

All of the packages are very similar, though manufacturers would claim differently. Each package has its own particular advantages and disadvantages but they all perform the same function, namely the production of a contoured plan.

It would not be possible nor desirable to describe the operation of all or even one of the packages completely. Some package manuals are hundreds of pages long. The following dissertation therefore describes the fundamental features of all packages and their effects on surveying procedure.

(b) Coding

On any radial positioning survey the data relating to a great number of points have to be recorded either manually or automatically using a data recorder. In manual recording, a sketch should always be made showing the point numbers and locations. The sketch enables the surveyor to join the various points correctly on the final plotted drawing.

When plotting by computer, however, the computer packages naturally cannot recognize a sketch, so the surveyor has to instruct the computer regarding the description of a point and how it is to be joined to other points. The points must be coded in some way such that the computer will recognize them. Thus any point of the survey will have a unique point number and a feature code, which may be a one-, two- or three-point code.

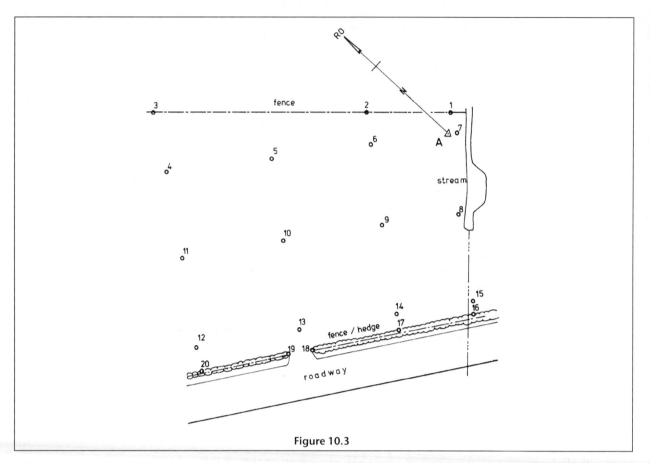

Figure 10.3

Table 10.5

Instrument station	A	Backsight station RO	Whole circle bearing A–RO = 00° 00′ 00″
Instrument height	1.35 m		
Easting	0.00 m		
Northing	0.00 m		
Reduced level	6.48 m		

Field data

Target station	Target height (m)	Slope length (m)	Whole circle bearing	Zenith angle	P (m)	ΔH (m)	ΔE (m)	ΔN (m)	Easting (m)	Northing (m)	Reduced level (m)	Target station
1	1.35	3.02	72° 7′ 30″	92° 6′ 35″	3.02	−0.11	2.87	0.93	2.87	0.93	6.37	1
2	1.35	11.10	324° 9′ 45″	89° 21′ 0″	11.10	0.13	−6.50	9.00	−6.50	9.00	6.61	2
3	2.35	30.02	318° 55′ 20″	87° 8′ 35″	29.98	1.50	−19.70	22.60	−19.70	22.60	6.98	3
4	1.35	29.36	306° 49′ 50″	89° 21′ 50″	29.36	0.33	−23.50	17.60	−23.50	17.60	6.81	4
5	1.35	19.34	307° 13′ 30″	90° 4′ 16″	19.34	−0.02	−15.40	11.70	−15.40	11.70	6.46	5
6	0.35	9.35	310° 12′ 0″	96° 13′ 30″	9.29	−1.01	−7.10	6.00	−7.10	6.00	6.47	6
7	1.35	0.85	69° 26′ 40″	90° 16′ 5″								
8	1.35	9.59	210° 51′ 40″	95° 13′ 40″								
9	1.35	13.41	260° 7′ 9″	92° 32′ 20″								
10	2.00	21.63	279° 35′ 35″	89° 28′ 50″								
11	1.00	30.89	287° 8′ 40″	91° 20′ 20″								
12	1.35	35.37	272° 26′ 0″	92° 43′ 45″								
13	1.35	27.59	261° 1′ 0″	93° 33′ 40″								
14	1.35	21.66	240° 56′ 45″	93° 19′ 10″								
15	1.35	19.67	213° 22′ 0″	93° 9′ 35″								
16	1.35	21.25	209° 8′ 37″	93° 15′ 0″								
17	1.35	23.38	238° 29′ 20″	93° 23′ 35″								
18	1.35	28.16	256° 24′ 35″	94° 14′ 35″								
19	1.35	30.11	260° 13′ 25″	93° 58′ 10″								
20	1.35	36.55	269° 31′ 45″	92° 58′ 10″								

Point number

Every point on the survey is given a reference to differentiate it from any other point on the survey. The reference may be alphabetic or numeric or alphanumeric. Where a particular package accepts only numeric characters, it is usual to separate the survey stations from the points of detail by allocating numbers (say 1 to 100) to the traverse stations and numbers from 101 onwards to the detail points.

Feature codes

In a feature coding system there are three types of code, namely, point codes, control codes and control codes with parameters:

1. *Point code*   A point code is related specifically to the point being observed and gives a description of the point. Codes are kept simple and restricted as far as possible to about four or five letters. Typical codes of a few common points of detail are shown in Table 10.6.

Table 10.6

Feature	Code
Boundary	BDY
Bank	BANK
Building	BLD
Fence	FCE
Footpath	FP
Hedge	HEDGE
Instrument station	IS
Kerb	KERB
Lamp post	LP
Manhole	MH
Pond	POND
Wall	WALL
Spot height	SH
Tree	TREE

Each package has its own feature code library which must be used in the field when recording point data; otherwise the package will not compute the data.

2. *Control code*   (a) Control codes control the way in which point codes are implemented. In general, in any computer package, points with the same code are joined in sequence by straight lines unless a control code instructs otherwise. Thus three points coded as 101 FCE, 102 FCE and 103 FCE will be computed and plotted by the software and joined by straight lines of a certain line type, as defined in the package library. If another point on the fence is observed later in the survey, e.g. point 109, the latter point will be joined to point 103 by a straight line unless prevented from being joined by the insertion of a control code.

(b) It is therefore usual on a radial point survey to observe all points that have the same code before observing points of a different code. This practice is called 'stringing' and is particularly advantageous when using a data recorder. Recorders automatically increment the point numbers by adding one but retain the previous feature code until the surveyor changes it. Thus, when observing points along a fence or a kerb, the observer simply sights the reflector held at the first point of the string, enters the point number and code and then triggers the recorder. The slope distance, zenith angle and horizontal angle are recorded automatically. At the second and subsequent points of a string, the observer sights the reflector and triggers the recorder. The point number is increased by one, the code is retained and the three measurement parameters are recorded automatically.

EXAMPLE

2 Figure 10.4 shows a new town parking area and public toilet block with boundaries comprising various hedges and fences. The area is to be radially surveyed from station 1 using line 1–2 as a reference line. Spot heights are to be observed at every point. Make a list of the coding required to enter points 101–106 into a computer package.

Answer

Point number	Code
101	FCE SH
102	FCE SH
103	FCE SH HEDGE
104	FCE SH HEDGE
105	FCE SH HEDGE
106	FCE SH

(c) Whenever a new string of points of the same feature commences, a control code must be added to prevent the string from joining to the previous string. The control code is usually the word START or ST. Thus in Fig. 10.4 point 111 is the start of a new fence and must be coded as 111 FCE ST. If the code START had been omitted, the package would have drawn a fence between point 106 and point 111. Alternatively, string numbers can be used to start new features.

EXAMPLE

3 Using Fig. 10.4, make a list of the coding required to enter points 107–119 into a computer package. Spot heights are required at every point.

Answer

Point number	Code
107	WALL SH
108	WALL SH

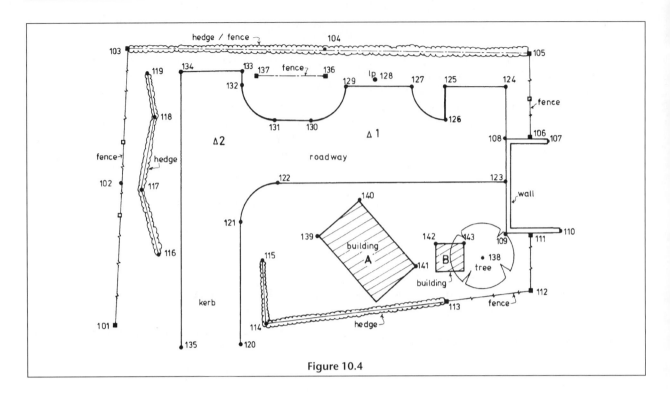

Figure 10.4

109	WALL SH
110	WALL SH
111	FCE START SH
112	FCE SH
113	FCE HEDGE SH
114	HEDGE SH
115	HEDGE SH
116	HEDGE START SH
117	HEDGE SH
118	HEDGE SH
119	HEDGE SH

(d) It has already been pointed out that, in a computer package, a feature code, e.g. 'kerb', will be joined to the previous occurrence of the point having the code 'kerb' by a straight line. Frequently, however, kerbs are not straight; hence, a control code must be introduced to indicate the start and finish of a curve. As always, the feature code libraries of different packages vary and so the following control codes are only examples, but the principles are the same.

In Fig. 10.4 point 120 is the first occurrence of the feature 'kerb' and is coded as 120 KERB. If a spot height is required the code becomes 120 KERB SH. Point 121 is the start of a curve of unknown radius. Line 120–121 is tangential to the curve. A typical control code is STCV, meaning 'start a smooth curve tangentially'; likewise point 122 is the end of the curve and line 122–123 is again tangential to the curve.

A typical control code to finish the curve is ENDCV or FINCV, meaning 'end the curve tangentially'. Some packages require a point about halfway around the curve, in order to be able to draw the curve smoothly.

EXAMPLE

4 Using Fig. 10.4, make a list of the coding required to enter points 120–123 into a computer package. Spot heights are required at every point.

Answer

Point number	Code
120	KERB SH
121	KERB STCV SH
122	KERB ENDCV SH
123	KERB SH

(e) In Fig. 10.4 the kerb between points 126 and 127 is again a smooth curve but is not tangential to any line. It therefore requires a different code from the previous curve code. A typical coding for point 126 is NEWCV and for point 127 is ENDONCV. These codes start and finish a curve, without being tangential to a specified tangent line.

EXAMPLE

5 Using Fig. 10.4, make a list of the coding required to enter points 123–128 into a computer package. Spot heights are required at all points.

Answer

Point number	Code
123	KERB SH
124	KERB SH
125	KERB SH
126	KERB NEWCV SH
127	KERB ENDONCV SH
128	LP SH

Exercise 10.2

1 Using Fig. 10.4, make a list of the coding required to enter points 128–137 into a computer package. Spot heights are required at all points.

(f) In the introduction to coding it was pointed out that surveyors should use a diagram when recording points manually so that they know which points are to be joined when they plot the survey. These relationships can be completely defined through the coding system, although it is advisable to keep diagrams until complete proficiency in coding is gained.

There are numerous 'join' commands in coding. Figure 10.5 illustrates typical situations where these commands are used. In Fig. 10.5(a) a four-sided enclosure is being surveyed. If the code FCE is used only, point 104 will not close back to point 101. The control code CLOSE is used with point 104, which is coded as 104 FCE CLOSE in order to effect a closure on point 101. Other 'join' commands include JCS, meaning 'join to closest point of same code', which is useful when the theodolite/EDM has to be moved to another set-up point and the continuation of a string is carried out from the second point (Fig. 10.5(b)); JC meaning

'join to the closest point regardless of code' (Fig. 10.5(c)); JP meaning 'join to the previous point regardless of code' (Fig. 10.5(d)).

3. *Control code with parameter* (a) A control code with a parameter allows the surveyor to pass additional information to the plot. It is really a control code with a value attached to it.

(b) The branch spread of a tree can be entered with the control code, SIZE, followed by the diameter, or radius in some packages, of the branches. Thus the tree in Fig. 10.4 is coded as 138 TREE SIZE 10.

(c) From any one instrument position it is impossible to observe the four corners of a building. The problem can be overcome by observing three corners of the building and adding a control code, usually in the form of a note, which will close the rectangle.

EXAMPLE

6 In Fig. 10.6(a) only three corner points, 101–103, can be observed from station 1. List the coding required to draw the building on the plot.

Answer

Point number	Code
101	BLD
102	BLD
103	BLD CLSRECT

The code CLSRECT will form a rectangle, depicting the building. A new point is coordinated in the data-

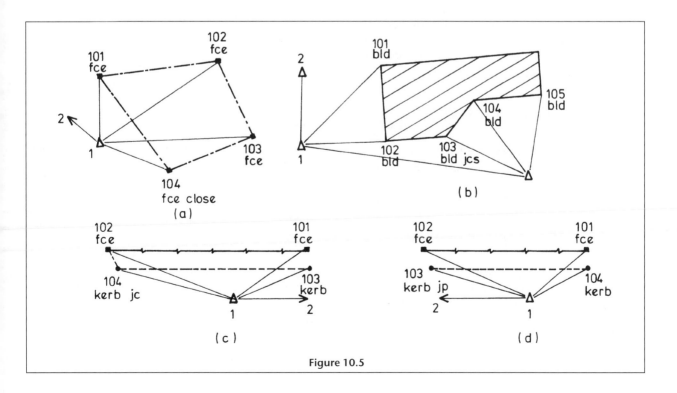

Figure 10.5

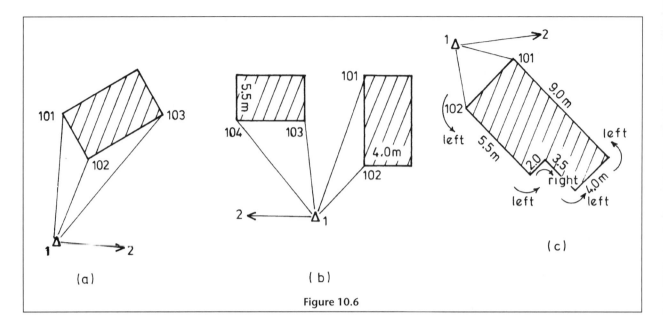

Figure 10.6

bases of the computer package, so that the new sides are parallel and equal in length to the sides computed from the three observation points. Unfortunately, any errors in the survey of the three points will be transferred to the new fourth point. This particular type of coding produces a parallelogram rather than a true rectangle. This situation can be avoided by observing only two corners and using a tape to measure the other side.

In Fig.10.6(b) points 101 and 102 form the baseline of the first building. The sides perpendicular to the baseline measure 4.0 metres. The control code is RECT followed by –4.0. The dimension is entered as a negative because the measured side makes a left turn from the base line.

Points 103 and 104 form the base line of the second building. The other side of the building measures 5.5 m and from the base line the measured side turns right. The 5.5 m parameter is added to the control code RECT as a positive dimension.

EXAMPLE

7 In Fig. 10.6(b) two buildings have been observed from a survey point, by observing only two corners of each building. List the coding required to draw the buildings on the plot.

Answer

Point number	Code
101	BLD
102	BLD RECT –4.0
103	BLD START
104	BLD RECT 5.5

Note: Point 103 begins a new string and requires the control code START.

Finally, in Fig. 10.6(c) a multi-sided building is to be surveyed. Some form of coordinate option is used as a control code. One solution, typical of most packages, is to enter a control code in the form of a 'distance' note followed by all the measured sizes of the building. Positive sizes are used for right turns, negative for left turns.

EXAMPLE

8 List the coding required to draw the building shown in Fig. 10.6(c).

Answer

Point number	Code
101	BLD
102	BLD DIST –5.5 –2.0 3.5
	–4.0 –9.0

Exercise 10.3

1 Using Fig. 10.4, list the coding required to enter points 138–143 into a computer package, given that building A is rectangular, building B is 3.0 metres square and the tree has a 7.0 m branch spread.

4. Radial survey with automatic total station

It is not possible to cover every type of instrument and software package, but a generic account of common procedures is given in this section.

(a) Fieldwork

The instrument is set up on a known station, a reference station is observed and detail points recorded. In

Figure 10.7

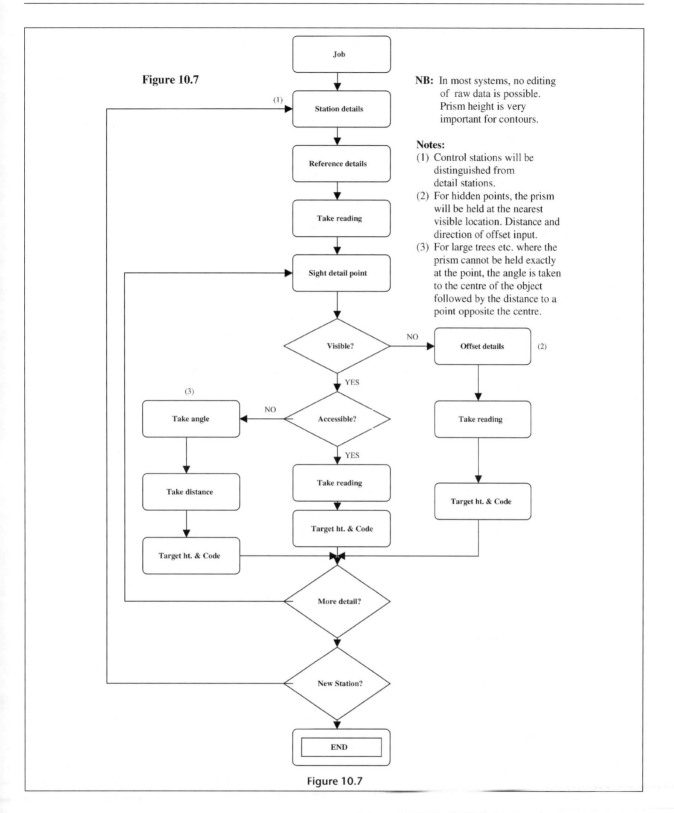

NB: In most systems, no editing of raw data is possible. Prism height is very important for contours.

Notes:
(1) Control stations will be distinguished from detail stations.
(2) For hidden points, the prism will be held at the nearest visible location. Distance and direction of offset input.
(3) For large trees etc. where the prism cannot be held exactly at the point, the angle is taken to the centre of the object followed by the distance to a point opposite the centre.

Figure 10.7

older instruments, a separate data recorder is used. Modern instruments have a large keyboard and screen and storage is internal. In robotic mode a remote control keyboard is used. The flow diagram in Fig. 10.7 illustrates the general procedure.

(b) Computer processing

A summary chart (Fig. 10.8) of the common steps to produce a contour plan from the raw field data is shown on page 200. The illustrations shown on pages 201 and 202 are taken from Trimble Terramodel software.

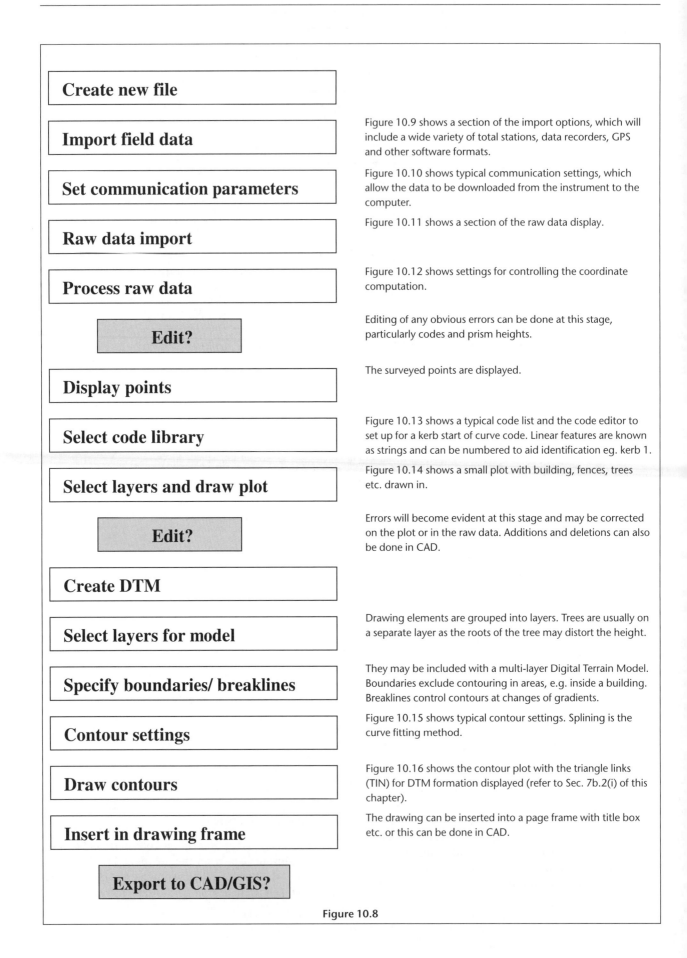

Create new file

Import field data

Set communication parameters

Raw data import

Process raw data

Edit?

Display points

Select code library

Select layers and draw plot

Edit?

Create DTM

Select layers for model

Specify boundaries/ breaklines

Contour settings

Draw contours

Insert in drawing frame

Export to CAD/GIS?

Figure 10.9 shows a section of the import options, which will include a wide variety of total stations, data recorders, GPS and other software formats.

Figure 10.10 shows typical communication settings, which allow the data to be downloaded from the instrument to the computer.

Figure 10.11 shows a section of the raw data display.

Figure 10.12 shows settings for controlling the coordinate computation.

Editing of any obvious errors can be done at this stage, particularly codes and prism heights.

The surveyed points are displayed.

Figure 10.13 shows a typical code list and the code editor to set up for a kerb start of curve code. Linear features are known as strings and can be numbered to aid identification eg. kerb 1.

Figure 10.14 shows a small plot with building, fences, trees etc. drawn in.

Errors will become evident at this stage and may be corrected on the plot or in the raw data. Additions and deletions can also be done in CAD.

Drawing elements are grouped into layers. Trees are usually on a separate layer as the roots of the tree may distort the height.

They may be included with a multi-layer Digital Terrain Model. Boundaries exclude contouring in areas, e.g. inside a building. Breaklines control contours at changes of gradients.

Figure 10.15 shows typical contour settings. Splining is the curve fitting method.

Figure 10.16 shows the contour plot with the triangle links (TIN) for DTM formation displayed (refer to Sec. 7b.2(i) of this chapter).

The drawing can be inserted into a page frame with title box etc. or this can be done in CAD.

Figure 10.8

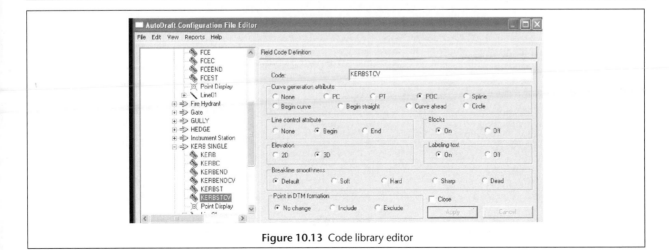

Figure 10.9 import

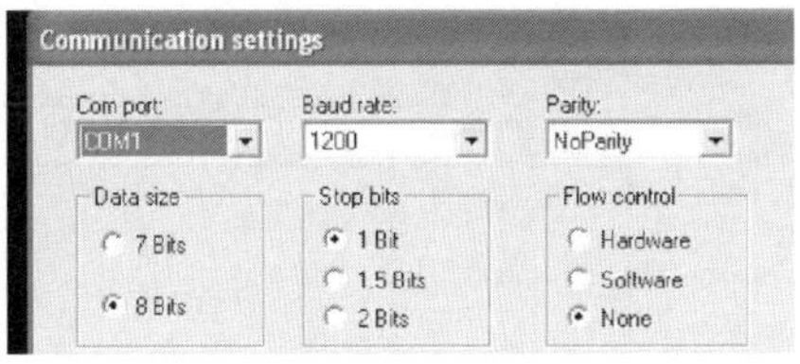

Figure 10.10 Com settings import

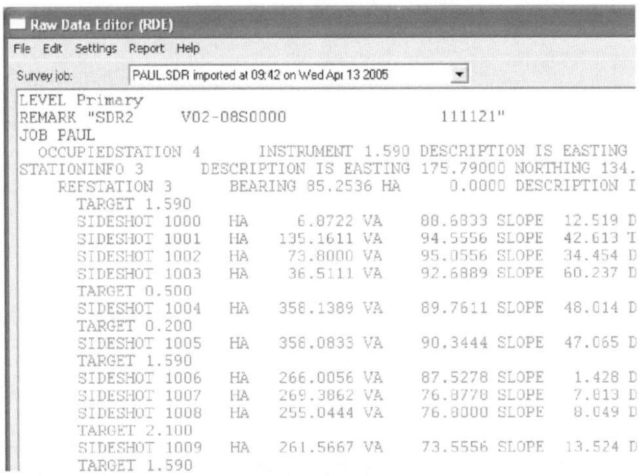

Figure 10.11 Raw data

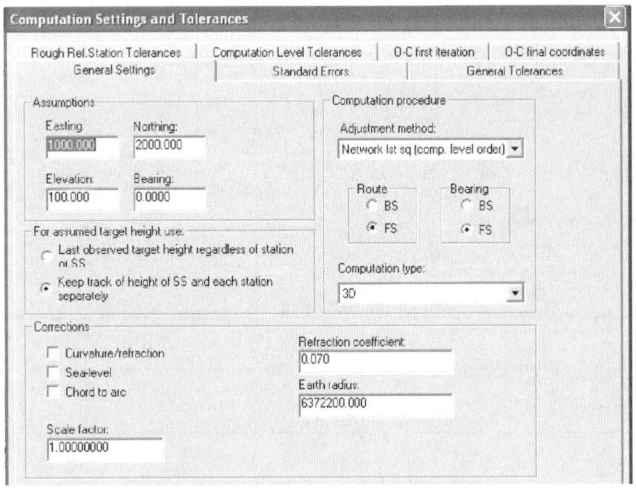

Figure 10.12 Computation settings

Figure 10.13 Code library editor

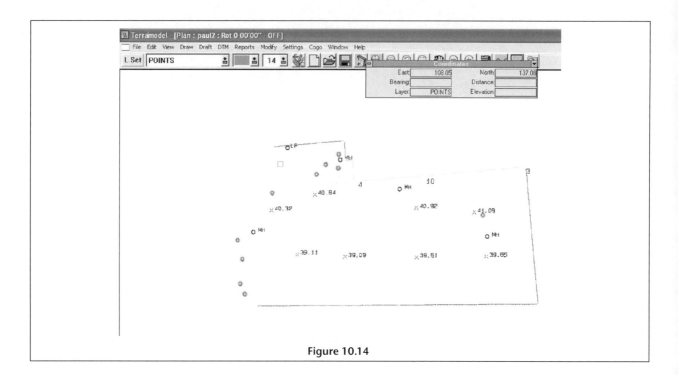

Figure 10.14

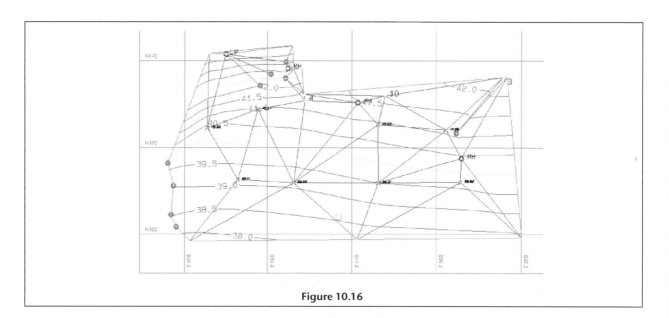

Figure 10.15

Figure 10.16

(c) Plotting

In preparation for plotting, a border and title box may be added round the detail plot. The detail drawing will be in layers, and a new layer should be used for the border and perhaps also for the text. Figure 10.17 shows typical layers in a survey plot.

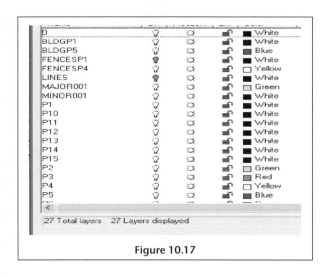

Figure 10.17

The title box can contain the following details:

Name of survey:	use large bold type
Job no:	smaller bold type
Scale:	smaller bold type
Date of survey:	small type
Surveyed by:	name plus company logo
Key:	list of symbols used
Coordinate system:	local or OS grid
Level datum:	local or OSDN
North point:	usually in top corner
Drawing no:	usually sequential.

A common place for the title box on site plans is on the right-hand side of the drawing as in Fig. 10.18. This allows the sheet to be folded so that the title is at the front. A company will usually have a standard sheet layout into which any plot can be inserted.

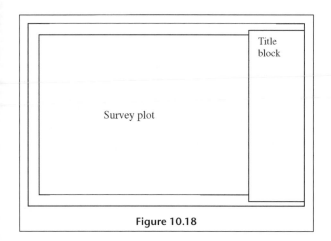

Figure 10.18

When the plotted area is very tight to fit the paper size, the title block may be positioned in any spare space in the plot as in Fig. 10.19.

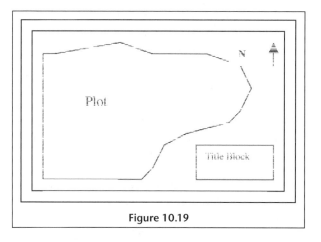

Figure 10.19

Traditionally plotters are of two types, *vector and raster*.

Vector plotters have a pen which is driven along a line by 'x' and 'y' motors. The paper can be fixed to a drum or a flat bed. This type of plotter is slow and is no longer used.

Raster plotters plot pixels, which are so small that the image appears to be continuous. Laser, electrostatic, thermal and inkjet types have been manufactured. Large format inkjet plotters are almost universally used now for plotting site plans. Figure 10.20 shows a Hewlett-Packard Designjet plotter.

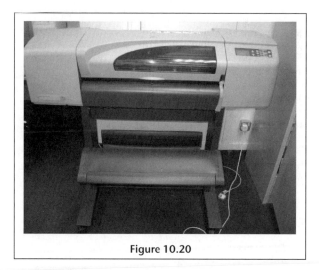

Figure 10.20

The print head sprays fine droplets of ink onto the paper, which steps in the 'y' direction, and the print head scans in 'x'. Four to eight colours are common.

(d) Accuracy

As in all methods of land survey, the propagation of errors must be minimized. The measurement errors with total stations are very small, but care must be taken over a large area that they do not accumulate to a noticeable level.

Errors are likely to occur from the following sources:

centring the instrument over the peg,
plumbing the prism (pogo stick),
the height of instrument measurement and input,
the height of prism measurement and input.

The control traverse is usually carried out at the same time as the detail observations are being taken, with sight(s) to the next control station(s) being taken on completion of the detail at a point. The instrument is then moved to the next station and a reference sight taken back to the pogo stick at the previous station. Great care should be taken in these observations. Traverse networks should close as a check. Additional cross-check observation should be taken if possible, and the software will calculate a *least squares* best fit.

To form the DTM a close network of spot heights is required. In an area with a regular gradient, points can be paced out in a rough grid pattern. In less regular terrain, breaks of slope must also be picked up. Fig. 10.21(b) shows the resultant contouring if breaks of slope are not surveyed.

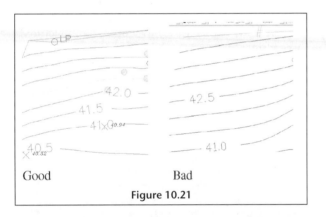

Good	Bad

Figure 10.21

For plotting, good quality paper should be used. Paper expands and contracts with moisture and heat, therefore for the best precision, plastic draughting film should be used.

5. GPS detail survey

(a) Fieldwork

Differential GPS in Real Time Kinematic (RTK) mode is used for detail survey, much in the same way as a robotic total station survey. The reference receiver is set up on the control station and the rover receiver and aerial are carried on a pole with a controller attached as in Fig. 10.22.

Initialization of the rover to establish its coordinates takes a few seconds at the commencement of the survey. It remains locked onto the satellites as the surveyor moves to the detail points. If lock is lost or the

Figure 10.22 (courtesy Topcon)

precision dilutes beyond tolerance a warning is given. The detail points are plotted on the controller screen as the points are fixed as in Fig. 10.23.

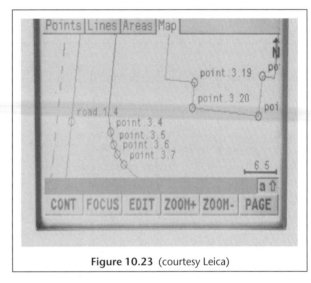

Figure 10.23 (courtesy Leica)

(b) Computer processing

If RTK GPS is being used the resultant detail file can be processed in the same software as used in the total station survey. If post processing of the data is required, this will be done in the GPS software before transferring to the detail software.

6. Photogrammetric detail survey

For national mapping and the production of contour plans for extensive ground areas, photogrammetric techniques of plotting the detail from aerial photography are more efficient and cost effective. This work is carried out by specialist companies.

The photography is taken from an aircraft with either a large format (23 × 23 cm) mapping camera or an aerial digital camera/scanner. It is flown in strips with about a 60 per cent fore and aft overlap to allow stereoscopic viewing, and about a 15 per cent lateral overlap of the strips. Figure 10.24 shows a Leica ADS40 digital sensor.

Figure 10.24 (courtesy Leica)

Much of the cost of photogrammetric mapping is in the cost of flying the photography, but if this is already available and up to date, then costs can be reduced. Digital imagery of the UK is available from 'Getmapping.com'.

Photogrammetric plotting machines have been developed which allows the viewing of the overlapping images as a 3D model. Detail and contours can be followed and recorded as a vector plot. Until the 1970s plotting machines were of the analogue type where the light rays from a ground point to the photograph were re-created by mechanical rods, which controlled the photograph viewing system. With conventional photography, analytical plotters are now used where the geometry of the viewing system is computer controlled. Digital imagery can be plotted in a high-end PC with a large screen, which with special software can display left and right stereopairs. Stereoscopic viewing is possible with special polarizing glasses which allow the left or right eye to see only the corresponding left or right image on the screen.

Figure 10.25 shows the plotting screen viewer.

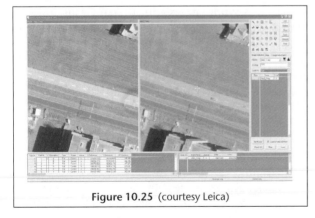

Figure 10.25 (courtesy Leica)

Plotting from a single photograph is also possible. The aerial photographic image has tilt and relief displacement. A *rectified* image has the distortion due to the tilt of the camera/scanner axis from the vertical removed. An ortho-rectified image has the tilt and the distortion due to ground relief removed. Rectification requires the image to be fitted to at least four ground control points. Ortho-rectification requires a digital terrain model of the area covered by the image.

7. Associated mapping software

(a) Computer aided draughting (CAD)

CAD is often the final stage of detail survey production. The draughting routines which are available in the survey software tend to be rather basic compared with software such as AutoCAD.

It is not the purpose of this book to give an in-depth knowledge of CAD, but some of the more useful commands for completing the drawing are detailed below. The examples are based on AutoCAD.

The screen view consists of a drawing area surrounded by menu headings at the top and bars of command icons, which can be arranged to preference. The command prompt window is at the bottom. An eye should be kept on this window as it is easy to lose track with the maze of commands, sub-commands and viewing options. **UNDO** and the escape button are much used by the learner.

Drawing commands

LINE	Draws a line from point to point. Use the *snap to end point* option to fill in missing sections on your plot.
ARC	Fits a curve between 3 points. Use *snap* to join to existing line.
CIRCLE	Draws a circle of specified radius. For concentric circles use *snap to centre point*. For a part of a circle, draw the full circle then **_TRIM_**.
RECTANGLE	Draws a rectangle square to drawing area by specifying diagonal corners. Use **_ROTATE_** to angle it. If only some corners of a building have been surveyed this command can be used to complete with right-angled corners.
HATCH	Fills a closed polygon with a pattern. A fill of dots is more pleasing on the eye than hatches and fancy patterns.
TEXT	Fits text into a box or aligned with a line. Simple fonts should be used.

Drawing layers

Drawing elements are separated into like features which are stored in layers such as fences and buildings. This facilitates the specifying of colours and line types. Layers can be switched of and on, locked and unlocked.

Figure 10.17 shows typical layers in a survey plot.

A template that sets up the standard layers and menus for a particular type of drawing can be saved.

View commands

ZOOM	Allows areas to be viewed at larger scale. *Zoom window, zoom previous* and *zoom dynamic* are most useful. Sometimes when a survey drawing is imported into CAD it does not appear on the screen. *Zoom all* will show all the drawing.
PAN	Moves the image in the drawing box.

Modifying commands

ERASE	Deletes a specified graphic element. The *window* option can be used to select a large area.
EXPLODE	Breaks a polyline into individual segments if only a section is to be erased.
TRIM	Erases a line overshoot.
EXTEND	Extends a line to meet another.
CHANGE	Changes the properties of an object.
MOVE	Moves an object.
OFFSET	Draws a line parallel to an existing line at a set distance. Useful for completing paths etc. where the width is known.

Insert commands

BLOCK	This is an alternative to the *COPY* command for inserting symbols in a drawing. A shape can be drawn and saved as a block.
WBLOCK	Saves the block as an external file for insertion into other drawings. This is useful for inserting a small local survey into an OS map. It can be fitted to existing detail.
INSERT	Drops an object into a drawing. Blocks or another drawing which has been cross-referenced can be inserted.

Layouts and paper space

Drawing is carried out in *model space*. A standard paper size *layout* can be saved and the drawing inserted into *view ports* in *paper space*.

Plotting

The scale of the drawing in the view port is expressed as the representative fraction plus the initials 'XP'. For plotting from model space, it must be remembered that AutoCAD drawing units are in millimetres and survey units are in metres, therefore the scale number has to be divided by 1000, e.g. 1:1250 becomes 1 = 1.25.

Autodesk MAP3D

MAP3D is a plug-in for AutoCAD which adds functionality for creating, editing and maintaining maps. It can read data from databases such as Access and Oracle and can handle a wide range of vector and raster data formats from other graphics and GIS software such as Microstation DGN, ArcGIS, Mapinfo, and OS Mastermap. In MAP3D all formats are transparent and held as Feature Data Objects (FDOs). Contours and digital terrain models can be formed, and map sheets can be set up for plotting a large area.

The software also incorporates tools for *Raster Design* allowing vectors to be created from photographs.

Functionality is much improved for the non-CAD user.

Autodesk CIVIL 3D

This is an improved version of LAND DESKTOP. It is a plug-in which provides an environment for the layout of road, rail and pipe networks. It produces a 3D model along the whole corridor width of the construction. Any surface can be defined and sections drawn. Changes such as altering a gradient are mirrored in all drawing elements. These design changes can be viewed by other team members sharing the same data.

(b) Digital Terrain Model (DTM)

A digital terrain model is a three-dimensional representation of the ground surface in a computer readable format. It is synonymous with Digital Ground Model (DGM), Digital Elevation Model (DEM) and Digital Height Model (DHM).

The terrain is represented by a set of points with known coordinates (x, y, z). These can be collected by a number of methods.

1. *Data acquisition*
(i)   *Ground survey*: the points are collected as part of a detail survey and are spaced depending on the terrain.
(ii) *Photogrammetry*: the points are recorded usually in a grid pattern with the spacing varying according to terrain. Random points may also be taken. Buildings are omitted.
(iii) *LIDAR*: this is another airborne method, where the aircraft or helicopter has a laser scanner which projects a swathe across the flight path measuring the distance to the ground. The position of the aircraft is monitored by GPS and corrected ground

heights recorded. The data is stored in a grid pattern with a very fine resolution.

(iv) *Synthetic Aperture Radar (SAR)*: This is similar to LIDAR, only using a radar beam. It is less accurate, but the radar beam can penetrate cloud.

(v) *Digital contours*: If a contour map is available, then the digitized contours can form the terrain model such as the Ordnance Survey *Profile*, which is derived from the 1/10 000 map contours. A grid is interpolated from the contours.

2. Modelling methods

(i) *Triangular irregular network (TIN)*: This method is used with data from ground survey. It is the most accurate of methods as the points used for modelling have been directly surveyed on the ground.

The software creates a network of triangles joining the surveyed points. In the search algorithm, the shape of the triangles is chosen as to be as near equilateral as possible. Figure 10.16 shows a triangulation pattern for contouring. The process starts with a base line on one of the boundaries and the software carries out a clockwise radial sweep to find the nearest neighbouring point to form a triangle. The radius is incremented until a neighbour is found, then a new baseline is chosen. Triangle formation is constrained by setting boundaries, within which triangulation can take place, and break lines at changes of slope. Triangle conditions can also be set such as minimum and maximum angles and maximum side length. Contours are interpolated and smoothed by a variety of methods of splining.

(ii) *Grid based methods*: This technique is used with aerial surveys and is generally for much more extensive areas. If the grid is interpolated from contours, it is much less accurate.

The grid of height points is analysed by the software and a best fit polynomial expression for the ground profile is calculated. The points can be analysed globally or in patches.

3. Model display

(i) *Wire frame*: This is the simplest method of 3D display, using a regular grid pattern, which is distorted by the viewing angle of the model as shown in Fig. 10.26.

(ii) *Shading*: Shading or rendering can be applied to each model cell by analysing its slope and orientation. Two forms are common:

Layer shading where a different colour shade is applied to separate relief intervals. A spectral range from blue–green–orange–red is commonly used. However, psychologically, the brighter layer should appear closer to the eye, so a progression from dull blue/green to brighter yellow and even white would give a better impression of height. Fig. 10.27(a) shows an example of this.

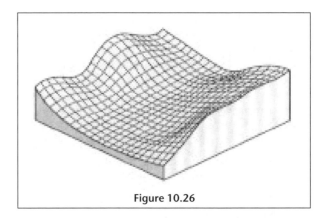

Figure 10.26

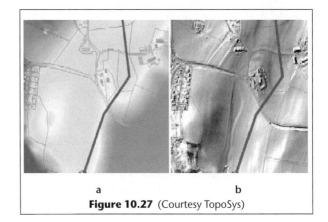

a   b
Figure 10.27 (Courtesy TopoSys)

Relief shading: An imaginary light source is set usually from the top left corner of the model, and the slopes opposite this direction are shaded, giving the 3D impression as in Fig. 10.27(b).

The model can be displayed in plan or oblique view.

(iii) *Draping*: An aerial photograph or map can be draped over the model to give surface detail as shown in Fig. 10.28.

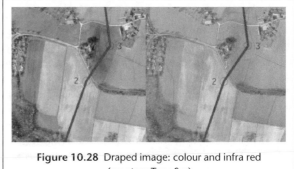

Figure 10.28 Draped image: colour and infra red
(courtesy TopoSys)

4. Applications of terrain models In conventional maps and plans, the third dimension has always been difficult to interpret for the less expert user. With DTMs there can be almost virtual reality terrain. The presentation of the data in 3D with 'fly throughs' is most impressive.

The scale of the model varies with the application.

(i) *Large scale*: Applications are mainly in the construction of roads, railways, dams, opencast mining and landscaping.

Figure 10.29 shows an opencast site.

Figure 10.29 (courtesy Trimble)

(ii) *Medium and small scale*: Applications tend to fall into two categories, *Terrain visualization* and *Terrain analysis*. Visualization is very big with the military for ground and air operations. Civilian uses include environmental impact studies for the construction of roads and buildings, route surveys for power lines and environmental monitoring. Terrain analysis is used by telecom companies for optimizing the coverage of radio masts and by insurance companies for flood risk assessment.

DTMs with national coverage such as *OS Profile* and *NEXTMap* are useful for these applications.

(c) Geographic information systems (GIS)

A Geographic Information System (GIS) or, as it is sometimes known in the land surveying field, a Land Information System (LIS), is the linking of the map detail with other spatial information. In simple terms it uses an intelligent map.

A common requirement in surveys of development sites is to carry out an inventory of the major trees. In the normal survey processing software a limited amount of attribute information can be collected by field coding. 'TREE1 1.5,4.5,26' could indicate a limited number of types of trees with a specific circle symbol; 1.5 m is the bowl diameter, 4.5 m is the spread and 26 the number on the tree. However, other information such as species, condition and height are also required. This is normally typed up in Word or Excel as a report. In a GIS this information would be linked (geocoded) to the position of the tree on the plot. Click on the tree and the data would be displayed. A photograph could even be linked. A simple query could be to select

trees with a bowl diameter greater than 0.5 m and a height of 5 m. A browser table of the trees selected would result and their positions highlighted on the map.

Similar applications include inventories of street furniture and services. GPS is widely used for this type of application. A hand-held or back-pack receiver is used in differential RTK mode. The map is displayed on the receiver screen or on a rugged tablet computer. The surveyor stands for a few seconds at the object and its position is recorded on the map. Other information is then added. An interesting variation on this is the 'Fast Map GPS' system where the GPS is attached to a vehicle and street furniture is fixed by intersection from still frames from a video camera which is pointed to the pavement from the vehicle.

The graphical objects in the plan have a reference number. This is used to link to the database tables in an operation termed *geocoding*. When completed, the database information is then referenced to the graphical objects in the plan.

GIS software is being developed as options in the standard surveying packages, but for sophisticated applications large systems dominate. *ArcGIS* from ESRI is widely used in large-scale applications with local authorities and survey companies. *MapInfo* is popular in marketing and with health boards.

8. Answers

Exercise 10.1

1 Table 10.7 and Fig. 10.30 [See page 209]

Exercise 10.2

1 *Point number*	*Code*
128	LP SH
129	KERB NEWCV SH
130	KERB ENDCV SH
131	KERB STCV SH
132	KERB ENDCV SH
133	KERB SH
134	KERB SH
135	KERB SH
136	FCE START SH
137	FCE SH

Exercise 10.3

1 *Point number*	*Code*
138	TREE Size 7
139	BLD
140	BLD
141	BLD
	Note CLS RECT
142	BLD START
143	BLD RECT 3.0

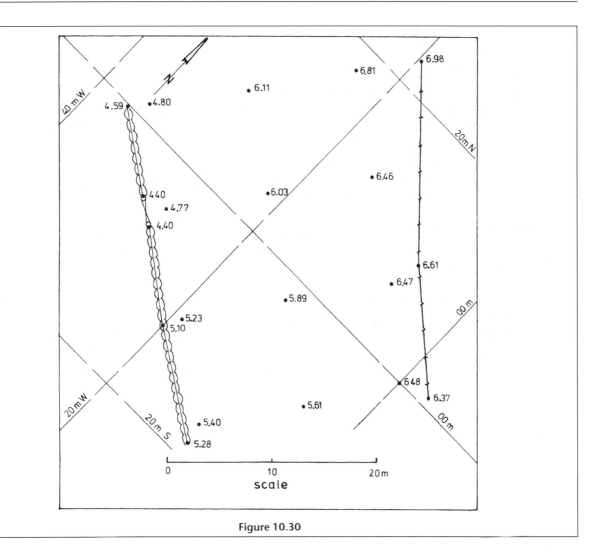

Figure 10.30

Table 10.7

Target station	Target height (m)	Slope length (m)	Whole circle bearing	Zenith angle	P (m)	ΔH (m)	ΔE (m)	ΔN (m)	Easting (m)	Northing (m)	Reduced level (m)	Target station
1	1.35	3.02	72° 7′ 30″	92° 6′ 35″	3.02	−0.11	2.87	0.93	2.87	0.93	6.37	1
2	1.35	11.10	324° 9′ 45″	89° 21′ 0″	11.10	0.13	−6.50	9.00	−6.50	9.00	6.61	2
3	2.35	30.02	318° 55′ 20″	87° 8′ 35″	29.98	1.50	−19.70	22.60	−19.70	22.60	6.98	3
4	1.35	29.36	306° 49′ 50″	89° 21′ 50″	29.36	0.33	−23.50	17.60	−23.50	17.60	6.81	4
5	1.35	19.34	307° 13′ 30″	90° 4′ 16″	19.24	−0.02	−15.40	11.70	−15.40	11.70	6.46	5
6	0.35	9.35	310° 12′ 0″	96° 13′ 30″	9.29	−1.01	−7.10	6.00	−7.10	6.00	6.47	6
7	1.35	0.85	69° 26′ 40″	90° 16′ 5″	0.85	0.00	0.80	0.30	0.80	0.30	6.48	7
8	1.35	9.59	210° 51′ 40″	95° 13′ 40″	9.55	−0.87	−4.90	−8.20	−4.90	−8.20	5.61	8
9	1.35	13.41	260° 7′ 9″	92° 32′ 20″	13.40	−0.59	−13.20	−2.30	−13.20	−2.30	5.89	9
10	2.00	21.63	279° 35′ 35″	89° 28′ 50″	21.63	0.20	−21.33	3.60	−21.33	3.60	6.03	10
11	1.00	30.89	287° 8′ 40″	91° 20′ 20″	30.89	−0.72	−29.51	9.10	−29.51	9.10	6.11	11
12	1.35	35.37	272° 26′ 0″	92° 43′ 45″	35.33	−1.68	−35.30	1.50	−35.30	1.50	4.80	12
13	1.35	27.59	261° 1′ 0″	93° 33′ 40″	27.54	−1.71	−27.20	−4.30	−27.20	−4.30	4.77	13
14	1.35	21.66	240° 56′ 45″	93° 19′ 10″	21.62	−1.25	−18.90	−10.50	−18.90	−10.50	5.23	14
15	1.35	19.67	213° 22′ 0″	93° 9′ 35″	19.64	−1.08	−10.80	−16.40	−10.80	−16.40	5.40	15
16	1.35	21.25	209° 8′ 37″	93° 15′ 0″	21.22	−1.20	−10.33	−18.53	−10.33	−18.53	5.28	16
17	1.35	23.38	238° 29′ 20″	93° 23′ 35″	23.34	−1.38	−19.90	−12.20	−19.90	−12.20	5.10	17
18	1.35	28.16	256° 24′ 35″	94° 14′ 35″	28.08	−2.08	−27.30	−6.60	−27.30	−6.60	4.40	18
19	1.35	30.11	260° 13′ 25″	93° 58′ 10″	30.04	−2.08	−29.60	−5.10	−29.60	−5.10	4.40	19
20	1.35	36.55	269° 31′ 45″	92° 58′ 10″	36.50	−1.89	−36.50	−0.30	−36.50	−0.30	4.59	20

Chapter summary

In this chapter, the following points are the most important:

- The three common methods of detail survey are radial positioning, GPS and photogrammetry.

- In radial detail survey a total station is set up at a known station and angles and distances measured and recorded to detail points, which are given codes to identify the feature. Coordinates are calculated, usually using a computer package. The feature codes are processed and a plot produced. Contours can be interpolated using an irregular network of triangles (TIN) joining spot heights.

- In GPS detail survey a roving receiver is first initialized to a reference receiver on a known station. The rover then moves to detail points, stopping for a few seconds to record the position and code of the point. In real-time kinematic (RTK) mode coordinates are processed and corrected in a few seconds.

- In photogrammetric detail survey strips of overlapping stereoscopic aerial photographs are flown using either a film camera or a digital sensor. A digital map is produced in a special plotting machine by tracing from a 3D image. Alternatively, the aerial photograph can be corrected for relief and tilt displacement and an orthophotograph produced.

- The drawing commands available in survey detail processing computer packages tend to be limited; consequently the drawing file is often exported to a CAD package for the final presentation or to link with other applications.

- A Digital Terrain Model (DTM) is a 3D representation of the ground surface in computer readable form. It is the basis for terrain visualization applications and area and volume calculations in civil engineering.

- A Geographic Information System (GIS) is essentially an intelligent map. Each feature has related information such as 'what', 'how much', 'when' and 'by whom' which can be associated with it from a database.

CHAPTER 11 **Curve ranging**

In this chapter you will learn about:

- circular geometry and the formulae for calculating the various elements of circular curves
- a variety of problems concerning the location of circular curves and the mathematical methods of solving them
- a selection of methods of setting out, on the ground, curves of small radius
- the tangential angles method and the coordinate method of setting out, on the ground, curves of large radius
- the method of overcoming obstructions which hinder setting out
- the geometry, computations and methods of setting out vertical curves
- the geometry, computations and methods of setting out transition curves

---

In construction surveying, curves have to be set out on the ground for a variety of purposes. A curve may form the major part of a roadway, it may form a kerb line at a junction or may be the shape of an ornamental rose bed in a town centre. Obviously different techniques would be required in the setting out of the curves mentioned above, but in all of them a few geometrical theorems are fundamental and it is wise to begin the study of curves by recalling those theorems.

1. Curve geometry

In Fig. 11.1, A, B and C are three points on the circumference of a circle.

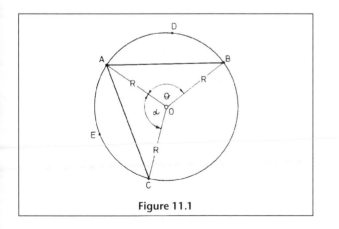

Figure 11.1

1. AB and AC are chords of the circle subtending angles θ and a respectively at the centre O. ADB and AEC are arcs of the circle. Their lengths are

$$2\pi R\left(\frac{\theta}{360}\right)^{\circ} \text{ and } 2\pi R\left(\frac{a}{360}\right)^{\circ} \text{ respectively}$$

More conveniently their lengths are $R\theta$ and Ra respectively, where θ and a are expressed in radians.

2. In Fig. 11.2, lines ABC and ADE are tangents to the circle at B and D respectively. AB = AD and angles ABO and ADO are right angles.

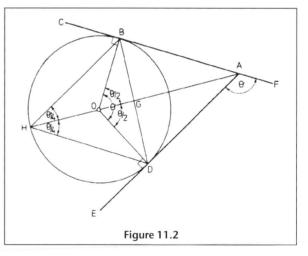

Figure 11.2

3. Since the opposite angles of a cyclic quadrilateral are supplementary, the figure ABOD must be cyclic as angles ABO and ODA together make 180°.
4. The exterior angle of a cyclic quadrilateral equals the interior opposite angle; therefore angle FAD = angle BOD = θ.
5. Join OA, the perpendicular bisector of chord BD. Angle OGB is therefore a right angle and angle BOG = $\theta/2$.

Angle ABG + angle GBO = 90°
and angle BOG + angle GBO = 90°
Therefore angle ABG = angle BOG = $\theta/2$

i.e. the angle ABG between the tangent AB and chord BD equals half the angle BOD at the centre.
6. Produce AO to the circumference at H and join HB. Angle BOG is the exterior angle of triangle BOH.

Therefore angle BOG = angle OHB + angle OBH

However, angles OHB and OBH are equal since triangle BOH is isosceles.

Therefore angle OHB = $\frac{1}{2}$ angle BOG
$= \theta/4$
Similarly angle OHD = $\theta/4$
Therefore angle BHD = $\theta/2$

i.e. the angle BHD at the circumference subtended by the chord BD equals half the angle BOD at the centre subtended by the same chord
Also the angle ABD between the tangent and chord equals the angle BHD at the circumference.

2. Curve elements

In Fig. 11.3, the centre lines AI and BI of two straight roadways, called simply the straights, meet at a point I called the intersection point. The roadways may actually exist on the ground or may simply be proposals on a roadway development plan. In either case, the two straights deviate by the angle θ, which is called the deviation angle. Alternatively, the angle may be called the deflection angle or intersection angle.

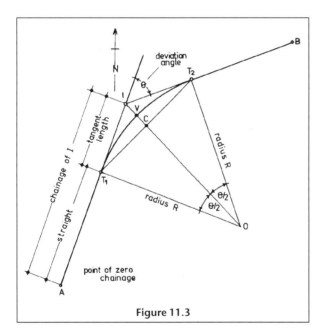

Figure 11.3

Clearly it is desirable to avoid having a junction at I, so the straights are joined by a circular curve of radius R.

The straights are tangential to the curve at the tangent points T_1 and T_2 and lengths IT_1 and IT_2, known as the tangent lengths, are equal. Before setting the curve on the ground, the exact location of the tangent points must be known.

If the two straights are existing roadways, then, in order to locate the tangent points, a theodolite is set at point I and the deviation angle θ is measured together with the lengths of the lines AI and IB.

If the roadway scheme exists only on a development plan, the angle θ and the distances AI and IB must be measured by protractor and scale rule or by calculation from the coordinates of A, I and B.

In either case, station A is the start of the curve calculation and is therefore the *point of zero chainage*. The chainage of point I is the distance AI.

The radius R is usually a multiple of 50 metres and is supplied by the architect or designer. Knowing only the deviation angle and radius, the tangent lengths and curve length are derived thus:

Angle IT_1O = angle OT_2I = 90°

Therefore IT_1OT_2 is a cyclic quadrilateral and

angle $T_1OT_2 = \theta$

Join I to O.

Angle T_1OI = angle $IOT_2 = \theta/2$

(a) Tangent lengths IT_1 and IT_2

In triangle IT_1O,

$$\frac{IT_1}{R} = \tan \theta/2$$

Therefore $IT_1 = R \tan \theta/2$
Chainage of T_1 = chainage of I – IT_1

(b) Length of curve T_1T_2

Curve $T_1T_2 = R \times \theta$ radians
Chainage of T_2 = chainage T_1 + curve length

(The chainage of the second tangent point is *always* derived via the curve.) This information enables the beginning and end of the curve to be located.

EXAMPLE

1 In Fig. 11.3, the whole circle bearings and lengths of AI and IB are:

Line	WCB	Length (m)
AI	20°	450.30
IB	70°	275.00

The radius of the curve joining the straights is 300 m. Calculate the chainages of the tangent points.

Answer

(a) Deviation angle θ = 70° − 20°

$= 50°$

(b) Chainage 1 = 450.30 m

(c) Tangent length $IT_1 = R \tan \theta/2$

$= 300 \tan 25°$

$= 139.89$ m

(d) Chainage $T_1 = 450.30 − 139.89$

$= 310.41$ m

(e) Curve length $= R \times \theta$ radians

$= 300 \times 0.872\ 66$

$= 261.80$ m

(f) Chainage $T_2 = 310.41 + 261.80$

$= 572.21$ m

Exercise 11.1

1 Two straight roadways AB and BC meet at junction B. The junction is to be replaced by a circular curve of 300 metres radius, which is to be tangential to straights AB and BC. The coordinates of points A, B and C are as follows:

Point	Easting	Northing
A	0.000	0.000
B	+859.230	+151.505
C	+1423.046	−53.707

Calculate:

(a) the lengths of the straight roadways AB and BC

(b) the deviation angle between straights AB and BC

(c) the lengths of the tangents to the straights

(d) the length of the curve joining the straights

(e) the chainages of the tangent points, assuming station A is the origin of the survey.

Other curve elements are frequently required and are calculated from the values of R and θ.

(c) Long chord $T_1 T_2$ (Fig. 11.3)

The long chord is the straight line joining T_1 and T_2. The line IO is the perpendicular bisector of T_1T_2 at C. In triangle T_1CO,

$$\frac{T_1C}{R} = \sin \theta/2$$

Therefore $T_1C = R \sin \theta/2$

and $T_1T_2 = 2R \sin \theta/2$

(d) Major offset CV (Fig. 11.3)

Frequently called the mid-ordinate or versine, the length CV is the greatest offset from the long chord to the curve:

$CV = R − OC$

In triangle T_1CO,

$$\frac{CO}{R} = \cos \theta/2$$

Therefore $CO = R \cos \theta/2$

and $CV = R − R \cos \theta/2$

$= R(1 − \cos \theta/2)$

(e) External distance VI (Fig. 11.3)

The length of VI is the shortest distance from the intersection point to the curve:

$VI = IO − R$

In triangle IT_1O,

$$\frac{IO}{R} = \sec \theta/2$$

Therefore $IO = R \sec \theta/2$

and $VI = R \sec \theta/2 − R$

$= R(\sec \theta/2 − 1)$

EXAMPLE

2 Two straights AI and IB deviate to the left by 80° 36′. They are to be joined by a circular curve such that the shortest distance between the curve and intersection point is 25.3 m (Fig. 11.4). Calculate:

(a) the radius of the curve,

(b) the lengths of the long chord and major offset.

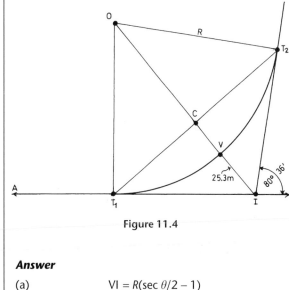

Figure 11.4

Answer

(a) $VI = R(\sec \theta/2 − 1)$

i.e. $25.3 = R(\sec 40° 18′ −1)$

$= R(1.311\ 186 − 1)$

Therefore $R = \dfrac{25.3}{0.311\ 186}$ m

$= 81.30$ m

(b) Long chord $T_1T_2 = 2R \sin \theta/2$
$$= 162.60 \sin 40° \, 18'$$
$$= 162.60 \times 0.646\,790$$
$$= 105.17 \text{ m}$$

Major offset VC $= R(1 - \cos \theta/2)$
$$= 81.30(1 - 0.762\,668)$$
$$= 81.30 \times 0.237\,332$$
$$= 19.30 \text{ m}$$

Exercise 11.2

Two straights XY and YZ deviate to the right by 47° 09′ 20″. They are to be joined by a circular curve of 50 metres radius. Calculate:

(a) the length of the long chord joining the tangent points
(b) the major offset CV
(c) the shortest distance YV from the curve to the intersection point
(d) the tangent lengths IT_1 and IT_2
(e) the length of the curve.

3. Designation of curves

In the United Kingdom, curves are designated by the length of the radius. Since there is generally some scope in the choice, the radius is usually a multiple of 50 metres.

The curve can also be designated by the degree of curvature which is defined as the number of degrees subtended at the centre by an arc 100 m long. The degree of curvature is given as a number of whole degrees. In Fig. 11.5, the angle $\theta = 5°$; i.e. the degree of curvature is 5°. The relationship between radius and degree of curvature is as follows:

Arc length $AB = R \times \theta$ radians

$$= R \times \theta \times \frac{\pi}{180} \ (\theta \text{ in degrees})$$

Therefore $R = \dfrac{100 \times 180}{\theta \times \pi}$ m

$$= \frac{5729.8}{\theta} \text{ m}$$

$$= 1145.96 \text{ m}$$

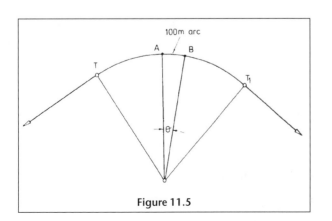

Figure 11.5

4. Problems in curve location

(a) Inaccessible intersection point

It happens frequently on site that the intersection point cannot be occupied because of some obstacle. In Fig. 11.6 the intersection point I is in a built-up area.

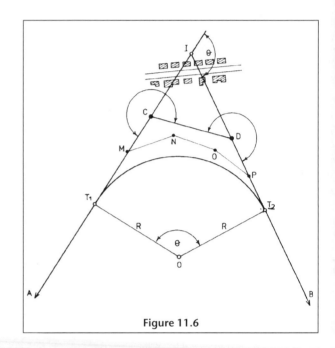

Figure 11.6

The relative position of the straights AI and IB must be obtained by traversing between them. The best conditions for surface taping determine the route of the traverse. In Fig. 11.6 the route AMNOP might provide the best conditions in which the traverse coordinates would be calculated and the position of station I deduced from a solution of triangle IMP.

The simplest traverse between the straights is one straight line between two stations C and D. Angles ACD and CDB are measured together with distances AC and CD. The chainage of C is therefore known.

EXAMPLE

3 In Fig. 11.6 the following data were derived from traverse ACDB.

Angle ACD 252° 15′ 00″   AC = 559.28 m
Angle CDB 227° 25′ 00″   CD = 256.50 m

Calculate the chainage of the tangent point T_1 if the straights are to be joined by a 300 m radius curve.

Solution

In triangle ICD,

Angle C = 72° 15′ 00″
Angle D = 47° 25′ 00″

Therefore angle I = 60° 20′ 00″

By sine rule $\dfrac{IC}{\sin 47° 25'} = \dfrac{CD}{\sin 60° 20'}$

Therefore IC $= \dfrac{256.50 \times 0.738\ 259}{0.868\ 920}$

$= 217.35$ m

Chainage of I $= AC + 217.35$

$= 776.63$ m

Deviation angle $\theta = 180° 00' 00''$
$\qquad\qquad -60° 20' 00''$
$\qquad\qquad = 119° 40' 00''$

Tangent length $IT_1 = R \tan \theta/2$
$\qquad\qquad = 300 \tan 59° 50'$
$\qquad\qquad = 300 \times 1.720\ 474$
$\qquad\qquad = 516.142$ m

Therefore chainage of T_1
$\qquad\qquad = $ chainage I $- 516.142$
$\qquad\qquad = 260.49$ m

Exercise 11.3

1 A horizontal curve of 100 metres radius is to be set out between two straights AX and XB, but the intersection point X is inaccessible. In order to overcome the problem, two points P and Q were selected on lines AX and BX respectively and the following information was recorded:

Length PQ = 55.00 m
Angle QPA = 151° 20'
Angle BQP = 143° 10'

Determine:

(a) the values of the three angles of triangle PXQ and hence the lengths PX and QX, using the sine formula
(b) the tangent lengths of the curve
(c) the distances of A and B from P and Q respectively
(d) the length of the curve AB.

(b) Curve tangential to three straights

In Fig. 11.7 three straights are to be joined by a circular curve, the radius of which is unknown and has to be calculated. One condition must be fulfilled, namely that each straight be a tangent to the curve of radius R.

First consider straights AB and BC only. BT_1 and BT_2 are equal tangent lengths deviating by angle θ. Therefore $BT_1 = BT_2 = R \tan \theta/2$.

Considering straights BC and CD only, CT_2 and CT_3 are equal tangent lengths deviating by angle a. Therefore $CT_2 = CT_3 = R \tan a/2$.

The length BC is known and is also equal to $(BT_2 + CT_2)$.

Therefore BC $= (BT_2 + CT_2)$
$\qquad\qquad = R \tan \theta/2 + R \tan a/2$
Hence $R = BC \div (\tan \theta/2 + \tan a/2)$

EXAMPLE

4 The following data refer to Fig. 11.7:

Straight	Bearing	Distance (m)
AB	34°	735.70
BC	74°	210.50
CD	124°	640.40

Calculate:

(a) the radius of the curve joining the straights,
(b) the length of curve.

Answer

(a) $\qquad$ Angle $\theta = 74° - 34° = 40°$
$\qquad\qquad$ Angle $a = 124° - 74° = 50°$
$\qquad\qquad\quad R = BC \div (\tan \theta/2 + \tan a/2)$ as before
$\qquad\qquad\qquad = 210.50 \div (\tan 20° + \tan 25°)$
$\qquad\qquad\qquad = 210.50 \div 0.830\ 277\ 9$
$\qquad\qquad\qquad = 253.53$ m

(b) Angle $T_1OT_3 = (40° + 50°) = 90°$ Therefore curve
$\qquad\qquad\qquad = \pi/2 \times R$ metres
$\qquad\qquad\qquad = 398.245$ m

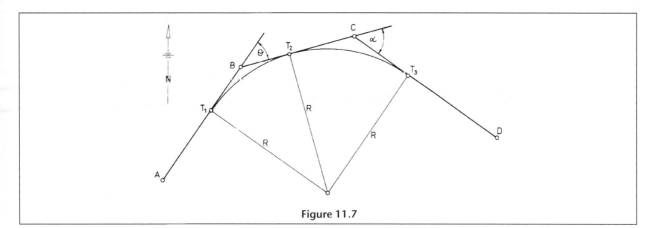

Figure 11.7

Exercise 11.4

1. A circular curve is to be set out so that it is tangential to three straight lines AB, BC and CD. The known bearings and lengths are:

Line	WCB	Length (m)
AB	37° 12′	–
BC	91° 02′	312.70
CD	147° 14′	–

(a) Calculate the radius of the curve
(b) Making the necessary calculations, describe how the initial tangent point on line AB would be located on site.

(c) Curve passing through three known points

In Fig. 11.8, P, Q and R are three points whose coordinates are known. They are to be joined by a curve of unknown radius, the length of which is required.

The circle that passes through the points is the circumscribing circle of triangle PQR. Therefore,

Angle QPR (at the circumference)
$= \frac{1}{2}$ angle QOR (at the centre)
$=$ angle SOR (SO bisects QR)

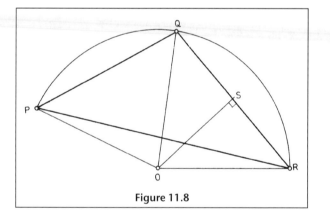

Figure 11.8

Angle QPR and SR can be calculated from the coordinates.

Therefore OR = SR cosec SÔR

A simpler solution is provided by the sine rule, which states that

$$\frac{p}{\sin P} = \frac{q}{\sin Q} = \frac{r}{\sin R}$$
$$= 2 \times \text{radius of circumscribing circle}$$

Side QR $= p$ and QP̂R can be found from the coordinates.

EXAMPLE

5 In Fig. 11.8 the following are the known coordinates of points P, Q and R:

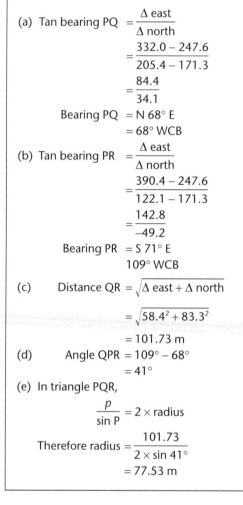

Point		
P	247.6 E	171.3 N
Q	332.0 E	205.4 N
R	390.4 E	122.1 N

Calculate the radius of the curve that passes through all three points.

Solution

(a) Tan bearing PQ $= \dfrac{\Delta \text{ east}}{\Delta \text{ north}}$

$= \dfrac{332.0 - 247.6}{205.4 - 171.3}$

$= \dfrac{84.4}{34.1}$

Bearing PQ $=$ N 68° E
$= 68°$ WCB

(b) Tan bearing PR $= \dfrac{\Delta \text{ east}}{\Delta \text{ north}}$

$= \dfrac{390.4 - 247.6}{122.1 - 171.3}$

$= \dfrac{142.8}{-49.2}$

Bearing PR $=$ S 71° E
109° WCB

(c) Distance QR $= \sqrt{\Delta \text{ east} + \Delta \text{ north}}$

$= \sqrt{58.4^2 + 83.3^2}$

$= 101.73$ m

(d) Angle QPR $= 109° - 68°$
$= 41°$

(e) In triangle PQR,

$\dfrac{p}{\sin P} = 2 \times \text{radius}$

Therefore radius $= \dfrac{101.73}{2 \times \sin 41°}$

$= 77.53$ m

Exercise 11.5

1 Figure 11.9 shows three manholes A, B and C, which are part of a town centre development scheme. The centre is to be landscaped such that the manholes will lie within a circular grassed area. The edge of the grassed

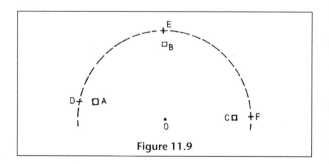

Figure 11.9

area is to be 1 metre clear of the manhole centres. The coordinates of the manholes are as follows:

Manhole	Easting	Northing
A	10.0	10.0
B	15.0	12.5
C	19.0	8.0

Calculate:
(a) the radius of the circle of the required grass area,
(b) the coordinates of the centre of the circle.

5. Setting out curves (calculations)

The purpose of curve calculations is to enable the curve to be set out in its predetermined position on the ground. The centre line of the curve is positioned by a series of pegs, set at intervals chosen by the surveyor.

Where curves are long and of large radius (over 100 m), a theodolite must be used to obtain the desired accuracy of setting out. Very often, however, curves of small radius have to be set out on site and setting out by tape is usually the chosen option. *Numerous methods are available but the following three methods are the most commonly used on site* because of their simplicity, accuracy and speed.

(a) Small radius curves

Method 1: Finding the centre
In Fig. 11.10 kerbs have to be laid at the roadway junction. Consider the right-hand curve. The deviation angle a is measured from the plan and the tangent lengths IT_1 and IT_2 $(= R \tan a/2)$ calculated. The procedure for setting the curve is then as follows:

1. From I, measure back along the straights the distances IT_1 and IT_2.

2. Hammer in pegs at these points and mark the exact positions of T_1 and T_2 by nails.
3. Hook a steel tape over each nail and mark the centre O at the point where the tapes intersect when reading R. Hammer in a peg and mark the centre exactly with a nail.
4. Any point on the curve is established by hooking the tape over the peg O and swinging the radius. This method is widely used where the radius of curvature is less than 30 m.

Method 2: Offsets from the tangent
When the deviation angle is small (less than 50°) the length of the curve short and the centre inaccessible, the curve can be set out by measuring offsets from the tangent. In the left-hand curve of Fig. 11.10, y is an offset from the tangent at a distance x metres from tangent point T_1. In the figure, the line AB is drawn parallel to the tangent until it cuts the radius. The length $AT_1 = y$ and the length $AO = (R - y)$.
In triangle OAB,

$$OA = \sqrt{OB^2 - AB^2} \text{ (by Pythagoras)}$$

$$\text{i.e. } (R - y) = \sqrt{R^2 - x^2}$$

$$\text{Therefore } y = R - \sqrt{R^2 - x^2}$$

Thus the offset y can be calculated for any distance x along the tangent and can be set by eye or by optical square.

EXAMPLE

6 Given that the deviation angle $\theta = 50°$ and the radius $R = 60$ m, calculate the offsets from the tangent at 5 metre intervals (Fig. 11.10).

(a) Tangent lengths $IT_1 = IT_2 = 60 \tan 25° = 27.98$ m

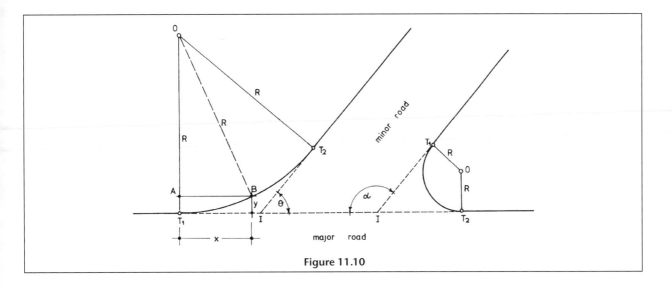

Figure 11.10

(b) Offsets at

$$5 \text{ m} = 60 - \sqrt{60^2 - 5^2} = 0.210 \text{ m}$$

$$10 \text{ m} = 60 - \sqrt{60^2 - 10^2} = 0.840 \text{ m}$$

$$15 \text{ m} = 60 - \sqrt{60^2 - 15^2} = 1.905 \text{ m}$$

$$20 \text{ m} = 60 - \sqrt{60^2 - 20^2} = 3.430 \text{ m}$$

$$25 \text{ m} = 60 - \sqrt{60^2 - 25^2} = 5.456 \text{ m}$$

$$27.98 \text{ m} = 60 - \sqrt{60^2 - 27.98^2} = 6.923 \text{ m}$$

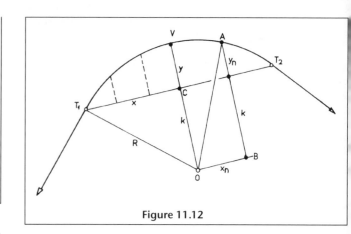

Figure 11.12

The procedure for setting out the curve is then as follows:

1. From I, measure back along the straights the distances IT_1 and IT_2 and drive in pegs to establish the exact positions of T_1 and T_2.
2. Establish pegs at 5 m intervals along the straights between I and the tangent points.
3. Using an optical square set out the appropriate offsets y at right angles to the tangents and drive in a peg at each point.

✍ Exercise 11.6

1. In Fig. 11.11, the deviation angle between the minor road and major road is 30° and the radius of the curve is 30.00 m. Calculate:

(a) the tangent lengths IT_1 and IT_2,
(b) the length IE of the extension to the left-hand straight.
(c) the lengths of the offsets at 2 metre intervals, required to set out the curve, from tangent point T_1 to tangent point T_2.

Method 3: Offsets from the long chord
This method is suitable for curves up to 100 m radius. The curve is established by measuring offsets y at right angles to the long chord T_1T_2 at selected distances from the tangent points (Fig. 11.12).

VC is the major offset y at the mid-point C of the long chord and OC is constant, k.

$$\text{Major offset } y = (R - k)$$

$$\text{In triangle } OT_1C, k = \sqrt{R^2 - x^2}$$

$$\text{Therefore } y = R - \sqrt{R^2 - x^2}$$

$$\text{Any other offset } y_n = (AB - k)$$

$$\text{In triangle ABO, } AB = \sqrt{R^2 - x_n^2}$$

$$\text{Therefore } y_n = \sqrt{R^2 - x_n^2} - k$$

Any offset y can be calculated for any distance along the long chord, and can be set out by eye or prism square.

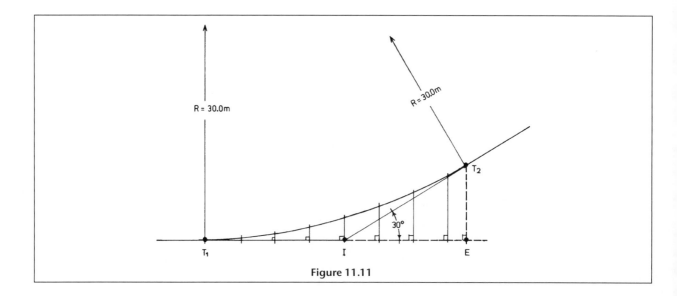

Figure 11.11

EXAMPLE

7 A roadway kerb in Fig. 11.12 has a radius of curvature of 40 m. The length of the long chord is 60 m. Calculate the offsets from the chord at 10 m intervals.

Answer

Major offset $= R - \sqrt{R^2 - x^2}$

$\quad = 40 - \sqrt{1600 - 900}$

$\quad = 40 - 26.46$

$\quad = 13.54$ m

$\quad k = 26.46$ m

Offset y_{10} ($x = 20$ m from C)

$\quad y_{10} = \sqrt{R^2 - x^2} - k$

$\qquad = \sqrt{1600 - 400} - 26.46$

$\qquad = 34.64 - 26.46$

$\qquad = 8.18$ m

Offset y_{20}($x = 10$ m from C)

$\quad y_{20} = \sqrt{R^2 - x^2} - k$

$\qquad = \sqrt{1600 - 100} - 26.46$

$\qquad = 38.73 - 26.46$

$\qquad = 12.27$ m

The offsets at 40 m and 50 m from T_1 are the same lengths as the offsets at 20 and 10 m respectively.

The procedure for setting out the curve is as follows:

1. Locate T_1 and T_2 and measure the distance between them. It should equal 60 metres.
2. At 10 m intervals along the long chord drive in pegs.
3. Set out the offsets at right angles to the long chord using a prism square and drive in pegs to mark the curve.

Exercise 11.7

1 A curved roadway kerb of radius 100 metres is to be set out using the method of offsets from the long chord. Given that the length of the long chord is 60 metres, calculate the lengths of the offsets at 10 metre intervals along the long chord.

(b) Curve composition

In setting out large radius curves, pegs are set at running chainage intervals of 5 m, 10 m or 20 m around the curve from the zero chainage point of the survey. It would therefore be unlikely that either tangent point of the curve would occur at an exact chainage interval.

In Fig. 11.13, straights AI and IB deviate by 13° at intersection point I, where the chainage is 171.574 metres.

Tangent lengths IT_1 and $IT_2 = 400 \tan 6.5°$	
	$= 45.574$ m
Therefore chainage T_1	$= 171.574 - 45.574$
	$= 126.000$ m
Curve length	$= 2\pi R \times 13/360$
	$= 90.757$ m
Therefore chainage T_2	$= 126.000 + 90.757$
	$= 216.757$ m

The last peg on the straight, measured at 20 metre intervals from A, occurs at chainage 120 m; therefore the first peg on the curve, at chainage 140 m, lies at a distance of $(140 - 126)$ m $= 14$ m from tangent point T_1. This short chord is called the initial sub-chord. Thereafter, pegs placed at standard chord intervals of 20 metres occur at chainages 160, 180 and 200 m. The final tangent point T_2 is reached at 216.75 m; therefore the final chord is $(216.757 - 200.000)$ m $= 16.757$ m. This short chord is called the final subchord.

Summarizing, the chord composition is derived as follows:

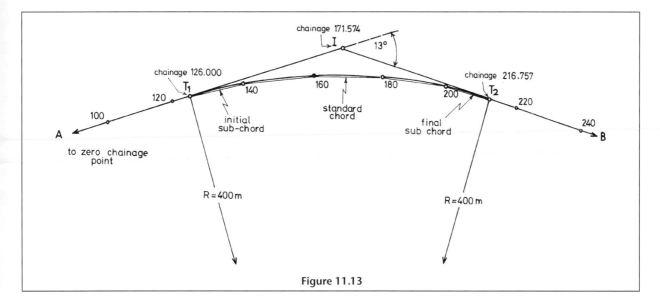

Figure 11.13

Chainage $T_1 = 126.000$ m
Chainage of first peg on curve $= 140.000$ m
Therefore initial sub-chord $= (140.000$
$- 126.000)$
$= 14.000$ m
Chainage of last peg on curve $= 200.000$ m
Therefore number of standard chords
$= (200 - 140)/20 = 3$
Chainage $T_2 = 216.757$ m
Therefore final sub-chord $= (216.757$
$- 200.000)$
$= 16.757$ m

In setting out large radius curves, the chords must be almost equal to the arcs that they subtend. An accuracy of about 1 part in 10 000 is obtainable, provided the chord length does not exceed one twentieth of the length of the radius, i.e. $c < R/20$.

(c) Setting out large radius curves

Method 1. Setting out by tangential angles
This is the traditional method of setting out curves and involves the use of tape and theodolite. In Fig. 11.14, the tangent point T_1 has been established along the left-hand straight as previously described. T_1B is the initial sub-chord c_1, which is shorter than the equal length standard chords $BC(c_2)$ and $CD(c_3)$ DT_2 is the

final sub-chord c_4, which is of course shorter than the standard chords.

Deflection angles Angles a_1, a_2, a_3 and a_4 are the angles by which the curve deflects to the right or left. They are called *deflection* angles, though the terms chord angle and tangential angles are commonly used. In this book they are called deflection angles. Their values must be calculated in order to set out the curve.

Procedure Assuming for a moment that the deflection angles are known, the curve is set out as follows:

1. Set the theodolite at T and sight intersection point I on zero degrees.
2. Release the upper clamp and set the theodolite to read a_1 degrees.
3. Holding the end of a tape at T, line in the tape with the theodolite and drive in a peg B at a distance of c_1 metres from T.
4. Set the theodolite to read $(a_1 + a_2)$ degrees.
5. Hold the rear end of the tape at B and with the tape reading c_2 metres, i.e. a standard chord length, swing the forward end until it is intersected by the line of sight of the theodolite. This is the point C on the curve.
6. Set the theodolite to read $(a_1 + a_2 + a_3)$ degrees. Repeat operation 5 to establish point D on the curve.

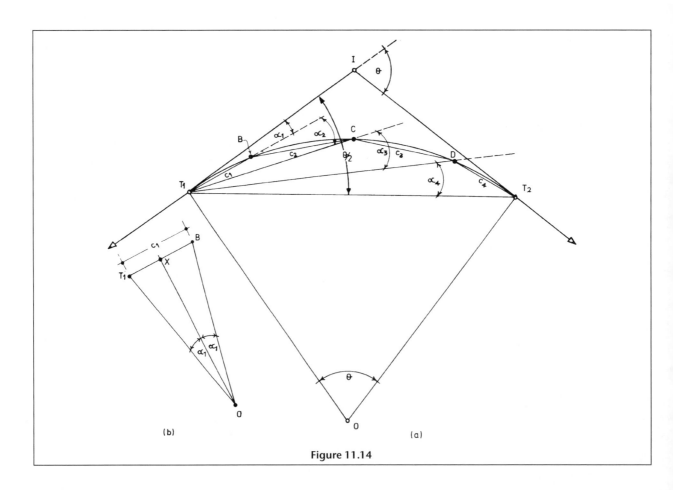

Figure 11.14

In most cases this operation will be repeated several more times to establish a number of pegs on the curve at one standard chord interval. In this example peg D is the last standard chord.

7. Set the theodolite to read $(a_1 + a_2 + a_3 + a_4) = \theta/2$ degrees.
8. Holding the rear end of the tape at D, swing the forward end until the reading of c_4 metres is intersected by the line of sight of the theodolite. This establishes tangent point T_2.

Calculation of deflection angles   Angle IT_1B is the angle between tangent T_1I and chord T_1B. Angle T_1OB is the angle at the centre subtended by chord T_1B.

Therefore angle $IT_1B = \frac{1}{2}$ angle $T_1OB = a$

In Fig. 11.14(b), OX is the perpendicular bisector of chord T_1B. Therefore angle $T_1OX =$ angle $XOB = a$.
In triangle T_1OX,

$$\sin T_1OX = \frac{T_1X}{T_1O}$$
$$= \frac{c_1/2}{R}$$
$$= \frac{c_1}{2R}$$

The value of any deflection angle (a_1, a_2, a_3 and a_4) can similarly be found and the formula can be written in general terms:

$$\sin a = \frac{c}{2R}$$

When c is less then one-twentieth of R, an accurate value of a can be determined thus:

$\sin a = a$ radians (since a is always small)

Therefore a radians $= \dfrac{c}{2R}$

Hence a degrees $= \dfrac{c}{2R} \times \dfrac{180}{\pi}$

and a minutes $= \dfrac{c}{2R} \times \dfrac{180}{\pi} \times 60$

i.e. $a = \left(\dfrac{c}{R} \times 1718.9\right)$ minutes

In Fig. 11.14(a),

Standard-chord angle $a_2 = \left(\dfrac{c_2}{R} \times 1718.9\right)$ minutes

or $\sin a_2 = \dfrac{c_2}{2R}$

Initial sub-chord angle $a_1 = \left(\dfrac{c_1}{R} \times 1718.9\right)$ minutes

or $\sin a_1 = \dfrac{c_1}{2R}$

From these calculations it can be seen that angles a_1 and a_2 are proportional to their chord lengths and the most convenient way to calculate deflection angles is firstly to calculate the standard chord angle and then by proportion to calculate the initial and final subchord angles:

Final sub-chord angle $= a_2 \times \dfrac{c_4}{c_2}$

Tangential angles   The reading to which the theodolite is set is the tangential angle, so called because the angle being set out is measured from the tangent. Other titles include total angle, setting out angle, etc.

Any tangential angle to a point on a curve is simply the summation of the deflection angles to that point. Thus, the tangential angle to point C $= a_1 + a_2$ and to point D $= a_1 + a_2 + a_3$.

EXAMPLE

8. Two straights AI and IB have bearings of 80° and 110° respectively. They are to be joined by a circular curve of 300 metres radius. The chainage of intersection point I is 872.485 m (Fig. 11.15). Calculate the data for setting out the curve by 20 m standard chords.

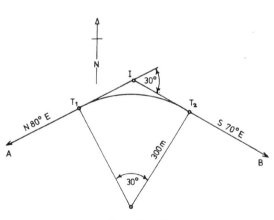

Figure 11.15

Answer

(a)   Tangent length = 300 tan 15°
           = 80.385 m
(b)   Chainage of T_1 = (872.485 − 80.385) m
           = 792.100 m
(c)   Curve length = 300 × θ radians
           = 157.080 m
(d)   Chainage of T_2 = 949.180 m

(e) Number of chords

Initial sub-chord $= (800.00 - 792.10)$

$= 7.90$ m

7 full chords of 20 m $= (7 \times 20.0)$

$= 140.00$ m

Sub-total $= 7.90 + 140.00$

$= 147.90$ m

Therefore final sub-chord

$= (157.08 - 147.90)$

$= 9.18$ m

Total $= 147.90 + 9.18$

$= 157.08$ m

$=$ curve length

(f) Tangential angle for standard 20 m chord

$= \left(\dfrac{20}{300} \times 1718.9 \right)$ minutes

$= 114.593$ minutes

$= 01° \, 54' \, 36''$

(g) Sub-chord angles

Initial $= \dfrac{114.593 \times 7.90}{20.0}$ minutes

$= 45.264$ minutes

$= 00° \, 45' \, 16''$

Final $= \left(\dfrac{114.593 \times 9.18}{20.0} \right)$ minutes

$= 52.598$ minutes

$= 00° \, 52' \, 36''$

For this and every other example, the setting-out information is presented in tabular fashion as in Table 11.1. Note that there is a discrepancy of 4″ in the final tangential angle in Table 11.1 due to the rounding off of the angles to whole seconds.

When curves are to turn to the left the tangential angles must be subtracted from 360°. For example, if the straights in Example 9 had deviated to the left by 30° the final tangential reading would have been

$(360° \, 00' \, 00'' - 15° \, 00' \, 00'') = 345° \, 00' \, 00''$

Exercise 11.8

1 Two straight roadways, AB and BC, intersecting at B, have bearings of 5° 30′ and 353° 30′ respectively. They are to be joined by a circular curve of 500 metres radius. The curve is to be set out by the method of tangential angles. Pegs are to be set out at intervals of 20 metres through chainage, from the point A. Given that the chainage of the point B is 842.75 m from A, calculate:

(a) the chainages of the tangent points of the curve,
(b) the chainages of the setting out pegs along the curve,
(c) the tangential angles required to set out the curve from the initial tangent point.

Method 2: Setting out by coordinates

When calculating setting-out distances, whether for curved roadways or otherwise, the plan distance is always obtained and has to be set out. However, it is often very difficult to measure distances horizontally; hence slope distances have frequently to be determined or inaccuracies in setting the pegs must occur. Furthermore, measuring by tape is slow, labour intensive and often uncomfortable.

Setting out points using total station instruments has therefore become standard practice on construction sites, since these instruments enable horizontal distances to be set out without difficulty.

Setting out by EDM methods requires that the coordinates of every proposed point be determined, usually by calculation. The coordinates are compared with those of the setting-out survey point and the horizontal distance and bearing between the two are computed.

On site, the EDM instrument is switched to the horizontal distance mode (tracking mode) and set on the correct bearing. The reflector is then aligned at the

Table 11.1

Chord number	Length (m)	Chainage (m)	Deflection angle	Tangential angle
(T₁)		792.10		
1	7.90	800.00	00° 45′ 16″	00° 45′ 16″
2	20.00	820.00	01° 54′ 36″	02° 39′ 52″
3	20.00	840.00	01° 54′ 36″	04° 34′ 28″
4	20.00	860.00	01° 54′ 36″	06° 29′ 04″
5	20.00	880.00	01° 54′ 36″	08° 23′ 40″
6	20.00	900.00	01° 54′ 36″	10° 18′ 16″
7	20.00	920.00	01° 54′ 36″	12° 12′ 52″
8	20.00	940.00	01° 54′ 36″	14° 07′ 28″
9(T₂)	9.18	949.18	00° 52′ 36″	15° 00′ 00″
	157.08	157.08	15° 00′ 04″	

required horizontal distance and no slope correction is ever necessary.

Coordinate calculations Figure 11.16 shows a curve of 572.96 m radius, connecting two straights AI and IB which have bearings of 40° 00′ 00″ and 44° 30′ 00″ respectively. Point A is the starting point (zero chainage) and has known coordinates of 105.26 m east and 352.15 m north. Based on the theory of Sec. 2 (curve elements) and Sec. 5(c) (setting out large radius curves), the data required to set out the curve are as shown in Table 11.2.

The curve is to be set out from a nearby survey traverse station S (with coordinates 148.50 m east, 370.01 m north) by the method of coordinates, there-

fore the coordinates of all chainage points of the roadway scheme must first be calculated.

Several methods are available to compute those coordinates, of which the easiest to understand is the following:

1. Calculate the coordinates of points A, T_1, I, T_2 and B as a traverse. The coordinates are computed in Table 11.3(a).
2. Calculate the coordinates of points T_1, O and T_2 as a traverse.

$$\begin{aligned}
\text{Forward bearing AT}_1 &= 40°\ 00′\ 00″ \\
\text{Back bearing T}_1\text{A} &= 220°\ 00′\ 00″ \\
-\text{angle AT}_1\text{O} &= \underline{-90°\ 00′\ 00″} \\
\text{Therefore bearing T}_1\text{O} &= \underline{\underline{130°\ 00′\ 00″}}
\end{aligned}$$

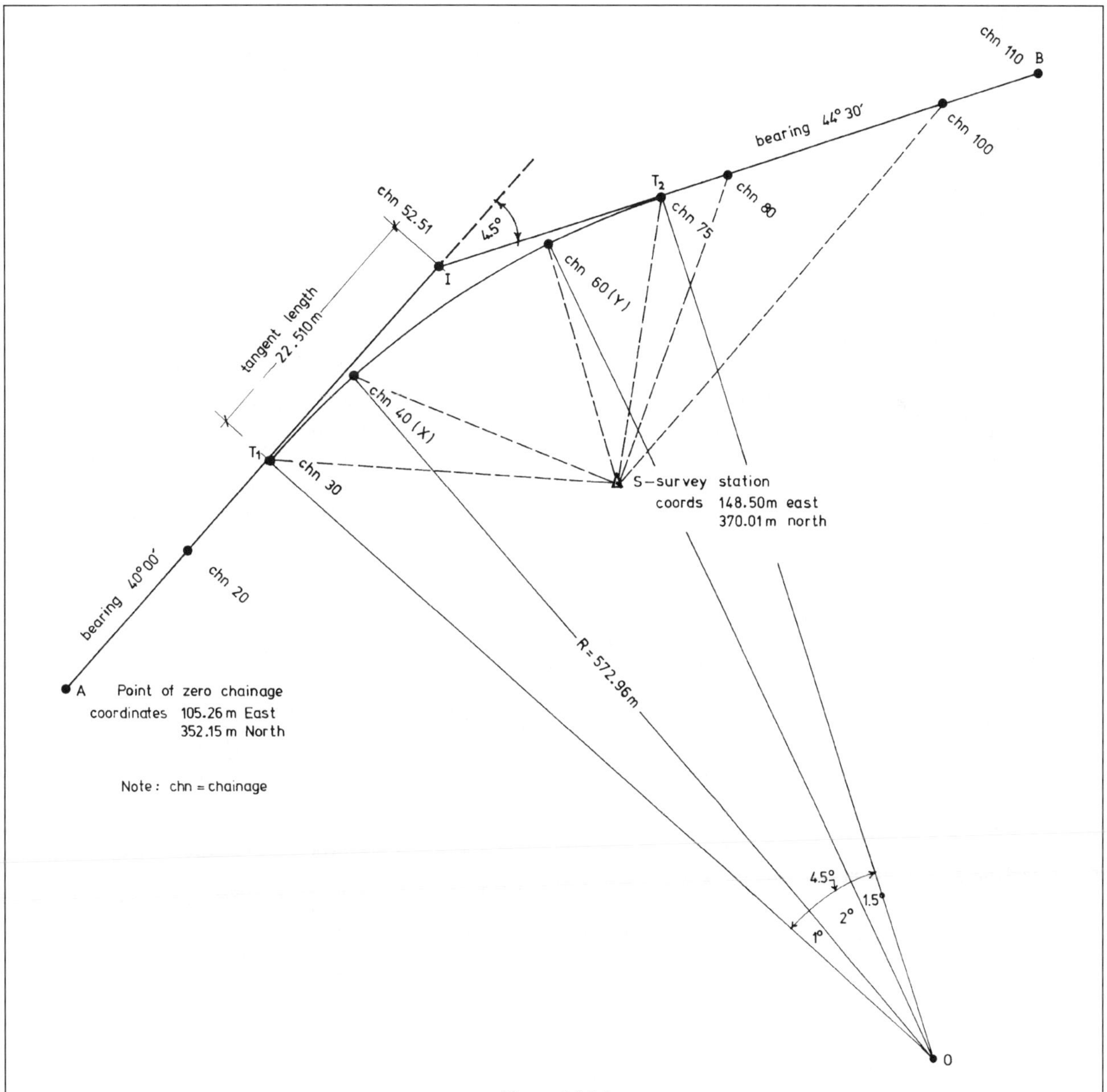

Figure 11.16

Back bearing $OT_1 = 310° 00' 00''$
+ angle T_1OT_2 $+4° 30' 00''$
Therefore forward bearing $OT_2 = \overline{314° 30' 00''}$

Table 11.2

1. Point A – point of zero chainage
    Coordinate 105.26 m east
    352.15 m north
2. Radius of curve = 572.96 m
3. Deviation angle = 4.5°
4. Chainage of intersection point = 52.51 m
5. Tangent length = 572.96 tan 2.25° = 22.51 m
6. Chainage T_1 = 52.51 – 22.51 m = 30.0 m
7. Curve length = $2\pi R \times (4.5/360)$ = 45.0 m
8. Chainage T_2 = 30 + 45 m = 75.0 m
9. Peg chainages
    0.0 m   Start
    20.0 m   On left-hand straight
    30.0 m   Tangent point T_1
    40.0 m   On curve (X)
    60.0 m   On curve (Y)
    75.0 m   Tangent point T_2
    80.0 m   On right-hand straight
    100 m   On right hand straight
    110 m   Point B
10. Curve composition
    Initial sub-chord = 10.0 m
    Standard chord = 20.0 m
    Final sub-chord = 15.0 m
11. Deflection angles
    Initial sub-chord = $\sin^{-1} 10/2R = 00° 30' 00''$
    Standard chord = $\sin^{-1} 20/2R = 01° 00' 00''$
    Final sub-chord = $\sin^{-1} 15/2R = 00° 45' 00''$
12. Centre angles (= 2 × deflection angles)
    $T_1OX = 01° 00' 00''$
    $XOY = 02° 00' 00''$
    $YOT_2 = \underline{01° 30' 00''}$
    Total = 04° 30' 00''

The coordinates of points T_1, O and T_2 are computed in Table 11.3(b). The coordinates of point T_2 in Table 11.3(b) check with those in Table 11.3(a).

3. Calculate the coordinates of the remaining curve chainage points, namely X (chainage 40 m) and Y (chainage 60 m).

Bearing $OT_1 = 310° 00' 00''$
+ angle T_1OX $+1° 00' 00''$
Forward bearing $OX = 311° 00' 00''$
+ angle XOY $+2° 00' 00''$
Forward bearing $OY = 313° 00' 00''$
+ angle YOT_2 $+1° 30' 00''$
Forward bearing $OT_2 = 314° 30' 00''$
(Checks with previous result)

The coordinates of points X and Y are computed in Tables 11.3(c) and 11.3(d) respectively.

4. Calculate the bearings and distances that will be required to set out the various chainage points from station S. Table 11.4 shows the relevant calculation, based, of course, on the formulae:

Bearing between two points = $\tan^{-1}$ (Δ east/Δ north) and

distance between two points = $\sqrt{\Delta \text{ east}^2 + \Delta \text{ north}^2}$

Field procedure It must be pointed out, at this juncture, that the mechanics of setting an electromagnetic distance measuring instrument along a certain line on a specific bearing varies with the type of instrument. Only the general principle is described here.

1. The EDM instrument (total station) is set up, centred and levelled at survey station S, and the bearing to point A (247° 33' 26'') is set on the instrument (Table 11.4).
2. A sight is taken to point A and the instrument

Table 11.3

Line	Length	Whole circle bearing	Δ east	Δ north	Easting	Northing	Station
(a)					105.260	352.150	A
A–Ch20	20.00	40° 0' 0''	12.856	15.321	118.116	367.471	Ch20
Ch20–T1	10.00	40° 0' 0''	6.428	7.660	124.544	375.131	T1
T1–1	22.51	40° 0' 0''	14.469	17.244	139.013	392.375	1
l–T2	22.51	44° 30' 0''	15.777	16.055	154.790	408.430	T2
T2–Ch80	5.00	44° 30' 0''	3.505	3.566	158.295	411.997	Ch80
80–100	20.00	44° 30' 0''	14.018	14.265	172.313	426.262	Ch100
Ch100–B	10.00	44° 30' 0''	7.009	7.133	179.322	433.395	Ch110(B)
(b)					124.544	375.131	T1
T1–O	572.96	130° 0' 0''	438.913	–368.292	563.457	6.839	O
O–T2	572.96	314° 30' 0''	–408.664	401.593	154.793	408.432	T2
(c)					563.457	6.839	O
O–X	572.96	311° 0' 0''	–432.418	375.896	131.039	382.735	X
(d)					563.457	6.839	O
O–Y	572.96	313° 0' 0''	–419.036	390.758	144.421	397.597	Y

Table 11.4

Ref Station	Easting	Northing				
S	148.500	370.010				

Station	Easting	Northing	Difference in eastings	Difference in northings	Set-out distance	Set-out bearing
A	105.260	352.150	−43.240	−17.860	46.783	247° 33′ 26.1″
Ch20	118.116	367.471	−30.384	−2.539	30.490	265° 13′ 23.7″
T1	124.544	375.131	−23.956	5.121	24.497	282° 3′ 58.8″
Ch40 (X)	131.039	382.735	−17.461	12.725	21.606	306° 5′ 00.0″
Ch60 (Y)	144.421	397.597	−4.079	27.587	27.887	351° 35′ 21.2″
T2	154.790	408.430	6.290	38.420	38.931	9° 17′ 52.1″
Ch80	158.295	411.997	9.795	41.987	43.114	13° 7′ 53.3″
Ch100	172.311	426.262	23.811	56.252	61.084	22° 56′ 33.1″
Ch110 (B)	179.320	433.395	30.820	63.385	70.481	25° 55′ 50.3″

clamped, reading 247° 33′ 26″. The instrument is set to tracking mode and the horizontal distance to point A is measured as a check. The distance should be 46.783 metres.

3. The bearing and horizontal distance to the first setting out point, chainage 20 m, are keyed into the instrument. From Table 11.4 the relevant data are: bearing, 265° 13′ 24″ and distance, 30.490 m.

4. The survey assistant will, by this time, have gone to the approximate position of the point to be set out, and will hold the reflector vertically.

5. A sight is taken to the reflector and as soon as contact is made, the difference in bearing and distance between the setting-out point and the actual position of the reflector are displayed on the keypad of the instrument as dHA and dHD.

6. The instrument is then rotated and the reflector moved until dHA and dHD become zero.

7. The assistant marks the reflector position, inserts a peg, re-checks the complete operation and when satisfied that it is correct, moves to the next setting out location, namely tangent point T_1.

8. The next bearing and distance are keyed in and the procedure repeated. All remaining pegs are set out in this manner.

Setting out by coordinates has several advantages but has one major disadvantage which must be guarded against.

The advantages stem mainly from the ease with which the design coordinates are obtained from software packages. These coordinates are error free, provided the correct input data has been used. The surveyor requires practically no knowledge of curve calculations and the points can therefore be set out by relatively inexperienced assistants, thus offering financial savings. The setting-out station is also remote from the centre line and the instrument is set up clear of the heavy road-making plant, thus offering safety and convenience.

The disadvantage is that there is no check on the setting out of the design coordinates and errors may go unnoticed until later. The points must therefore be set out from a second survey point by computing a second set of data (bearings and distances) and setting them out as described.

Exercise 11.9

1 In Fig. 11.17, two straight roadways, PQ and QR, have bearings of 05° 00′ 00″ and 357° 20′ 00″ respectively and intersect at point Q (chainage 78.793 m).

They are to be joined by a curve of 429.718 metres radius.

(a) Using Table 11.5, calculate all of the curve elements.

(b) Calculate the coordinates of all setting out pegs, from chainage 40 m on the left-hand straight to chainage 140 m (point R) on the right-hand straight, at through chainages of 20 metres.

(c) Given that all of the points are to be set out from tangent point T_1, calculate the bearings and distances required to set out the pegs in their correct locations.

6. Obstructions

Often it is impossible to set out all the points on the curve from the tangent point because of some obstruction. In Fig. 11.18, the third point E on the curve has been set out by the tangential angle $(a_1 + a_2 + a_2) =$ say 5°.

Owing to the trees, point F cannot be seen from T_1. In such a case the theodolite is removed to E and a sight taken back to the tangent point T_1 with the horizontal circle of the theodolite reading $180° − (a_1 + a_2 + a_2) = 175°$ (say).

If a tangent is drawn to the circle at E, angle XT_1E equals angle T_1EX, i.e. angle $T_1EX = 5°$. The theodolite is then set to read $(175° +$ angle $T_1EX) = (175° + 5°) = 180°$, in which case the line of sight is along tangent EX. The telescope is swung through a further 180° to

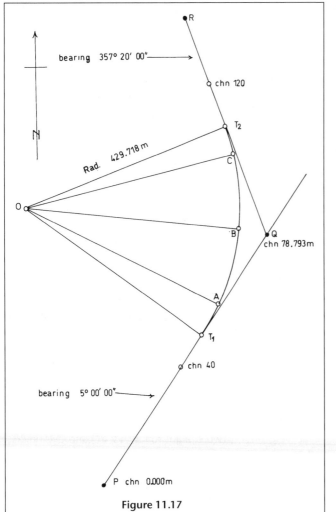

Figure 11.17

Table 11.5

1. Point P – point of zero chainage
   Coordinates: 00.000 m east
   00.000 m north
2. Radius of curve = 429.718 m
3. Deviation angle = 7° 40′ 00″
4. Chainage of intersection point = 78.793 m
5. Tangent length =
6. Chainage T_1 =
7. Curve length =
8. Chainage T_2 =
9. Peg chainages
   40.000 m on left-hand straight
   m tangent point T_1
   m on curve
   m on curve
   m on curve
   m tangent point T_2
   120.000 m on right-hand straight
   140.000 m on right hand straight
10. Curve composition
    Initial sub-chord =
    Two standard chords =
    Final sub-chord =
11. Deflection angles
    Initial sub-chord =
    Standard chord =
    Final sub-chord =
12. Centre angles (= 2 × deflection angels)
    T_1OA =
    AOB =
    BOC =
    COT_2 =

point along the continuation of the tangent EY and the horizontal circle reads zero.

Point E is then treated as being a tangent point and F is set out by setting the circle to read a_2 degrees and measuring the standard chord length EF.

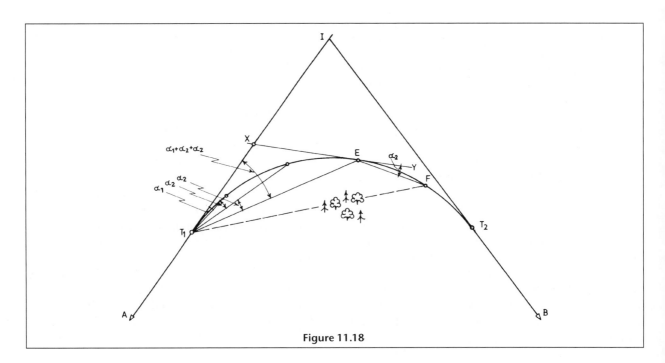

Figure 11.18

7. Vertical curves

(a) Summit and valley curves

Whenever roads or railways change gradient, a vertical curve is required to take traffic smoothly from one gradient to the other. When the two gradients form a hill, the curve is called a summit curve and when the gradients form a valley, a sag or valley curve is produced (Fig. 11.19).

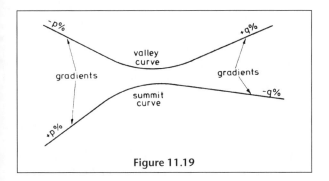

Figure 11.19

(b) Percentage gradients

The gradients are expressed as percentages. A gradient of 1 in 50 is a 2 per cent gradient, i.e. the gradient rises or falls by 2 units in 100 units. Similarly a gradient of 1 in 200 is a 0.5 per cent gradient.

In vertical curve calculations, the left-hand gradient is p per cent and the right-hand gradient is q per cent.

EXAMPLE

9 A downhill gradient of 1 in 40 is to be connected to an uphill gradient of 1 in 18, by a vertical curve. Calculate the values of the gradients as percentages.

Answer

$$\text{Gradient 1 in 40 (uphill)} = +1/40$$
$$\text{Percentage grade} = (1/40) \times 100$$
$$= +2.5 \text{ per cent}$$
$$\text{Gradient 1 in 18 (downhill)} = 1/18$$
$$\text{Percentage grade} = (-1/18) \times 100$$
$$= -5.556 \text{ per cent}$$

Since the change of gradient from slope to curve is required to be smooth and gradual, parabolic curves are chosen. This form of curve is flat near the tangent point and calculations are reasonably simple. The form of the curve is $y = ax^2 + bx + c$, where

> y = reduced level of any point on the curve,
> x = distance to that point measured from the start of the curve,
> a = multiplying coefficient, which will be derived in Sec. 7(e),
> b = value of the left-hand gradient,
> c = reduced level of the first point on the curve.

(c) Properties of the parabola

1. The distance between the points T and T_1 as measured along (a) the curve TT_1, (b) the tangents TIT_1 and (c) the chord TT_1 are so close in length that they are considered equal (Fig. 11.20).
2. The intersection point I is treated as being midway between the points T and T_1; thus the lengths IT and IT_1 are equal. The curve is in fact often called an equal tangent parabola.
3. The height IV is called the correction in gradient and equals the height VC. In other words, the parabola bisects the length CI.

(d) Setting-out data

In setting out a vertical curve on the ground, the objective is to place large pegs at the required intervals along the line of the proposed roadway (Fig. 11.21) and to nail a cross-piece to each peg at a certain height

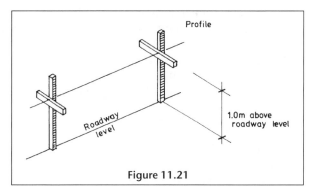

Figure 11.21

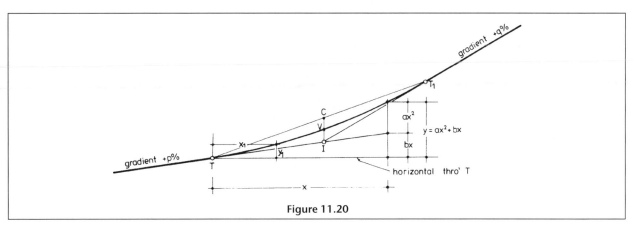

Figure 11.20

(usually 1.0 m), above the proposed road level. These pegs are called profiles and the erection of these profiles is the standard method of setting out proposed levels on any construction site. A full explanation is given in Chapter 12.

The following information is required for any setting out calculations:

1. The gradients of the slopes and the reduced level of one chainage point, preferably the intersection point.
2. The length of the curve. The length of the curve is dependent upon: (a) the gradient of the straights; (b) the sight distance required to stop or overtake.
(a) Generally the steeper the approach gradients, the greater will be the centrifugal effect caused by the change of gradient from the slope to the curve. The centrifugal effect has to be introduced in a controlled progressive fashion. This is achieved by the use of parabolic curves. The Dept of Transport has introduced design standards which restrict the length and gradient of vertical curves on various classes of roadway. As an example, the absolute maximum gradient on a single carriageway is 8 per cent or 1 in 12.5.
(b) The sight distance is the length required by a vehicle to stop from the moment a driver sees an obstruction over the brow of a summit curve. There is no problem in a sag curve since visibility is usually unrestricted. The 'stopping sight distance' as defined by the Dept of Transport includes thinking, braking, stopping and safety margin distances.

The 'full overtaking sight distance' is rather different in that it requires a much longer distance to overtake than to stop. Obviously it is dangerous to overtake on the summit curve of a single carriageway but there is usually no problem with sag curves or, in either case, with dual carriageways.

The Dept of Transport design standards recommend the use of 'K' values when determining the length of vertical curves. These values are published in tables from which the user extracts the appropriate K value from the design speed of the road. This value is multiplied by the positive algebraic difference 'A' between the two gradients forming the vertical curve to produce the length of the curve KA. As an example, for a roadway where the design speed is 70 kph, the full overtaking sight distance, taken from the tables, is 200 metres. If the curve is to connect an upward gradient of + 2 per cent to a downward gradient of –1 per cent, the algebraic difference is A = 2–(–1) = 3 and the length KA of the curve becomes 200 × 3 = 600 metres. The length may be exceeded if required.

The tables give other criteria for lengths of curve, namely desirable minimum and absolute minimum K values.

Understandably, the lengths and gradients of vertical curves will be decided by the roadway design team and *this textbook will simply assume arbitrary values throughout.*

(e) Calculation of data

In Fig. 11.22(a) a gradient of +1 per cent (i.e. 1 in 100) meets a gradient of +4 per cent (i.e. 1 in 25) at intersection point I, the chainage and reduced level of which are 500 and 261.30 m respectively. A 100 m long vertical curve is to be inserted between the straights.

The levels on the roadway curved surface at 25 metre intervals are required. These levels are called corrected grade elevations or curve levels.

Step 1 Calculate the reduced levels of the initial tangent point T, the final tangent point T_1 and the intersection point I. In Figure 11.22(a),

$$IT = IT_1 = 100/2 = 50 \text{ m}$$
Reduced level I = 261.30 m (given)
Reduced level T = reduced level I – 1 per cent of 50
$$= 261.30 - (0.01 \times 50)$$
$$= 260.80 \text{ m}$$
Reduced level T_1 = Reduced level I + 4 per cent of 50
$$= 261.30 + (0.04 \times 50)$$
$$= 263.30 \text{ m}$$

Step 2 Calculate the tangent levels, i.e. the levels that would obtain on the left-hand gradient if it were extended above or in this case below the right-hand gradient, towards the final tangent point. Mathematically any tangent level is

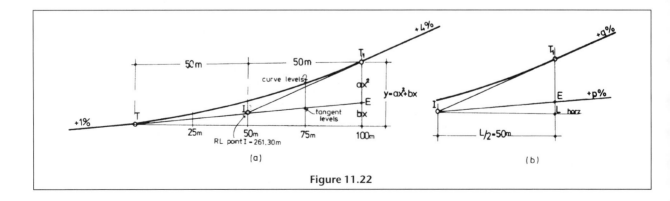

Figure 11.22

Tangent level = (reduced level T + bx)

where b = left-hand gradient (p per cent)

x = distance from T

Tangent level at 25 m = reduced level T + bx

= 260.80 + 1 per cent of 25 m

= 261.05

Tangent level 50 m (I) = 260.80 + 1 per cent of 50 m

= 261.30

Tangent level at 75 m = 260.80 + 1 per cent of 75 m

= 261.55

Tangent level 100 m (E)

= 260.80 m

+ 1 per cent of 100 m

= 261.80 m

Step 3 Calculate the grade corrections at the required chainage points. The grade correction is the value of ax^2 which, when added to the various tangent levels in step 2, will give the level on the curve.

The term x is, of course, the distance of the chainage point from the initial tangent point. The value of a is unknown and has to be found in order to calculate these grade corrections.

In Fig. 11.22(b) the reduced levels of T_1 and E are 263.30 and 261.80 respectively; therefore reduced level T_1 – reduced level E = 263.30 – 261.80 = 1.50 m. This value of 1.50 m is really the grade correction at point E, i.e. the value that is applied to tangent level E to produce the curve level T_1. Therefore

Grade correction 1.50 m = ax^2 (where x = 100 m)

= $a \times 100^2$

Therefore a = 1.50 ÷ 100^2

= 1.50×10^{-4}

Value of ax^2 at 25 m intervals from T:

Chainage	ax^2
25 m	$(1.50 \times 10^{-4}) \times 25^2 = 0.094$ m
50 m	$(1.50 \times 10^{-4}) \times 50^2 = 0.375$ m
75 m	$(1.50 \times 10^{-4}) \times 75^2 = 0.844$ m
100 m	$(1.50 \times 10^{-4}) \times 100^2 = 1.500$ m

The following is an alternative method of calculating the term a. In Fig. 11.22(b)

Reduced level T_1 = reduced level I + q per cent of $L/2$

$$= \text{RL I} + \frac{qL}{200}$$

Reduced level E = reduced level I + p per cent of $L/2$

$$= \text{RL I} + \frac{pL}{200}$$

Now ax^2 = reduced level T_1 – reduced level E

$$= \frac{qL}{200} - \frac{pL}{200}$$

$$= \frac{(q-p)L}{200}$$

Since $x = L$ at point T_1,

$$aL^2 = \frac{(q-p)L}{200}$$

Therefore $a = \dfrac{(q-p)}{200} \times \dfrac{L}{L^2}$

$$= \frac{q-p}{200L}$$

The formula applies in all situations where a is required. The proper sign convention for positive and negative gradients p or q must of course be used.

In this case, q = + 4 per cent, p = + 1 per cent and L = 100 m. Therefore

$$a = \frac{4-1}{200 \times 100} = \frac{3}{20\,000} = 1.5 \times 10^{-4}$$

Step 4 Calculate the curve level at the various chainage points. The curve level at any point is the algebraic addition of the tangent level (T + bx) and grade correction (ax^2):

i.e. curve level = tangent level + gradient correction

Therefore curve level 25 m = 261.05 + 0.094

= 261.144 m

Curve level 50 m = 261.30 + 0.375

= 261.675 m

75 m = 261.55 + 0.844

= 262.394 m

100 m = 261.80 + 1.500

= 263.300 m

In all examples the calculations are performed in tabular fashion, as shown in Table 11.6.

Table 11.6

Chainage (m)	Tangent level (m) T + bx	Grade correction (m) ax²	Curve level (m) T + ax² + bx
0(T)	260.80	0	260.800
25	261.05	0.094	261.144
50	261.30	0.375	261.675
75	261.55	0.844	262.394
100(E)	261.80	1.500	263.300

EXAMPLE

10 A rising gradient of 1 in 40 is to be connected to a falling gradient of 1 in 75 by means of a vertical parabolic curve 400 m in length. The reduced level of the intersection point of the gradients is 26.850 m above Ordnance Datum. Calculate:

(a) the reduced levels of the tangent points,
(b) the reduced levels at 50 m intervals along the curve.

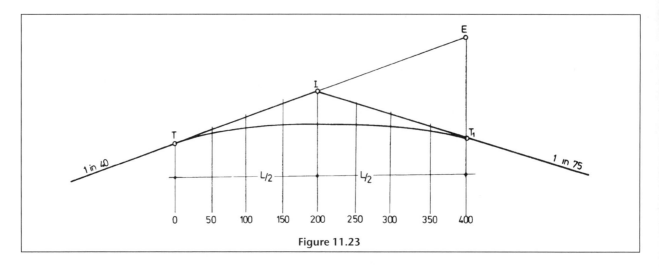

Figure 11.23

Solution

See Fig. 11.23 and Table 11.7

$$IT = IT_1 = 400/2 = 200 \text{ m}$$

Reduced level I (given) $= 26.850$ m

Reduced level T $= 26.850 - 200/40$
$$= 21.850 \text{ m}$$

Reduced level $T_1 = 26.850 - 200/75$
$$= 24.183 \text{ m}$$

Reduced level E $= 26.850 + 200/40$
$$= 31.850 \text{ m}$$

$ET_1 = ax^2$ at 400 m $= 24.183 - 31.850$
$$= -7.667$$

Therefore $a = -7.667/400^2$
$$= -4.792 \times 10^{-5}$$

$$\text{or } a = \frac{q - p}{200L}$$

$$p\% = 1/40 \times 100$$
$$= +2.5 \text{ per cent}$$
$$q\% = -1/75 \times 100$$
$$= -1.333 \text{ per cent}$$

Therefore

$$a = \frac{-1.333 - 2.5}{200 \times 400}$$
$$= -4.792 \times 10^{-5},$$
$$\text{as before}$$

Table 11.7

Chainage (m)	Tangent level (m) $T + bx$	Grade correction (m) ax^2	Curve level (m) $T + ax^2 + bx$
0(T)	21.850	0	21.850
50	23.100	−0.120	22.980
100	24.350	−0.479	23.871
150	25.600	−1.078	24.522
200	26.850	−1.917	24.933
250	28.100	−2.995	25.105
300	29.350	−4.313	25.037
350	30.600	−5.870	24.730
400(T_1)	31.850	−7.667	24.183

Exercise 11.10

1 A downward gradient of 1 in 50 is to be connected to an upward gradient of 1 in 35 by means of a 60 metre long vertical parabolic curve. Given that the reduced level of the intersection point of the gradients is 26.000 m, calculate the reduced levels on the curve at 25 metre intervals.

(f) Highest (or lowest) point on any curve

The highest or lowest point on any curve is the turning point, i.e. the position where the gradient of the tangent is zero.

The gradient of the tangent is found by differentiating y with respect to x in the equation $y = ax^2 + bx$.

$$\frac{dy}{dx} = 2ax + b$$

When $dy/dx = 0$

$$x = -b/2a$$

In Example 10,

$$b = +2.5 \text{ per cent} = +2.5 \times 10^{-2}$$
$$a = -4.792 \times 10^{-5}$$

$$\text{Therefore } x = \frac{-2.5 \times 10^{-2}}{2 \times -4.792 \times 10^{-5}}$$
$$= 10^3 \times 0.260 \, 85$$
$$= 260.851 \text{ m from T}$$

Level of highest point $= 21.850 + ax^2 + bx$
$$= 21.850 - 3.261 + 6.521$$
$$= 25.110 \text{ m}$$

Exercise 11.11

1 Using the figures of Exercise 11.10, calculate the chainage and reduced level of the lowest point on the curve.

Exercise 11.12 (Horizontal and Vertical curves)

1 A 40 m radius roadway kerb is to be set out at the junction of two roadways by the method of offsets from the long chord. Given that the length of the long chord is 30 m, calculate the lengths of the offsets from the chord at 5 m intervals.

2 Two straights AX and XB are to be connected by a 400 m radius circular curve. The bearings and lengths of the straights are as follows:

Straight	Whole circle bearing	Length (m)
AX	73° 10′	197.5
XB	81° 40′	–

Calculate:

(a) the intersection angle between the straights
(b) the tangent lengths
(c) the curve length
(d) the chainages of the tangent points
(e) the setting-out information to enable the curve to be set out at 20 m intervals of through chainage.
    (Present the information in the form of a setting out table.)

3 Figure 11.24 shows two straights AB and CD which are to be joined by a curve of 330 m radius. The intersection point I is inaccessible and a traverse ABCD produced the results shown in the figure. Calculate:

(a) the tangent lengths
(b) the chainages of the tangent points
(c) the setting-out information to set out the curve at even chainages of 20 metres.

4 Two straight roadways AI and IB are to be joined by a circular curve of radius 40 metres. The total coordinates of A, I and B are shown in Table 11.8 Calculate:

(a) the distance from A and B to the tangent points of the curve
(b) the *total* coordinates of each tangent point.

Table 11.8

Station	Northing	Easting
A	+75.38	+111.20
I	+154.60	+146.81
B	+128.70	+165.35

5 A vertical curve is to be used to connect a rising gradient of 1 in 60 with a falling gradient of 1 in 100; they intersect at a point having a reduced level 65.25 metres AOD. Given that the curve is to be 150 metres long calculate:

(a) the reduced levels of the tangent points
(b) the levels at 30 metre intervals along the curve and the depths of cutting required
(c) the distance from the tangent point on the 1 in 60 gradient to the highest point on the curve, and the reduced level of this point.

6 Two straights AB and BC having gradients of −1:40 and +1:50 respectively meet at point B at a level of 40.00 m AOD and chainage 1500.0 m in the direction AB. The gradients are to be joined by a vertical parabolic arc 240 m in length. Calculate the level and chainage of the lowest point on the curve.

8. Transition curves

(a) The need for transition curves

When road or rail vehicles travel in a straight line along a road or railway track, then have to change direction via a curve, there is a natural tendency for the vehicle to continue along the straight due to the centrifugal, or radial, force acting on the vehicle. The smaller the radius encountered, the greater is the force. This force causes road vehicles to skid and causes discomfort to passengers. The radial force must therefore be counteracted in some way.

In the study of mechanics, the radial force (P) acting on the vehicle (Fig. 11.25) can be shown to be equal to (mv^2/R) or (Wv^2/Rg) since $m = W/g$. The figure shows a

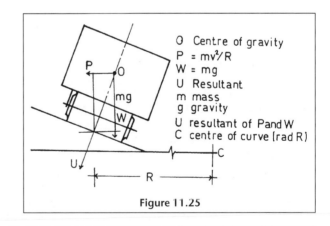

O Centre of gravity
P = mv^2/R
W = mg
U Resultant
m mass
g gravity
U resultant of P and W
C centre of curve (rad R)

Figure 11.25

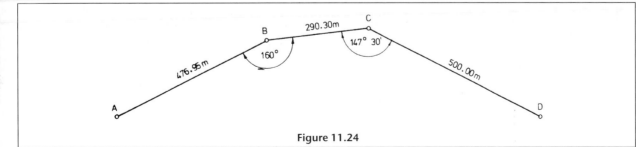

Figure 11.24

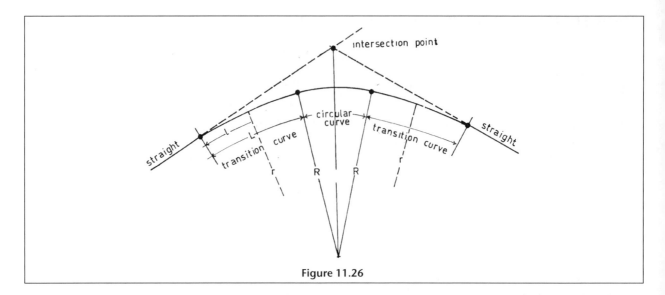

Figure 11.26

railway vehicle travelling on a circular curve of radius R under equilibrium conditions. The resultant U of the forces P and the weight W affecting the vehicle is acting centrally on the vehicle, which is therefore correctly balanced and passengers feel no discomfort.

When vehicles travel along a 'straight' where the radius is infinite, the force P acting on the vehicle is zero. On a circular curve, the radius R is constant, hence the radial force is greatest, therefore

$P = (mv^2/R)$

If the radial force acting on the vehicle could be made to increase gradually between the two extremes, the problem would be solved. The solutions are (i) to superelevate the outer side of the road or rail and/or (ii) introduce a transition curve between the straight and circular curve, which action also achieves a gradual change of direction from the straight to the circular curve (Fig. 11.26).

(b) The ideal transition curve

It has already been stated above that the purpose of a transition curve is to achieve gradual changes in (i) the direction of the road or track and (ii) the application of radial force from zero to maximum.

Recalling that $P = mv^2/r$ and that the radial force P is to increase at a constant rate, P must vary with time and since the speed v of the vehicle does not change, the distance l, measured from the beginning of the curve, must also vary with time.

Therefore $P \propto l \propto mv^2/r$
and since m and v are constants,
$l \propto l/r$
and so $rl =$ a constant (K).

At the junction of the transition curve (length L) and the beginning of the circular curve (radius R), RL must also $= K$.

The distance l at any point, multiplied by the radius

of curvature r at that point, must remain constant in order to produce the perfect transition curve.

(c) Superelevation

The effect of the radial force can be greatly reduced by banking the outer rail of a railway or outer side of a carriageway, i.e. the side furthest from the centre of curvature

Figure 11.27 shows a railway track (radius R) where the outer rail has been raised by the maximum superelevation amount E. The radial force acting upon the vehicle is as shown previously,

$(P = mv^2/R)$

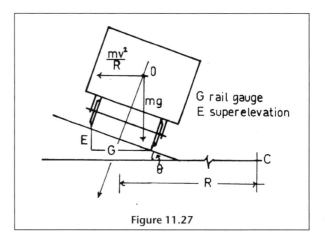

Figure 11.27

In the study of mechanics, given that the amount of cant is E and the rail gauge is G, the inward force due to the cant is (Emg/G). For equilibrium conditions, the two forces must balance,

$$\text{therefore } \frac{Emg}{G} = \frac{mv^2}{R}$$

$$E = \frac{Gv^2}{Rg}$$

Trigonometrically, $\tan \theta = E/G$

$$= \frac{Gv^2/Rg}{G}$$

$$= \frac{v^2}{Rg}$$

E X A M P L E

11 Calculate the required superelevation on a railway line where radius $R = 500$ m, the rail gauge is 1.52 m and the average vehicle speed is 72 kph.

Answer

$$72 \text{ kph} = (72 \times 1000) / (60 \times 60) \text{ m/sec}$$
$$= 20 \text{ m/sec}$$
$$E = Gv^2/Rg$$
$$= \frac{(1.52 \times 400)}{(500 \times 9.815)} \times 1000 \text{ mm}$$
$$= \underline{124 \text{ mm}}$$

(d) Maximum allowable superelevation

Example 11 shows the theoretical amount of super-elevation for the given design criteria. Dept of Transport rules, however, state that only 45 per cent of the theoretical amount should be applied and the steepest angle of cant should not exceed 4 degrees approximately.

The maximum amount of allowable cant in Example 11 should therefore be $(45\% \times 124)$ mm = 56 mm.

(e) Length of transition curve

Section (b) explained that the ideal transition curve is one where the term rl (radius × length at any point) remains constant throughout the length of the transition or, in other words, as the length increases the radius decreases in inverse proportion. Such a description applies exactly to a curve called the clothoid. The calculations of the clothoid are complex, however, involving infinite series, so engineers tend to favour the cubic parabola which is very similar to the clothoid but with the advantage that the calculations are much simpler. Having said that, roadway software design packages will always favour the clothoid since all calculations are simple on a computer.

The length of a transition curve has been determined in a variety of ways over the years, namely:

(i)   An arbitrary length was chosen.
(ii) The total length of the curve was simply divided into three equal parts, i.e. 1/3rd to each transition and 1/3rd to the circular portion.
(iii) An arbitrary gradient was used to 'run in' the cant, for example at a 1/300 gradient, supereleva-

---

tion of 100 mm requires a transition of 30 metres long.

In 1908, Mr WH Shortt, an eminent engineer of the L & SW Railway Co., realized that a very important factor which must always be taken into consideration is passenger comfort. He advanced the theory that a rate of gain of radial acceleration (c) of 0.3m/sec^2, in a second, would pass unnoticed by passengers and they would not be inconvenienced in any way. This is the accepted method of calculating the length of transition curves.

Rate of gain of radial acceleration (c)   Recalling from Sec. 8 (c) that the inward force due to cant is $F = Wv^2/Rg$, the 'centrifugal ratio' is given by F/W:

$$F/W = Wv^2/Rg$$
$$= v^2/Rg$$

For any constant speed v, the value of this centrifugal ratio is proportional to v^2/Rg. This term is known as the 'radial acceleration', which increases from zero at the beginning of the transition to v^2/Rg at the junction with the circular curve.

The time (t) taken over the complete transition L is $t = L/v$; therefore the rate of gain of radial acceleration (a) is:

$$a = \frac{\text{radial acceleration (max)}}{\text{time taken}}$$
$$= (v^2/R) / (L/v)$$
$$= (v^3/RL) \text{ m/sec}$$

Thus, the length of the transition curve L is

$$L = (v^3/aR) \text{ m}$$

E X A M P L E

12 Calculate the length of transition curve required for a rate of gain of radial acceleration of (0.305 m/sec^3) on a 500 m radius curve, if vehicles travel at an average speed of 72 kph, i.e. 20 m/sec.

Answer

$$L = v^3/aR$$
$$= 20^5/(0.305 \times 500)$$
$$= \underline{52.45 \text{ m}}$$

Exercise 11.13

1 A section of roadway being designed to accommodate vehicles travelling at an 85 percentile speed of 80 kph has a radius of 800 m. Given that the roadway is 6.00 m wide, calculate the Department of Transport's maximum allowable superelevation. (*Note*: 85 percentile speed is the speed not normally exceeded by 85 per cent of road vehicles.)

2 Calculate the length of transition curve required for a rate of gain of radial acceleration of (0.3 m/sec³) on a 350 m radius curve if vehicles travel at an average speed of 80 kpm.

(f) Curve composition

Throughout this section the emphasis has been and will be placed on composite curves, i.e. those curves composed of one central circular curve with equal length transition curves on entry to and exit from the circular curve.

It is possible to design a completely transitional curve comprising two transitions of equal length, the argument being that it is slightly safer than a composite curve. The Dept of Transport places limitations on this type of curve, however, and indeed it is sometimes not possible to place a wholly transitional curve between two straights. It is really a matter for the curve design team and is not required by a construction surveyor.

(g) The cubic parabola and 'shift'

It was pointed out in Sec. 8(e) that the ideal transition curve is the clothoid but because the calculations are somewhat complex, many engineers prefer the cubic parabola transition. The two curves are practically identical when the deviation angle is less than 12 degrees (Fig. 11.28). The cubic parabola cannot be used when the deviation angle exceeds that value.

The equation of the cubic parabola (derived from the formulae of the clothoid) is

$y = x^3/6RL$

where R = radius of the circular curve
L = length of the transition curve
x = distance along the straight from the start of the curve (Fig. 11.28)
y = offset to the curve at distance x.

In Fig. 11.29, two straights AI and IB are to be joined by a circular arc AB of radius R with cubic parabola transition curves at each end. In order to accommodate the transitions which start at radius infinity and end at radius R, it is necessary either to (a) move the circular curve inwards, or (b) move the straights outwards.

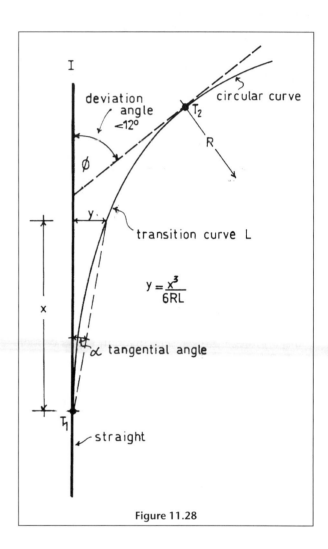

Figure 11.28

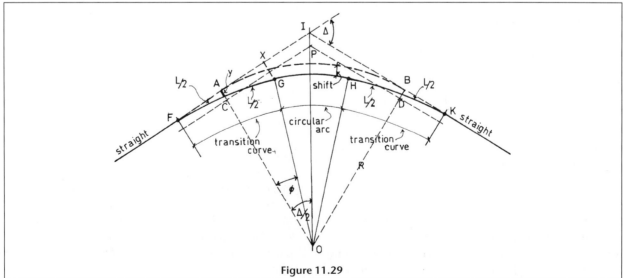

Figure 11.29

Since the latter course of action would alter the course of the road or railway, method (a) is always adopted. The amount by which the circular curve is moved inwards is known as the 'shift.' In Fig. 11.29, the curve AB is moved inwards to new position CD in order to accommodate the transitions FG and HK. The shift is therefore AC and since the radius of the curve CD = R, then the radius of curve AB = (R + shift).

The lengths of transitions FG and HK are both equal to L. For the small deflections to which the curves usually apply, curve length FG is very nearly equal to straight length FX and is indeed assumed to be so. The curve FG of length L is placed so that half its length is on each side of the original tangent point A of the circular curve. The new tangent point F at the start of the transition curve is therefore located at distance FA = (L/2) from A. Similarly, tangent point K lies at distance BK = (L/2) from B. Note that curve FG bisects shift AC.

Calculation of shift Let y = the offset to the curve at A which lies at distance L/2 from F, that is, $y = 1/2$ shift.
  Applying the formula $y = (x^3/6RL)$,

$$y = \frac{(L/2)^3}{6RL} = \frac{L^3/8}{6RL} = \frac{L^2}{48R}$$
$$\text{Hence shift AC} = 2y = \frac{L^2}{24R}$$

EXAMPLE

13 Two straights are to be joined by a circular curve of 600 m radius with cubic parabola transitions at both ends. The transitions are 75 m long.

(i) Calculate the lengths of the offsets required to set out the curve at 25 m intervals.
(ii) Calculate the shift, i.e. the amount by which the curve has to be moved to accommodate the transitions.

Answer

(i) $y = \dfrac{x^3}{6RL}$

Since (1/6RL is common for all calculations)
1/6RL = 1/(6 x 600 x 75) = 1/270 000

x	x^3	$x^3/270\,000$
25	15 626	0.058
50	125 000	0.463
75	421 875	1.563

(ii) Shift = $L^2/24R$
= $75^2/(24 \times 600)$
= 0.391 m

Setting out the transition (Fieldwork)
1. A theodolite is set up at tangent point T_1 and aligned to intersection point I (Fig. 11.28).
2. A steel tape and constant tension tension handle are used to measure the chainages (x distances) along the line T_1I. The tape must be accurately lined in along the straight using the theodolite. Any slope and temperature corrections should be made at this point, though it is unlikely that they will be necessary.
3. The y offsets are then set out at right angles to the straight from the measured chainage points.

(h) Tangent lengths

In Fig. 11.29, the deviation angle of the circular curve is Δ. The tangent points at the beginning and end of the transition occur at F and K, hence the tangent lengths are IF and IK.

AC = shift 'S'
AO = (Radius + shift) = (R + S)

In triangle AIO,

IA = AO tan $(\Delta/2)$
   = $(R + S)$ tan $(\Delta/2)$

Also AF   = 1/2 transition length = $L/2$
and IF   = (IA + AF)
therefore IF = $(R + S)$ tan $(\Delta/2) + L/2$

(i) Tangential angles and deviation angle

In triangle QFP of Fig. 11.30, point P is any point (offset y) on the transition curve at distance x from the tangent point F. The small angle QFP (= a) which is the angle formed by the straight and the chord to point P is called the 'tangential' angle.

(i) In Triangle QFP,

tan $a = y/x$
From Sec. (d), $y = x^3/6RL$
therefore, tan $x = x^3/6RLx$
       $= x^2/6RL$

At the junction point G, where $x = L$, the tangential angle will equal $L^2/6RL$, which is $L/6R$.
  At this point G, if a tangent is drawn, it will strike straight FI at point R, thus forming an angle IRG = ϕ with the straight. This angle is called the 'deviation' angle. The angle is really the gradient of the tangent to the cubic parabola $y = x^3/6RL$, or in the calculus parlance, is dy/dx.

(ii) So, when $y = x^3/6RL$

$\phi = \tan \phi = dy/dx = 3x^2/6RL$
       $= x^2/2RL$

From (i) above the *tangential* angle to point G is $x^2/6RL$ and from (ii) above the *deviation* angle is $x^2/2RL$,

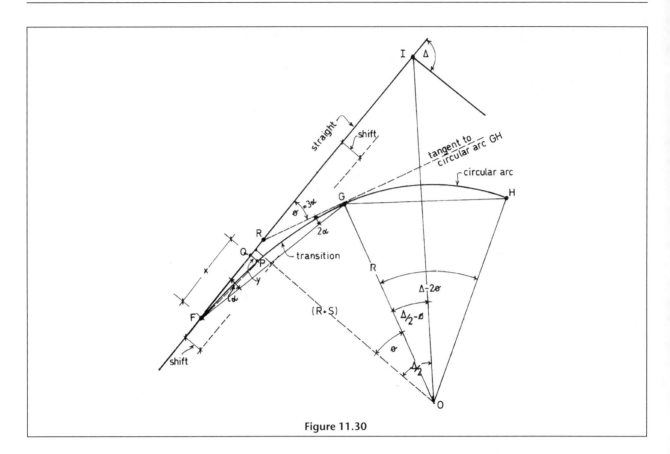

Figure 11.30

therefore the tangential angle to any point on the transition is 1/3rd the size of the deviation angle to that point.

Note: Since both tangential and deviation angles are small in transition curve calculations, the tangent of any angle equals (without error) the angle in radians.

Summarizing, therefore

$$\tan a = x^2/6RL$$
$$\text{therefore } a = x^2/6RL \text{ radians}$$
$$= (x^2/6RL) \times (180/\pi) \text{ degs}$$
$$= [(x^2/6RL) \times (180/\pi) \times 60] \text{ mins}$$
$$= (x^2/6RL) \times 3438 \text{ mins}$$
$$= 573 \, x^2/RL \text{ mins}$$

This is usually considered to be a shortcut in calculations but with a good hand calculator is hardly necessary.

(i) Length of circular curve

Consider Figs 11.30 and 11.31.
Angle GOI = $(\Delta/2 - \phi)$ degrees.
The angle GOH at the centre, subtended by the circular arc GH, equals $(2 \times$ angle GOI$)$.
Therefore angle GOH $= 2(\Delta/2 - \phi)$deg
$$= (\Delta - 2\phi)\text{deg}$$
Length of arc GH $= \dfrac{(\Delta-2\phi)}{360} \times 2\pi R$
$$\text{or} = R \times (\Delta - 2\phi) \text{ radians}$$

EXAMPLE

14 Two roadway straights meet at point I (chn 50 m) where they deviate to the right by 30 degrees. They are to be joined by a circular curve of 100 m radius, with 25 m transitions at either end.

(i) Compute all of the data required to set out the first transition curve and the circular curve by tangential angles with chords at 5 m intervals of through chainage.

(ii) Describe how the first chord of the circular curve is set out from the common tangent.
(HNC Topographic Studies)

Answer

(i) See Fig. 11.31.

1 Deviation angle $\Delta = 30° \, 00' \, 00''$

Curve radius $R = 100.000$ m

Chainage $I = 50.000$ m

Transitions $L = 25.000$ m

2 Shift $= L^2/24R = 25^2/(24 \times 100)$

$$= 0.260 \text{ m}$$

3 Tan length $= (R + S) \tan \Delta/2 + L/2$

$$= (100.260 \tan 15°) + 12.500$$

$$= 39.365 \text{ m}$$

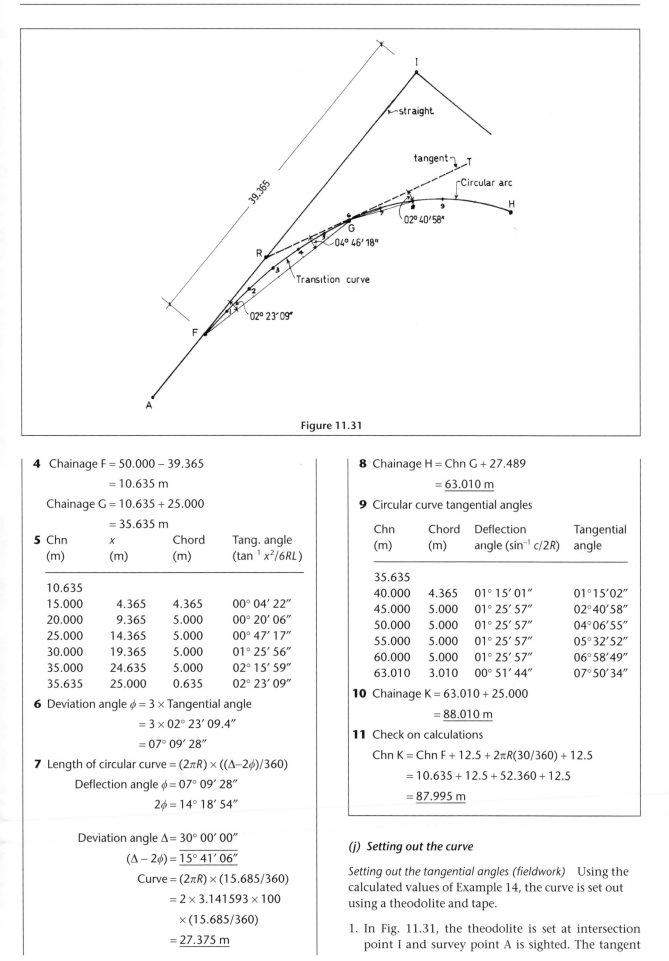

Figure 11.31

4 Chainage F = 50.000 − 39.365

$\qquad$ = 10.635 m

Chainage G = 10.635 + 25.000

$\qquad$ = 35.635 m

5

Chn (m)	x (m)	Chord (m)	Tang. angle ($\tan^{-1} x^2/6RL$)
10.635			
15.000	4.365	4.365	00° 04′ 22″
20.000	9.365	5.000	00° 20′ 06″
25.000	14.365	5.000	00° 47′ 17″
30.000	19.365	5.000	01° 25′ 56″
35.000	24.635	5.000	02° 15′ 59″
35.635	25.000	0.635	02° 23′ 09″

6 Deviation angle ϕ = 3 × Tangential angle

$\qquad$ = 3 × 02° 23′ 09.4″

$\qquad$ = 07° 09′ 28″

7 Length of circular curve = $(2\pi R) \times ((\Delta - 2\phi)/360)$

$\qquad$ Deflection angle ϕ = 07° 09′ 28″

$\qquad\qquad$ 2ϕ = 14° 18′ 54″

$\qquad$ Deviation angle Δ = 30° 00′ 00″

$\qquad\qquad$ $(\Delta - 2\phi)$ = $\overline{15° 41′ 06″}$

$\qquad$ Curve = $(2\pi R) \times (15.685/360)$

$\qquad\qquad$ = 2 × 3.141593 × 100

$\qquad\qquad$ × (15.685/360)

$\qquad\qquad$ = 27.375 m

8 Chainage H = Chn G + 27.489

$\qquad$ = 63.010 m

9 Circular curve tangential angles

Chn (m)	Chord (m)	Deflection angle ($\sin^{-1} c/2R$)	Tangential angle
35.635			
40.000	4.365	01° 15′ 01″	01°15′02″
45.000	5.000	01° 25′ 57″	02°40′58″
50.000	5.000	01° 25′ 57″	04°06′55″
55.000	5.000	01° 25′ 57″	05°32′52″
60.000	5.000	01° 25′ 57″	06°58′49″
63.010	3.010	00° 51′ 44″	07°50′34″

10 Chainage K = 63.010 + 25.000

$\qquad$ = 88.010 m

11 Check on calculations

$\qquad$ Chn K = Chn F + 12.5 + $2\pi R(30/360)$ + 12.5

$\qquad\qquad$ = 10.635 + 12.5 + 52.360 + 12.5

$\qquad\qquad$ = 87.995 m

(j) Setting out the curve

Setting out the tangential angles (fieldwork)   Using the calculated values of Example 14, the curve is set out using a theodolite and tape.

1. In Fig. 11.31, the theodolite is set at intersection point I and survey point A is sighted. The tangent

length 39.365 m is measured back from point I and a peg lined in to locate tangent point F.

2. The theodolite is relocated at F and point I sighted on zero degrees.

3. The circle is set to read 00° 04′ 22″ (tangential angle) and a peg is lined in on this bearing at chord length 4.365 m. This locates peg no. 1 (chainage point 15 m).

4. The circle reading is changed to read 00 20′ 06″; the tape zero is held on peg 1 and a second peg is lined in on this bearing at chord length 5.00 m. This locates peg no. 2 (chainage 20 m).

5. The procedure is repeated for the various tangential angles and chord lengths to locate pegs nos 3 to G (chainage 35.635 m). The tangential angle a to peg G (the common tangent point) is 02° 23′ 09″.

6. In order to set out the circular arc from point G, the direction of the common tangent GT has to be located as follows (Fig. 11.31):

7. The theodolite is set at the common tangent point (peg G) and the point F is sighted on circle reading (180° – 2a). This equates to (180° – 2 × 02° 23′ 09″) which equals 175° 13′ 42″. When the theodolite reads 180°, the telescope will point along the common tangent towards R and when set to 360°, it will point towards T, from which direction the circular curve is set out.

8. At peg G, the circle is set to read 01° 15′ 01″; the tape zero is held on peg G and a peg is aligned at distance 4.365 m. This locates peg no. 7 at chainage 40 m).

9. Peg no. 8 is set out at 5.000 m from peg no. 7 on reading 02° 40′ 58″ and the procedure is repeated for the various tangential angles and chord lengths to locate the remainder of the curve.

Setting out by coordinates Setting out curves by coordinates using total stations is standard practice on construction sites. The coordinates of each point on the curve have to be determined and set out from known survey points or free stations. A complete example of the setting-out method is given earlier in this chapter on page 222.

Exercise 11.14

1 Two railway straights with gauge 1.435 m intersect at an angle of 135°. They are to be connected by a curve of 370 m radius with cubic parabola transitions at either end.

The curve is to be designed for a maximum speed of 60 kph with a rate of gain of radial acceleration of 0.305 m/sec^3.

Calculate

(a) the required length of transition
(b) the maximum cant of the outer rail.
(c) the amount of shift required for the transition
(d) the lengths of the tangent points from the intersection of the straights.

(Royal Institution of Chartered Surveyors)

2 Two straight portions of a railway line intersecting at an angle of 155° are to be connected by two cubic parabola transition curves each 250 m long and a circular arc of 1000 m radius.

Calculate the necessary data for setting out the curve using chords of 50 m.

(Royal Institution of Chartered Surveyors)

3 Two straights having a deviation angle of 14°02′40″ are to be connected by a circular curve of 600 m radius with spiral transitions at each end.

Calculate

(a) the superelevation for equilibrium on the circular arc if the design speed is 70 kph, $g = 9.82$ m/sec^2 and gauge is 1.52 m.
(b) the lengths of the transitions and circular arc if cant is introduced at 1 in 600.
(c) all data required for setting out one of the spirals by means of deflection angles and 15 m chords.

(Royal Institution of Chartered Surveyors)

9. Answers

Exercise 11.1

1 (a) Figure 11.32

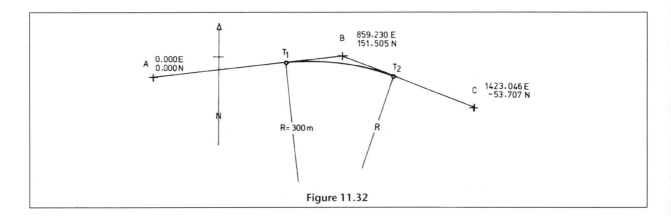

Figure 11.32

Point	Easting	Northing
A	0.000	0.000
B	859.230	151.505

	$\Delta E_{AB} = 859.230$	$\Delta N_{AB} = 151.505$

B	859.230	151.50
C	1423.046	−53.707

	$\Delta E_{BC} = 563.816$	$\Delta N_{BC} = -205.212$

$$\text{Length } AB = \sqrt{859.230^2 + 151.505^2}$$
$$= 872.485 \text{ m}$$
$$\text{Bearing } AB = \tan^{-1}(859.230/151.505)$$
$$= 80° \ 00' \ 00''$$

$$\text{Length } BC = \sqrt{563.816^2 + 205.212^2}$$
$$= 600.000 \text{ m}$$
$$\text{Bearing } BC = \tan^{-1}(563.816/{-205.212})$$
$$= 110° \ 00' \ 00''$$

(b) Deviation angle θ = bearing BC − bearing AB
$$= 110° \ 00' \ 00'' - 80° \ 00' \ 00''$$
$$= 30° \ 00' \ 00''$$

(c) Tangent length BA = BC = $R \tan \theta$
$$= 300.000 \tan(30° \ 00' \ 00''/2)$$
$$= 80.385 \text{ m}$$

(d) Curve length = $2\pi R \times 30/360$
$$= 2\pi 300/12$$
$$= 157.080 \text{ m}$$

(e) Length AT_1 = AB − BT_1 =

$$\begin{array}{r} 872.485 \\ - 80.385 \\ \hline 792.100 \text{ m} \end{array}$$

Curve length $T_1T_2 = + 157.080$

Therefore chainage $T_2 = \underline{949.180 \text{ m}}$

Exercise 11.2

1 $T_1T_2 = 2R \sin\theta/2$
$$= 2R \sin (47° \ 09' \ 20'')/2$$
$$= 100 \sin (23° \ 34' \ 40'')$$
$$= 40.000 \text{ m}$$

Major offset CV = $R (1 - \cos \theta/2)$
$$= 50 (1 - \cos 23° \ 34' \ 40'')$$
$$= 50 (1 - 0.916\ 518)$$
$$= 50 \times 0.083\ 482$$
$$= 4.174 \text{ m}$$

Shortest distance VY = $R (\sec \theta/2 - 1)$
$$= 50 \times (1.090\ 994 - 1)$$
$$= 4.550 \text{ m}$$

Tangent length $YT_1 = YT_2 = R \tan \theta/2$
$$= 50 \tan 23° \ 34' \ 40''$$
$$= 21.810 \text{ m}$$

Curve length $= 2\pi R \times (47° \ 09' \ 20''/360°)$
$$= 100\pi \times 0.130\ 988$$
$$= 41.151 \text{ m}$$

Exercise 11.3

1 Make a sketch from the given information (Fig. 11.33)

(a) Angle XPQ = 180° − QPA
$$= 28° \ 40'$$

Angle PQX = (180° − BQP)
$$= 36° \ 50'$$

Angle QXP = 180° − (28° 40' + 36° 50')
$$= 180° - 65° \ 30'$$
$$= 114° \ 30'$$

In triangle XQP,
$$XQ = \frac{\sin 28° \ 40' \times 55.0}{\sin 114° \ 30'}$$
$$= 28.995 \text{ m}$$

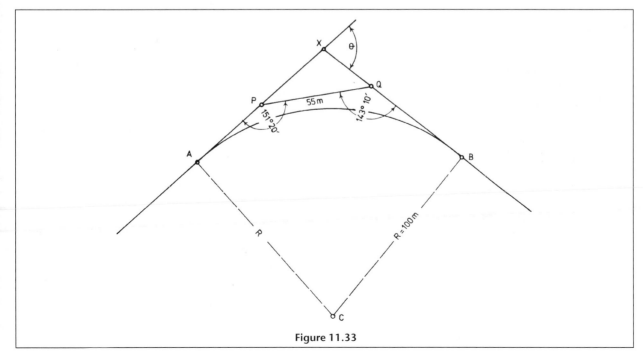

Figure 11.33

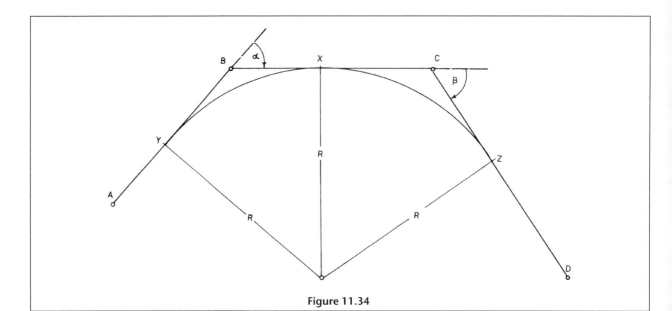

Figure 11.34

and XP $= \dfrac{\sin 36° \, 50' \times 55.0}{\sin 114° \, 30'}$

$= 36.324$ m

(b) Deviation angle $\theta = 180° - 114° \, 30'$

$= 65° \, 30'$

Therefore $\theta/2 = 32° \, 45'$

Tangent lengths XA and XB $= R \tan \theta/2$

$= 100 \tan 32° \, 45'$

$= 64.322$ m

(c)   AP $=$ AX $-$ PX

$= 64.322 - 36.324$

$= 27.998$ m

QB $=$ BX $-$ QX

$= 64.322 - 28.995$

$= 35.327$ m

(d) Curve $= 2\pi \times 100 \times (65.5/360)$

$= 114.319$ m

Exercise 11.4

1 (a) Figure 11.34

Bearing line AB $= 37° \, 12'$

Bearing line BC $= \underline{91° \, 02'}$

Therefore angle α $= 53° \, 50'$

and $\alpha/2$ $= 26° \, 55'$

Bearing line BC $= 91° \, 02'$

Bearing line CD $= \underline{147° \, 14'}$

Therefore angle β $= 56° \, 12'$

and $\beta/2$ $= 28° \, 06'$

BY $=$ BX $= R \tan \alpha/2 = R \tan 26° \, 55'$

XC $= R \tan \beta/2 = R \tan 28° \, 06'$

BC $=$ (BX $+$ XC)

$= R \tan 26° \, 55' + R \tan 28° \, 06'$

i.e. $312.7 = R \, (\tan 26° \, 55' + \tan 28° \, 06')$

Therefore $R = \dfrac{312.700}{\tan 26° \, 55' + \tan 28° \, 06'}$

$= \dfrac{312.700}{0.507 \, 694 \, 8 + 0.533 \, 950 \, 3}$

$= 300.20$ m

(b) BY $=$ BX $= R \tan 26° \, 55' = 152.41$ m

The theodolite is set over peg B and made ready for sighting. A backsight is taken to station A and the instrument is locked along the line. The tape is laid out along this line and is kept in line by constant sights through the theodolite. When the required distance of 152.41 m is reached a peg is hammered securely into the ground and a nail is lined on to the peg at the required distance BY.

Exercise 11.5

1 Figure 11.35

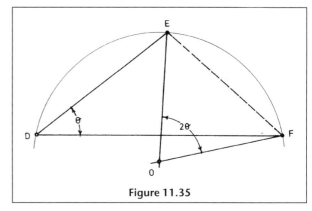

Figure 11.35

	Point	East (m)	North (m)
Existing coordinates	A	10.0	10.0
	B	15.0	12.5
	C	19.0	8.0
Proposed coordinates	D	9.0	10.0
	E	15.0	13.5
	F	20.0	8.0

Tan bearing DE = Δ easting/Δ northing
$$= (15 - 9)/(13.5 - 10)$$
$$= 6/3.5$$
Therefore bearing DE = 59° 44′ 37″

Tan bearing DF = Δ easting/Δ northing
$$= (20 - 9)/(8 - 10)$$
$$= 11/-2$$
Therefore bearing DF = 100° 18′ 17″

Distance EF $= \sqrt{\Delta \text{ east}^2 + \Delta \text{ north}^2}$

$$= \sqrt{(15 - 20)^2 + (13.5 - 8)^2}$$

$$= \sqrt{5^2 + 5.5^2}$$

$$= 7.43 \text{ m}$$

Angle EDF(θ) = 100° 18′ 17″ − 59° 44′ 37″
$$= 40° 33′ 40″$$

In ΔDEF,

$$\frac{EF}{\sin EDF} = 2R$$

Therefore $R = 7.43/(2 \sin 40° 33′ 40″)$
$$= 5.71 \text{ m}$$

Angle EOF = 2 × angle EDF
$$= 2\theta$$
$$= 81° 07′ 20″$$

Angles (FEO + OFE) = 180° − 81° 07′ 20″
$$= 98° 52′ 40″$$
Therefore angle FEO = 49° 26′ 20″

Tan bearing EF = Δ easting/Δ northing
$$= (20 - 15)/(8 - 13.5)$$
$$= 5/-5.5$$
Therefore bearing EF = 137° 43′ 34″

Bearing EO = bearing EF + angle FEO
$$= 137° 43′ 34″ + 49° 26′ 20″$$
$$= 187° 09′ 54″$$

Easting O = easting E
+ EO sin 187° 09′ 54″
$$= 15.0 + 5.71 \sin 187° 09′ 54″$$
$$= 15.0 - 0.71$$
$$= 14.29 \text{ m}$$

Northing O = northing E
+ EO cos 187° 09′ 54″
$$= 13.5$$
+ 5.71 cos 187° 09′ 54″
$$= 13.5 - 5.67$$
$$= 7.83 \text{ m}$$

Exercise 11.6

1 Tan length IT$_1$ = IT$_2$ = R tan θ/2
$$= 30 \tan 15°$$
$$= 8.038 \text{ m}$$

In ΔIET$_2$,

IE = IT$_2$ cos θ
$$= 8.038 \cos 30°$$
$$= 6.962 \text{ m}$$

T$_1$E = IT$_1$ + IE
$$= 8.038 + 6.962$$
$$= 15.000 \text{ m}$$

Offset at 2 m $= 30 - \sqrt{30^2 - 2^2}$ = 0.067 m

4 m $= 30 - \sqrt{30^2 - 4^2}$ = 0.268 m

6 m $= 30 - \sqrt{30^2 - 6^2}$ = 0.606 m

8 m $= 30 - \sqrt{30^2 - 8^2}$ = 1.086 m

10 m $= 30 - \sqrt{30^2 - 10^2}$ = 1.716 m

12 m $= 30 - \sqrt{30^2 - 12^2}$ = 2.504 m

14 m $= 30 - \sqrt{30^2 - 14^2}$ = 3.467 m

15 m $= 30 - \sqrt{30^2 - 15^2}$ = 4.019 m

Exercise 11.7

1 Make a sketch from the given information (Fig. 11.36)

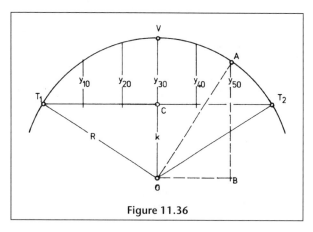

Figure 11.36

In ΔOTC,
OT$_1$ = R = 100.0 m
CT$_1$ = x = 30.0 m
By Pythogoras' theorem
$$OC = k = \sqrt{R^2 - x^2}$$

$$= \sqrt{100^2 - 30^2}$$

$$= 95.394 \text{ m}$$

Major offset VC = 100 − 94.394
$$= 4.61 \text{ m}$$

In ΔABO,
$$OB - x_n = 20 \text{ m}$$
$$OA = R = 100 \text{ m}$$

$$AB = \sqrt{R^2 - x_n^2}$$

Offset y_{50} (and any other offset)

$$= \sqrt{R^2 - x_n^2} - k$$

$$= \sqrt{100^2 - 20^2} - 95.394$$

$$= 97.980 - 95.394$$

$$= 2.59 \text{ m}$$

Also y_{10} = 2.59 m

Offset $y_{20} = y_{40} = \sqrt{R^2 - x_n^2 - k}$

$$= \sqrt{100^2 - 10^2} - 95.394$$
$$= 99.499 - 95.394$$
$$= 4.11 \text{ m}$$

$y_0 = 0.00$ m
$y_{10} = 2.59$ m
$y_{20} = 4.11$ m
$y_{30} = 4.61$ m
$y_{40} = 4.11$ m
$y_{50} = 2.59$ m
$y_{60} = 0.00$ m

Exercise 11.8

1 Table 11.9

Table 11.9

Point	Chainage	Tangential angle (left)	Theodolite reading (10″)
T$_1$	790.20		
1	800.00	00° 33′ 41″	359° 26′ 20″
2	820.00	01° 42′ 27″	358° 17′ 30″
3	840.00	02° 51′ 13″	357° 08′ 50″
4	860.00	03° 59′ 59″	356° 00′ 00″
5	880.00	05° 08′ 45″	354° 51′ 10″
6(T$_2$)	894.92	06° 00′ 03″	354° 00′ 00″

Exercise 11.9

1 (a) Table 11.10
   (b) Table 11.11 (page 243)
   (c) Table 11.12 (page 243)

Exercise 11.10

1 Table 11.13 (page 243)

Exercise 11.11

1 Chainage $x = \dfrac{-b}{2a}$

$$= \frac{+2.0 \times 10^{-2}}{2 \times 0.000\,404\,76}$$
$$= 24.710$$

Level $y = ax^2 + bx + c$
$$= 0.000\,404\,76 \times 24.710^2 + ((-2/100)$$
$$\times 24.710) + 26.600$$
$$= 0.247 + (-0.494) + 26.600$$
$$= 26.353 \text{ m}$$

Exercise 11.12

1 (a) Offset at 5m intervals = 1.65 m, 2.61 m, 2.92 m, 2.61 m, 1.65 m

2 (a) 08° 30′ (b) 29.73 m (c) 59.34 m
   (d) 167.77 m, 227.11 m

Table 11.10

1. Point P–point of zero chainage
   Coordinates: 00.000 m east
        00.000 m north
2. Radius of curve = 429.718 m
3. Deviation angle = 7° 40′ 00″
4. Chainage of intersection point = 78.793 m
5. Tangent length = 429.718 tan 3° 50′ 00″ = 28.793 m
6. Chainage T$_1$ = 78.793 − 28.793 = 50.000 m
7. Curve length = $2\pi R \times (7° 40'/360°)$ = 57.500 m
8. Chainage T$_2$ = 50.000 + 57.500 = 107.500 m
9. Peg chainages
   40.000 m on left-hand straight
   50.000 m tangent point T$_1$
   60.000 m on curve
   80.000 m on curve
   100.000 m on curve
   107.500 m tangent point T$_2$
   120.000 m on right-hand straight
   140.000 m on right hand straight
10. Curve composition
   Initial sub-chord = 60.0 − 50.0 m = 10.00 m
   Two standard chords = (20×2) m = 40.00 m
   Final sub-chord = 107.500 − 100.000 = 7.500 m
11. Deflection angles
   Initial sub-chord = $\sin^{-1}10/2R$ = 00° 40′ 00″
   Standard chord = $\sin^{-1}20/2R$ = 01° 20′ 00″
   Final sub-chord = $\sin^{-1}7.500/2R$ = 00° 30′ 00″
12. Centre angles (= 2 × deflection angles)
   T$_1$OA = 01° 20′ 00″
   AOB = 02° 40′ 00″
   BOC = 02° 40′ 00″
   COT$_2$ = <u>01° 00′ 00″</u>
   Total = <u>07° 40′ 00″</u>

(e)

Chord number	Length	Chainage	Tangential angle
T$_1$		167.77	
1	12.23	180.00	00° 52′ 33″
2	20.00	200.00	02° 18′ 30″
3	20.00	220.00	03° 44′ 27″
4(T$_2$)	7.11	227.11	04° 15′ 00″

3 (a) 162.74
   (b) 527.62, 830.00
   (c) Angles,   1 at 1° 04′ 30″ initial sub-chord
        14 at 1° 44′ 10″
        1 at 0° 52′ 10″ final sub-chord

4 On AI tangent point is between A and I:
   Distance A to TP = 17.29 m

   On BI tangent point is outwith BI:
   Distance B to TP = −37.71 m

   Coordinates   TP1 118.29 mE,   91.15 mN
          TP2 197.94 mE,   109.73 mN

5 (a)   Reduced level left-hand TP = 64.000 m AOD
       Reduced level right-hand TP = 64.500 m AOD

Table 11.11

Line	Length	Whole circle bearing	Δ east	Δ north	Easting	Northing	Station
					0.000	0.000	P
P/40	40.00	5° 0′ 0″	3.486	39.848	3.486	39.848	40
40/T1	10.00	5° 0′ 0″	0.872	9.962	4.358	49.810	T1 (50)
T1/Q	28.79	5° 0′ 0″	2.509	28.683	6.867	78.493	Q
Q/T2	28.79	357° 20′ 0″	−1.340	28.762	5.528	107.255	T2 (107.50)
T2/120	12.50	357° 20′ 0″	−0.582	12.486	4.946	119.741	120
T20/140	20.00	357° 20′ 0″	−0.931	19.978	4.015	139.719	140 (R)
					0.000	0.000	P
P/T1	50.00	5° 0′ 0″	4.358	49.810	4.358	49.810	T1
T1/O	429.72	275° 0′ 0″	−428.083	37.452	−423.725	87.262	O
O/T2	429.72	87° 20′ 0″	429.253	19.993	5.528	107.255	T2
					−423.724	87.262	O
O/60	429.718	93° 40′ 0″	428.838	−27.481	5.114	59.781	60
					−423.724	87.262	O
O/80	429.718	91° 0′ 0″	429.653	−7.500	5.929	79.762	80
					−423.724	87.262	O
O/100	429.718	88° 20′ 0″	429.536	12.498	5.812	99.760	100
					−423.724	87.262	O
O/T2	429.718	87° 20′ 0″	429.253	19.993	5.529	107.255	T2

Table 11.12

Ref. Station	Easting ***	Northing ***
T1	4.358	49.810

Station	Easting	Northing	Difference in eastings	Difference in northings	Set-out distance	Set-out bearing
40	3.486	39.848	−0.872	−9.962	10.000	185° 00′ 00″
T1 (50)	4.358	49.810	0.000	0.000	0.000	– – –
60	5.114	59.781	0.756	9.971	10.000	4° 20′ 13″
80	5.929	79.762	1.571	29.952	29.993	3° 0′ 10°
100	5.812	99.760	1.454	49.950	49.971	1° 40′ 3″
T2 (107.50)	5.528	107.255	1.170	57.445	57.457	1° 10′ 1″
120	4.946	119.741	0.588	69.931	69.934	0° 28′ 55″
140	4.015	139.719	−0.343	89.909	89.910	359° 46′ 54″

Table 11.13

1. Left-hand grade-1 in (p)   −50.000   (p)% = −2.000
2. Right-hand grade 1 in (q)   35.000   (q)% = 2.857
3. Length of curve (l)   60.000   a = 0.000 404 76
4. Chainage initial tangent point   0.000
5. Reduced level initial tangent point   26.600

Chainage	Chainage interval	Tangent level	Grade correction	Curve level	Remarks
0.000		26.600		26.600	Start of curve
10.000	10.000	26.400	0.040	26.440	
20.000	10.000	26.200	0.162	26.362	
30.000	10.000	26.000	0.364	26.364	
40.000	10.000	25.800	0.648	26.448	
50.000	10.000	25.600	1.012	26.612	
60.000	10.000	25.400	1.457	26.857	End of curve

5 (b)

Chainage	Curve level	Cut
0	64	0
30	64.42	0.08
60	64.68	0.32
90	64.78	0.32
120	64.72	0.08
150	64.50	0

(c) Highest point, chainage 93.757, reduced level
    64.781 m AOD

6 Chainage 1513.3 m
   Reduced level 41.34 m AOD

Exercise 11.13

1 $E = 45\%(Gv^2/Rg)$
$= 0.45 \times (6 \times 22.222^2 / 800 \times 9.815)$
$= 170$ mm

2 $L = v^3/aR$
$= 22.222^3/(0.3 \times 350)$
$= 104.514$ m

Exercise 11.14

1 Speed $v = 60$ kph $= 16.67$ m/sec
   Length of transition
$= v^3/aR = 16.667^3/(0.305 \times 370)$
$= 41.0$ m

$$\text{Cant} = Gv^2/Rg = \frac{(1.435 \times 16.667^2)}{(370 \times 9.815)}$$

$= 110$ mm

Shift $= L^2/24R = 41^2/(24 \times 370)$
$= 0.190$ m

Deviation angle $= 180° - 135° = 45°$

Tangent length $= (R + S) \tan 22° 30' + 20.5$
$= 370.19 \times 0.4142 + 20.5$
$= 173.83$ m

2 Shift $= L^2/24R = 250^2/(24 \times 1000)$
$= 2.61$ m

Deviation angle $= 180° - 155° = 25°$
1/2 Dev angle $= 12° 30''$

Tan length $= (R + S) \tan \Delta/2 + L/2$
$= 1002.61 \tan 12° 30' + 125.00$
$= 347.27$ m

Tan angles $= \tan^{-1}(x^2/6RL) = 50^2/(6 \times 1000 \times 250)$
$= 00° 05' 44'', 00° 2 2' 50'', 00° 51' 34''$
$01° 31' 39'', 02° 23' 09''$

3 Speed $v = (70 \times 1000)/3600 = 19.44$ m/sec
Cant $= Gv^2/Rg = 97.5$ mm
Length of cant $= 0.0975 \times 600 = 58.5$ m
For transition
Deviation angle $\phi = \tan^{-1}(L/2R)$
$= \tan^{-1}(58.5/2 \times 600)$
$= 02° 47' 36''$

For circular arc
1/2 (Dev angle Δ) $= 1/2(14° 02' 40'')$
$= 07° 01' 20''$

$(\Delta/2 - \phi) = (07° 01' 20'' - 02° 47' 36'')$
$= 04° 13' 44''$

Length of circular arc $= 2\pi R \times (04° 13' 44''/360°)$
$= 88.57$ m

Tangential angles $= \tan^{-1}(x^2/6RL)$
At 15 m $= 03' 36''$
30 m $= 14' 42''$
45 m $= 33' 36''$
58.5 m $= 55' 52''$

Chapter summary:

In this chapter, the following are the most important points:

- In curve surveying, the circular curve elements of radius, length, arc, chord and tangent are inter-related and connected by various formulae requiring knowledge of the mathematics of circular geometry and the relationship of radian measure to sexagesimal degree measure:

  1 radian $= (180/\pi)$ degrees

- Three main types of curve are used in setting out construction works. Simple circular curves as used on roadways may be long on main roads or short at road junctions. Parabolic vertical curves are required where roads are constructed on gradients. They provide a smooth transition from one gradient to another. Horizontal transition curves are also parabolic and are required on railways and roads where a circular curve is to join to a straight length of track or carriageway. These curves again provide a smooth and safe passage where otherwise there would be a sudden change of direction.

- In all circular curve situations, two straights (e.g. roads, railways) meet at an intersection point I. The deviation angle θ between them is measured from a plan or in the field. A circular curve of radius R is used to make the deviation from one straight to the other. The curve begins and ends on tangent points T_1 and T_2, situated at equal distances from I. These lengths are called tangent lengths and are calculated from the formula:

  $IT_1 = IT_2 = R \tan \theta/2$

  The chainages (Chn) of tangent points T_1 and T_2 are:

  Chn T1 = Chn I − IT1
  Chn T2 = Chn T1 + length of curve

The length of the curve is $2\pi R \times (\theta/360)$

These are the basic calculations for all circular curves. However, various problems may complicate these simple calculations. Examples with solutions are detailed in pages 212 to 216.

- Very short curves occur at roadway junctions and are set out using a variety of methods, namely:

  1. by finding the centre of the circle and swinging a tape of known radius through the required arc.
  2. by measuring offsets from the tangent lengths, detailed in pages 217–218, or from the long chord on pages 218–219. Since the formulae are many and varied, they are not listed here but are used in the text in the aforesaid pages.

- Large radius curves are set out using two methods, namely:

  1. by tangential angles where the chainage points of the centre line of the curve are set out by a series of chords which progressively diverge from T1 until T2 is reached. This is the traditional method which is self checking, in that failure to finish on T2 indicates an error of calculation or setting out. The calculations are too numerous and complex to be abbreviated in summary form. The reader is referred to pages 220–222 for a full explanation and numerical examples.
  2. The centre line chainage points are set out independently from a remote control station by EDM methods. Again the calculation are numerous and complex so complete details are given in pages 222–225. This method has advantages in that it is speedy and the surveyor is removed from the area of road making operations. The disadvantage is that the stake-out points are unchecked and errors may pass unnoticed until later. The points must therefore be set out independently from a second control station.

- In roadway vertical curves, two gradients meet at an intersection point I. The reduced level of point I and the values of the two gradients are measured from a plan or in the field. A parabolic curve is used to smooth the change of gradient between the left and right hand gradients. The curve begins and finishes on tangent points T and T1, situated at equal distances from I and the curve is an equal tangent parabola of the form

$$y = (ax^2 + bx + c)$$

where y = the reduced level of any stakeout point on the curve at distance x from the start of the curve, b = the value of the left-hand gradient and a = the multiplying coefficient derived from the two gradient values.

- The length of the curve is dependent on the two gradients and the sight distance required by a vehicle when stopping or overtaking. The length is taken from Dept of Transport design standards publications.

- Since the calculations are of an advanced nature, the reader is referred to pages 227–230 for complete explanations and numerical examples.

- Transition curves are required on most roads and railways wherever a curve leaves a straight and thereby would make a sudden change of direction. Vehicles travelling along the straight would tend to continue on that path and passengers would feel discomfort due to the centrifugal, or radial, force acting on the vehicle. The transition curve smoothes the change between the straight and curve. The ideal curve is a clothoid but is usually calculated as a cubic parabola in construction circles.

- Transition curves are used where two straights meet at an intersection point I and deviate by an angle of Δ degrees by means of a circular curve of radius R. Transitions are placed at both positions where the curve joins the straights.

- The length L of the curve is calculated from the radius R of the circular curve, the speed v of the vehicles and the rate a at which radial acceleration is gained along the curve, therefore $L = v^3/aR$

- The transition curve, being inserted between a straight and a circular curve, has to be accommodated by moving the curve inwards by an amount known as the shift S and the curve is placed astride the initial tangent point such that half of its length is on each side of the tangent point. The shift is computed from the formula $S = L^2/24R$.

- The tangent lengths from I to the start and end of the transitions are equal and are calculated thus:

tan length $= (R + S) \tan \Delta/2 + L/2$

- The curve may be set out at intervals by offsets y from the tangent using the formula $y = x^3/6RL$ where x is the distance from the tangent point.

- The centre line of the curve may also be set out by tangential angles or by coordinates, in the manner of simple circular curves. The methods are complex and best understood by reference to the text at pages 235–238, where explanations and numerical examples are given.

CHAPTER 12 **Setting out construction works**

In this chapter you will learn about:

- the equipment used in the horizontal control of setting out construction works and the process of setting a stakeout peg on a specified bearing and distance
- the setting out procedure and associated calculations used in staking out a roadway centre line as a traverse and by coordinates
- the procedure and associated calculations used in staking out building lines and small buildings with the use of basic instruments
- the procedure for setting out large buildings using theodolites and precise linear measuring techniques
- the calculations to establish the positions of free stations for setting out by coordinate methods
- the use of laser and GPS instruments in setting out and in checking verticality
- the equipment used in the vertical control of setting out construction works and the process of setting a stakeout peg at a specified level
- the process and calculations used in setting out floor, drainage and roadway levels, using basic equipment and rotating laser instruments
- the use of laser and GPS equipment in large-scale earthworks

---

1. Setting out – principles

Setting out any construction work, e.g. a proposed roadway, is in reality the reverse of surveying an actual roadway. In the surveying process, angles and distances are measured in the field; coordinates are calculated from the data and a plan is drawn of the results.

Before setting out any construction works, the development plan is studied carefully and coordinates of selected points on the works are obtained by some means. They may be scaled from the plan, read from a CAD drawing or may be computed from design sizes. Bearings and distances to these points from known survey stations are calculated and those data used to set out the proposed works.

Figure 12.1 is a copy of the development plan of the proposed construction works at the site shown in Fig. 6.1 (page 100). The development shows two large houses, Nos 10 and 11 Lochview Road, and a private roadway R1–R4 serving them. A storm water drain S1–S4 and a foul sewer F1–F4 provide site drainage. The plan also shows the traverse stations A to F which were used to survey the site. The coordinates of those points are shown in Table 12.1.

On any construction site, it is general practice to

Table 12.1

Survey Point	Easting (m)	Northing (m)
A	100.000	100.000
B	65.621	134.379
C	48.497	113.013
D	87.930	84.898
E	105.348	71.216
F	116.336	85.572

construct firstly the roads and sewers in order to provide (a) access to the site and (b) main drainage to all buildings.

2. Equipment for setting out

All roads, buildings, sewers and other construction works are set out using standard surveying equipment, comprising:

1. *Steel tapes* Steel tapes must always be used for setting out purposes as they are not subject to the same degree of stretching as are Fibron tapes.

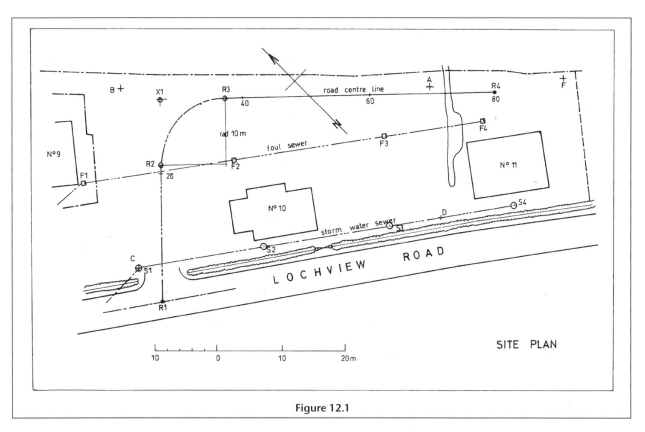

Figure 12.1

2. *Levels*   Automatic and digital levels are most commonly used. Optical levels still find favour but produce less reliable results.

3. *Theodolites*   A wide range of theodolites is available, but total stations and laser instruments are the preferred option because of their reliability and ease of use.

4. *GPS equipment*   A wide range of GPS equipment is in use and this form of setting out will probably supersede all other methods. It has disadvantages amongst high-rise buildings and sufficient satellite coverage may not be available when required.

5. *Laser instruments*   Line lasers, rotating lasers and laser plummets have revolutionized setting out procedures and greatly increase accuracy. They are, however, expensive to purchase and maintain.

6. *Pegs*   Pegs are either wooden 50 mm × 50 mm × 500 mm stakes for use in soft ground or 25 mm × 25 mm × 300 mm angle irons for hard standing.

Pegs should be colour coded with paint and the code should be used throughout the duration of the contract to avoid confusion. Any code will suffice but in general centre line pegs are white (Fig. 12.2), offset pegs are yellow and level pegs are blue.

In setting out roads and sewers, the centre line pegs (white) are first established. Naturally they are removed during excavation, so offset pegs (yellow) are positioned 3–5 m to the right and left of centre lines (Fig. 12.2).

7. *Profiles*   A profile is a wooden stake to which a cross-piece, painted in contrasting coloured stripes, is nailed (Fig. 12.2). For sewer work, goalpost-type profiles, called sight rails, are preferable. The profiles are erected over the offset pegs in order to remain clear of the excavations.

8. *Travellers*   Travellers are really mobile profile boards used in conjunction with sight rails. The length of the traveller equals the sight rail level – sewer invert level. The length should be kept to multiples of 0.25 metres and travellers are usually about 2 metres long.

9. *Corner profiles*   During the construction of buildings, the pegs denoting the corners of the buildings are always removed during the construction work. The corner positions have, therefore, to be removed some distance back from the excavations on to corner profiles (see Fig. 12.6). These are constructed from stout wooden stakes 50 mm × 50 mm, on to which wooden boards 250 mm × 25 mm × 1.00 m long are securely nailed.

3. Setting out a peg on a specified distance and bearing

(a) Setting out on level ground

In order to set out the roads, buildings and sewers shown on the development plan (Fig. 12.1), a total of some forty to fifty pegs need to be accurately placed on the ground in their proposed positions. Fortunately, every peg is set out in exactly the same manner. Physically setting a peg in the ground in its proposed position

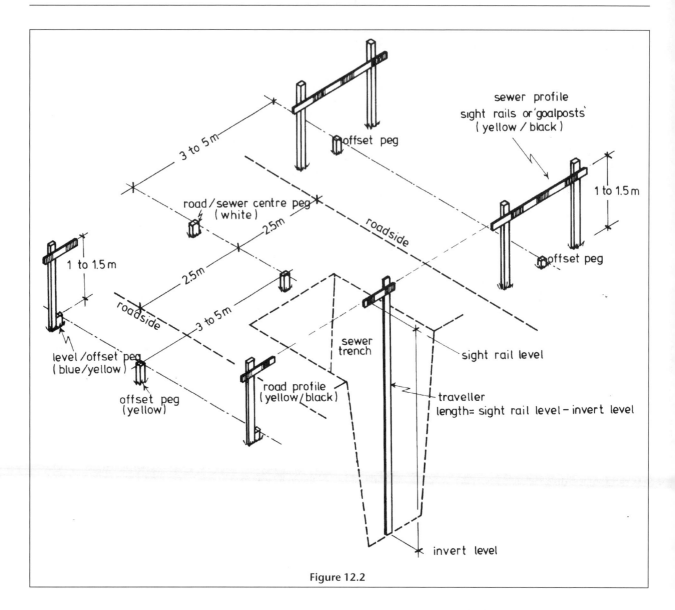

Figure 12.2

is not an easy task, therefore the work is described in detail in the next section.

Procedure

In Fig. 12.3 (a), a peg C is to be set out from a survey line AB at an angle of 65° 30' and a distance of 10.25 m. A surveyor and preferably two assistants are required to set out pegs.

1. The theodolite is set over station B and correctly levelled and centred. On face left, a backsight is taken to station A with the theodolite reading zero degrees (the method varies with the type of theodolite).
2. The horizontal circle is set to read 65° 30'; thus the theodolite is pointing along the line BC.
3. The end of the tape is held against the nail in peg B and laid out approximately along the line BC by the assistants.
4. The 10.25 m reading on the tape is held against the *side* of the peg C (Fig. 12.3 (b)); the tape is tightened

and slowly swung in an arc until the surveyor sees it clearly through the telescope of the theodolite.

5. The peg is carefully moved, on the observer's instructions, until the *bottom, front* edge of the peg is accurately bisected. The peg is then hammered home.
6. The tape is again held at peg B, by assistant 1, while assistant 2 tightens it and marks a pencil line across the peg C at distance 10.25 m.
7. A pencil is held vertically on this line by an assistant and is moved slowly along the line until the surveyor sees it bisected by the line of sight through the theodolite. The assistant marks this point on the peg.
8. The distance of 10.25 m is checked and the operation is repeated on face right. If all is well, the two positions of point C should coincide or differ by a very few millimetres. The mean is accepted and a nail hammered into the peg to denote point C.

It is not good practice to hook the end of the tape over the nail at peg B when setting out the distance, as

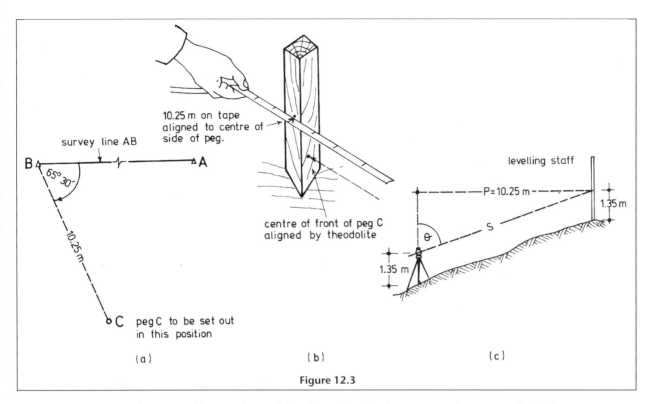

survey line AB

10.25 m on tape
aligned to centre of
side of peg.

B ⊿ 65° 30'

10.25 m

C   peg C to be set out
in this position

centre of front of peg C
aligned by theodolite

levelling staff

P = 10.25 m

1.35 m

S

θ

1.35 m

(a)   (b)   (c)

Figure 12.3

excessive tension on the tape will move the nail head or even move the peg.

(b) Setting out on sloping ground

In all setting out operations, the horizontal distance is required, but frequently, because of ground undulations, it will be necessary to set out the slope distance. The setting-out procedure is very similar to that of Sec. 3(a).

Procedure
1. The theodolite is set up at station B, backsighted to A reading zero and foresighted along the line BC reading 65° 30'. The instrument height is measured and noted. Let the height be 1.35 m.
2. The tape is stretched out in the proposed direction of line BC and a levelling staff held vertically at distance 10.25 m.
3. The instrument height (1.35 m) is read on the staff and the angle of inclination (vertical or zenith angle) is noted. Let the zenith angle be 84° 15'.
4. The distance to be set out on the slope will have to be increased to be the equivalent of 10.25 m of plan length, as follows (Fig. 12.3(c)):

Plan length P/slope length S
        = sin zenith angle
Slope length S = plan length P/sin zenith angle
        S = 10.250/sin 84° 15'
        = 10.302 m

5. The procedure in setting out peg C then follows exactly the procedure detailed in Sec. 3(a), using the new setting out length of 10.302 m.

4. Setting out roadways – calculations

(a) Preparation of setting out data

The first step in setting out a roadway is to establish the centre line. In order to achieve this, the coordinates of all points of change of direction must be obtained from the development plan.

Figure 12.4 is a skeletal layout of the roadway R1–R4. The coordinates of (a) starting point R1, (b) finishing point R4 and (c) intersection point X1 must be scaled as accurately as possible from the plan or obtained from a CAD version of the plan (on a plan of scale 1:200, scaling accuracy is $1/4$ mm × 200 = 50 mm).

The coordinate values are shown in Table 12.2.

Table 12.2

Point	Easting (m)	Northing (m)
R1	47.00	106.70
X1	69.00	128.70
R4	107.00	91.80

(b) Methods of setting out

In Chapter 11, Setting out curves, two methods of setting out large radius roadway and railway curves were fully explained. Setting out straight roadways (or railways) uses exactly the same two methods.

1. *Setting out the roadway as a traverse.* Points C (survey Stn), R1 (start of road), X1 (intersection point), R4 (end of road) and F (survey point) are treated as points of a traverse survey and are set out by angles and

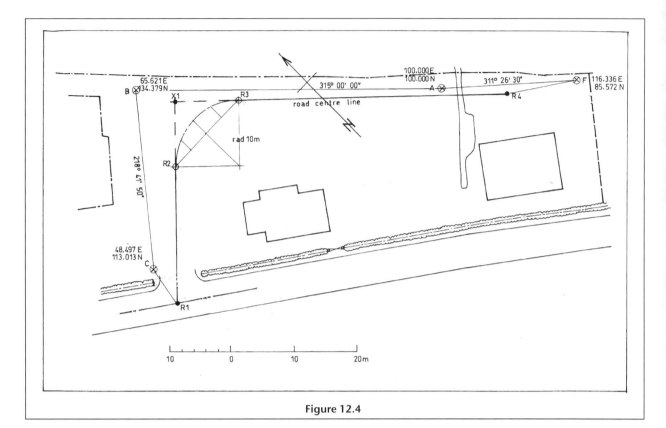

Figure 12.4

distances using a theodolite or total station. The chosen instrument is set up in turn on each of the points C, R1, X1 and R4 and the traverse is closed on point F.

This method takes time and assistants have to be skilled, particularly if distances are to be set out by tape. However, the method is self-checking in that the survey finishes on the known coordinates of point F and any errors are therefore detected.

2. *Setting out coordinates by radiation.* The coordinates of points R1, X1 and R4 are set out directly from some convenient survey point, probably B, using a total station instrument and prism.

This method is speedy and an unskilled assistant may be employed to handle the prism. The disadvantage is that the points are unchecked and errors are not noticed till later, resulting in more expenditure.

(c) Setting out the road as a traverse

1. *Calculation of data* The traverse, as already explained, comprises the points C, R1, X1, R4, F.

(i) The coordinates of points R1, X1 and R4 are shown in Table 12.2. (The coordinates of points C and F are already known.)

(ii) Calculate the bearing and distance of the lines C–R1, R1–X1, X1–R4 and R4–F. by the familiar formulae:

Bearing (any line) $= \tan^{-1} \Delta E/\Delta N$
Distance (any line) $= (\Delta E^2 + \Delta N^2)^{1/2}$

EXAMPLE

1 The coordinates of survey stn C and roadway point R1 are shown below. Calculate the bearing and distance C–R1.

Point	Easting (m)	Northing (m)
R1	47.000	106.700
C	48.497	113.013

$\Delta E_{C-R1} = -1.497$
$\Delta N_{C-R1} = -6.313$

Solution

Bearing C–R1 $= \tan^{-1} (-1.497 / -6.313)$
$= 193° \ 20' \ 24''$
Distance C–R1 $= (-1.497^2 + -6.313^2)^{1/2}$
$= 6.488$ m

Exercise 12.1

1 The coordinates of survey stn F and roadway point R4 are shown in Table 12.3. Calculate the bearing and distance R4–F.

Table 12.3

Point	Easting (m)	Northing (m)
F	116.336	85.572
R4	107.000	91.800

Table 12.4

Station	Easting (m)	Northing (m)	Line	Length (m)	Whole circle bearing
B	65.621	134.379			
C	48.497	113.013	B–C	27.395	218° 41′ 50″
R1	47.000	106.700	C–R1	6.488	193° 20′ 24″
X1	69.000	128.700	R1–X1	31.113	45° 00′ 00″
R4	107.000	91.800	X1–R4	52.968	134° 09′ 30″
F	116.336	85.572	R4–F	11.223	123° 42′ 25″
A	100.000	100.000	F–A	21.795	311° 26′ 30″

When all of the bearings and distances have been computed, a setting out table showing all data from station C through R1, X1, R4 to F is compiled as in Table 12.4.

The roadway may be set out perfectly adequately from the data shown in Table 12.4 but many surveyors prefer to set the theodolite to zero at each set-up point. The clockwise angle between the two relevant lines is therefore required. At any set-up point the bearing to the back station is subtracted from the bearing to the forward station to give the angle. The calculation is demonstrated in Example 2.

E X A M P L E

2 The whole circle bearings of all lines of the set-out traverse required to locate roadway R1–R4 are shown in Table 12.4. Calculate the clockwise angles between adjacent lines of the traverse

Answer

e.g. In Table 12.4 the bearing from C to R1 is 193° 20′ 24″ and the bearing from R1 to X1 is 45° 00′ 00″. The instrument station is R1, the back station is C and the forward station is X1.

Forward bearing C–R1 = 193° 20′ 24″
    therefore,
backbearing R1–C   = 193° 20′ 24″ − 180° 00′ 00″
                      = 13° 20′ 24″
Forward bearing R1–X1 = 45° 00′ 00″

Angle C–R1–X1   = 45° 00′ 00″ − 13° 20′ 24″
                      = 31° 39′ 36″

All other angles are computed similarly and are shown as Table 12.5.

---

2. Fieldwork in setting out One surveyor and two assistants are required to set out a roadway, using this method. Using Tables 12.4 and 12.5, roadway R1–X1–R4 is set out as follows.

1. The theodolite or total station is set up, face left, at survey station C and a backsight is taken to station B with the instrument reading zero degrees.

Table 12.5

Angle	Clockwise Value
B–C–R1	154° 38′ 34″
C–R1–X1	31° 39′ 36″ (Ex.2)
R1–X1–R4	269° 09′ 30″
X1–R4–F	169° 32′ 55″
R4–F–A	07° 44′ 05″

2. The horizontal circle is set to read 154° 38′ 34″; thus the theodolite now points towards station R1.
3. The assistants set the tape along this line C–R1 and measure 6.488 m horizontally, thus establishing point R1. (The mechanics of actually setting out the peg are described in Sec. 3(a) and Fig. 12.3.)
4. If the horizontal distance cannot be set out for any reason, the equivalent slope distance must be determined and the peg established in the manner of Sec. 3(b).
5. The theodolite is set to face right and the procedure of setting the peg is repeated. There may be a minor error in the peg position, in which case the mean position would be accepted.
6. The theodolite is removed to station R1, backsighted, face left, to station C and the setting out procedure described in steps 2 to 5 is repeated, with the next set of data.
7. The setting out is continued to reach peg F, whereupon the survey should check acceptably.

(d) Setting out by coordinates (radiation)

When calculating setting out distances, the plan distance is always produced. However, on the ground it is often difficult to measure distances horizontally; hence the ground slope along the line being set out has to be found and the plan distance amended. Furthermore, measuring by tape is labour intensive and often uncomfortable.

Setting out building and engineering works, using total station instruments, has therefore become standard practice on construction sites, since those instruments allow horizontal distances to be set out without difficulty.

Using EDM methods requires that the coordinates of every proposed point be determined, usually by calculation. The bearing and distance to these points, from some survey station, are then computed.

On site, the EDM instrument is switched to tracking mode, and set on the correct bearing. The target prism is then aligned at the correct horizontal distance and no slope correction is ever necessary.

1. Calculation of data   In the particular case of roadway R1–R4 (Fig. 12.1) the most convenient survey station for setting out the road would probably be station B. The bearing and distance to points R1, X1 and R4 from station B are therefore required using the familiar formulae:

$$\text{Bearing (any line)} = \tan^{-1} \Delta E / \Delta N$$
$$\text{Distance (any line)} = (\Delta E^2 + \Delta N^2)^{\frac{1}{2}}$$

After calculation of those bearings and distances, the roadway may be set out from the data. However, as was stated in the previous method, many surveyors prefer to set the theodolite to zero for the first sight; therefore the clockwise angles to the next and subsequent stations are required. These clockwise angles are simply the differences in bearings of the various lines.

The setting out data, both from bearings and from clockwise angles, are shown in Table 12.6.

Table 12.6

Line	Length	Bearing	Angle	Clockwise Reading
B–A	–	135° 00′ 00″	–	–
B–R4	59.373	135° 49′ 08″	A–B–R4	00° 49′ 08″
B–X1	6.608	149° 14′ 50″	R4–B–X1	13° 25′ 42″
B–R1	33.360	213° 55′ 50″	X1–B–R1	78° 06′ 42″

2. Fieldwork in setting out   Using Table 12.6 the roadway is set out as follows:

1. The total station instrument is set up, face left, at survey station B and a backsight is taken to station A with the instrument reading zero degrees.
2. Tracking mode is selected and the instrument is set to read (00° 49′ 08″ & 59.373 m). The assistant is directed towards point R4 by the surveyor (usually by radio), and adjusts the prism position until the point R4 has been accurately located, where a peg is hammered into the ground.
3. The input readings are changed to (13° 25′ 42″ & 6.608 m) and step 2 is repeated to establish point X1. Similarly point R1 is set out with the appropriate data of (78° 06′ 42″ & 33.360 m).

This method is obviously faster but the points are unchecked. It should be normal practice to repeat the setting out from another survey point; hence the time

taken in calculation and setting out is little different from method (a).

(e) Setting out the tangent points of the curve

1. Assuming that the intersection point X1 has been correctly set out, the next task is to locate the tangent points, R2 and R3, at the beginning and end of the roadway curve (Fig. 12.4). The curve elements, namely the tangent lengths, curve length and chainages of the tangent points, must firstly be computed.

EXAMPLE

3 The bearings and lengths of straights R1–X1 and X1–R4, taken from Table 12.4, are as follows:

Straight	Bearing	Length
R1–X1	45° 00′ 00″	31.113 m
X1–R4	134° 09′ 30″	52.968 m

Given that the curve joining the straights has a radius of 10.000 m, calculate

(a) the deviation angle,
(b) the tangent lengths,
(c) the curve length, and
(d) the chainages of the tangent points.

Answer

(a) Deviation angle θ = 134° 09′ 30″ – 45° 00′ 00″
　　　　　　　　　 = 89° 09′ 30″
(b) Tangent lengths　= $R \tan \theta/2$
　　　　　　　　　 = 10.000 × tan 44.579 167°
　　　　　　　　　 = 10.000 × 0.985 417
　　　　　　　　　 = 9.854 m
(c) Curve length　　 = $R \times \theta$ radians
　　　　　　　　　 = 10.000 × 1.556 106 5
　　　　　　　　　 = 15.561 m
(d) Chainage R2　　 = 31.113 – 9.854
　　　　　　　　　 = 21.259 m
　　Chainage R3　　 = 21.259 + 15.561
　　　　　　　　　 = 36.820 m

---

2. In the field, the tangent points R2 and R3 are established by measuring either forward or backward from the intersection point X thus: (a) The theodolite is set at point X, (b) a backsight is taken to point R1 and (c) peg R2 is aligned at a distance of 9.854 m from X.

Similarly the tangent point R3 is established from point X by (a) aligning the theodolite on to point R4 and (b) measuring 9.854 m forward from X towards R4.

(f) Setting out the curve

There are three methods of setting out a small radius curve, all of which are detailed in Chapter 11. The methods are (a) finding the centre of the curve and swinging an arc of 10.0 m radius from R2 to R3, (b) measuring offsets from the tangent lengths R2–X and R3–X and (c) measuring offsets from the long chord R2–R3. The choice of method depends upon whether there are any obstructions in the vicinity of the curve. In this case method (c) is the preferred choice. The student is advised to study Chapter 11, Example 7 before attempting Exercise 12.2.

Exercise 12.2

1 In Figure 12.4, the curve is to be set out by offsets from the long chord R2–R3. Calculate:

(a) the length of the long chord given that the radius is 10.00 m and the deviation angle R1–X–R4 is 89° 09′ 30″,
(b) the lengths of the offsets at the quarter points of the long chord.

(g) Offset pegs

Once the centre line of the roadway, including curve, has been pegged out on the ground, offset pegs, coloured yellow, are set at right angles to the left and right of the centre line, at distances such that they will not be disturbed by future construction work.

5. Setting out small buildings

The trend in modern dwelling-house building is towards timber-framed kits while large factory buildings and multi-storey buildings are always prefabricated. Little, if any, inaccuracy can be tolerated with either method of building.

The exact position that the building is to occupy on the ground is governed by the building line as defined by the Local Authority. Figures 12.1 and 12.5 show that two timber framed buildings are to be erected. The building line is parallel to Lochview Road at a distance of 7.0 metres from the centre line of the road. This building line must therefore be set out first of all.

(a) Setting out the building line – preparatory work – calculations

The building line may be set out simply by measuring two 7.0 m offsets to the north of the roadway centre line or by scaling the coordinates of two points on the line and setting them out from an existing survey station by theodolite. The latter is the more common method.

EXAMPLE

4 Figure 12.5 shows the position of two points BL1 and BL2 on the building line and two traverse stations C and D. The coordinates of the points C, BL1 and BL2 are as follows:

Point	Easting	Northing
C	48.497	113.013
BL1	55.600	109.300
BL2	104.800	74.200

The bearing of line CD, known from the survey, is 125° 29′ 17″. Calculate the bearing and distance from:

(a) survey point C to building line point BL1,
(b) point BL To BL2; hence calculate:
(c) the angle required at survey point C to set out point BL1,
(d) the angle required at BL1 to set out BL2.

Answer

(a)

$$\Delta E(C{-}BL1) = 7.103$$
$$\Delta N(C{-}BL1) = -3.713$$

Tan Bearing C–BL1	$= 7.103/{-}3.713$
	$= 117°\ 35'\ 51''$
Distance C–BL1	$= (7.103^2 + 3.713^2)^{1/2}$
	$= 8.015$ m

(b)

$$\Delta E(BL1{-}BL2) = 49.200$$
$$\Delta N(BL1{-}BL2) = 35.100$$

Tan Bearing BL1–BL2	$= 49.200/{-}35.100$
	$= 125°\ 30'\ 16''$
Distance BL1–BL2	$= (49.200 + 35.100)^{1/2}$
	$= 60.437$ m

(c)

Back bearing C–D	$= 125°\ 29'\ 17''$
Fwd Bearing C–BL1	$= 117°\ 35'\ 51''$
Angle D–C–BL1	$= \underline{352°\ 06'\ 34''}$

(d)

Back bearing BL1–C	$= 297°\ 35'\ 51''$
Fwd Bearing BL1–BL2	$= 125°\ 30'\ 16''$
Angle C–BL1–BL2	$= \underline{187°\ 54'\ 25''}$

(b) Setting out the building line – fieldwork

1. Set the theodolite over peg C and take a backsight reading to D on zero degrees.
2. Set the horizontal circle to read 352° 06′ 24″ and set out peg BL1 at distance 8.015 m.
3. Transfer the theodolite to point BL1 and take a backsight reading to C on zero degrees.
4. Set the horizontal circle to read 187° 54′ 25″ and set out peg BL2 at 60.437 m.

Alternatively, as in setting out roadways and curves, the points BL1 and BL2 can be set out from station C by coordinates, using a total station instrument.

(c) Setting out the building – fieldwork

In Figs 12.1 and 12.5, the two buildings fronting Lochview Road have different shapes but when setting out, each building is reduced to a basic rectangle, enabling checks to be easily applied.

Figure 12.6(a) shows the position of No. 10 Lochview Road and the relevant building line. The house may be set out from the building line using only a steel tape and some basic geometry or a theodolite and tape may be required.

The geometry usually consists of setting out a right angle. In the right-angled triangle ABC (Fig. 12.7) angle 'C' is 90 degrees and side 'c' is the hypotenuse. The relationship between the three sides is as follows:

Side a $= (2n + 1)$: Side b $= 2n(n + 1)$:
Side c $= 2n(n + 1) + 1$

When $n = 1$ the relationship becomes 3:4:5. This holds good for any right-angled triangle and any constant 'n'.

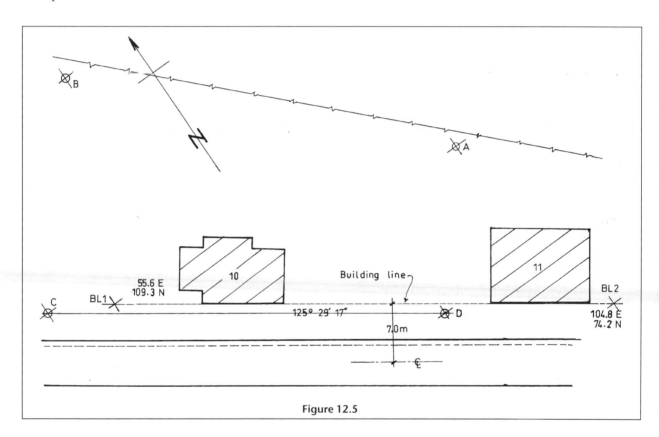

Figure 12.5

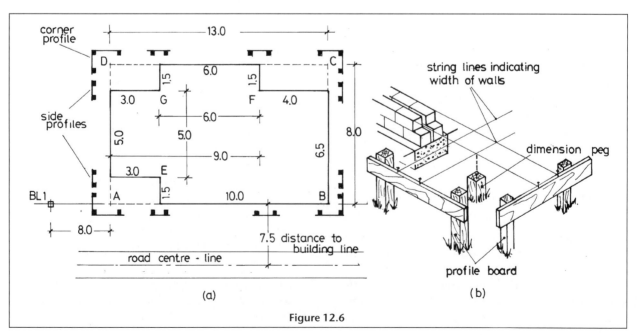

(a)   (b)

Figure 12.6

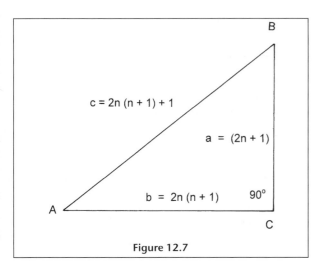

$c = 2n (n + 1) + 1$

$a = (2n + 1)$

$b = 2n (n + 1)$

$90°$

A

B

C

Figure 12.7

Procedure
1. From the plan, measure the distance between the building starting point BL1 and the corner of the house. The scaled dimension is 8.00 m.
2. Determine, from the plan, the dimensions of a basic rectangle to enclose the house. The dimensions are 13.0 m by 8.0 m.
3. Using a steel tape, set out the distance 8.00 m along the building line from point BL1, to establish corner A of the house. Mark the point A by a nail driven into a wooden peg.
4. Measure the distance AB (13.0 m) along the building line and establish a peg B. Mark the point by a nail.
5. Using a basic 3:4:5 right angle, measure the lengths AD and BC (8.0 m) and establish pegs at C and D.
6. Check the lengths of the diagonals AC and BD (15.264 m). The two measurements should be equal, thus proving that the building is square.

Although the method of setting out a right angle using a 3:4:5 triangle is theoretically sound, in practice it tends to lead to inaccuracies in positioning. By calculating the length of the diagonal of the rectangle and using two tapes, the setting out can be accomplished much more accurately and speedily as follows:

1. As before, measure the length AB and mark the positions A and B by nails driven into the wooden stakes.
2. Calculate the diagonal size of the rectangle using the theorem of Pythagoras:

$$AC = \sqrt{13.0^2 + 8.0^2} = 15.264 \text{ m}$$

3. Hold the zero of tape 1 against point A; hold the zero of tape 2 against point B and stretch them out in the direction of the point C.
4. At the intersection of 15.264 m of the first tape and 8.000 m of the second tape, mark point C on a peg.
5. Repeat for point D by measuring AD = 8.00 m with tape 1 and BD = 15.264 m with tape 2.
6. Check that DC = 13.000 m.

7. In order to establish peg G, measure a distance of 3.00 m from D (Fig. 13.6(a)) along line DC and 3.00 m from A along line AB. Insert nails into pegs at these points. Measure 1.5 m from D along line DA and 1.5 m from C along line CB. Insert nails into pegs at these points. Stretch tapes or builder's lines between the two sets of nails. The intersection of the lines is point G.
8. Similarly establish pegs E and F.
9. As a final check on the work, measure the dimensions EG, GF and FE. The distances should be EG = 5.00 m, GF = 6.00 m and EF = $\sqrt{5.0^2 + 6.0^2}$ = 7.810 metres.

Profile boards
During the excavation of the foundations, the pegs A, B, C and D will be destroyed and it is necessary to establish subsidiary marks on profile boards (Fig. 12.6(b)).

Profile boards are stout pieces of timber 150 mm by 25 mm cut to varying lengths. The boards are nailed to 50 mm square posts hammered into the ground, well clear of the foundations. Once the profiles have been established, builder's lines are strung between them and accurately plumbed above the pegs A, B, C and D, and nails are driven into the boards to hold the strings and mark the positions of the walls, foundations, etc.

Setting out on sloping ground
When setting out buildings on sloping ground, it must be remembered that the dimensions taken from the plan are horizontal lengths and consequently the tape must be held horizontally and the method of step taping used.

The diagonals must also be measured horizontally and in practice considerable difficulty is experienced in obtaining checks under such conditions. Besides, the method is laborious and time consuming.

When site conditions prove to be too difficult, a theodolite should be used to establish right angles. The instrument is set up over a peg and a plumb-bob is attached. Horizontal distances can usually be measured accurately from the plumb line. If this still proves to be too difficult, the method for setting out large buildings (following) should be used.

6. Setting out large buildings

In setting out large buildings, which are mainly prefabricated, accuracy is absolutely essential. The factory-built components cannot be altered on site and even though some allowance is made for fitting on site, faulty setting out causes loss of time and money in corrective work.

Accuracy in setting out can be achieved in a variety of ways, using a variety of instruments, of which the following is a reasonably acceptable selection.

(a) Method 1 – by tape and theodolite

This is the traditional method of setting out a large building. It is accurate and self-checking but is slow and requires a great deal of skill on the part of surveyor and assistants.

In setting out a building using this method, the surveyor must be aware of the effects of calibration, temperature and tension on taped measurements and the effects of instrumental errors on angular observations.

Linear measurements
Fabric-based tapes should never be used for setting out. Good-quality steel tapes are always employed and due allowance made for the following potential sources of error:

1. *Calibration* Calibration errors can be ignored when good-quality tapes are used. After long use, however, the tape should be tested against a standard. If the tape has been broken and repaired it should not be used for accurate setting out.

2. *Temperature* The length of a steel tape varies with temperature. The tape is the standard length at 20°C only. If left lying in direct sunlight the tape may reach an abnormally high temperature. In winter the temperature may well be at freezing point. In such cases, a correction has to be applied for differences in temperature but a fairly accurate estimation is obtained by allowing 1 mm per 10 metres per 10°C difference in temperature. For example, when using a 30 m steel tape on a winter day when the temperature is 0° an allowance of $- (1 \times 3 \times 2) = -6$ mm should be made for each tape length measured.

3. *Tension* Steel tapes should be used on the flat with a tension of 50 N. Without a spring balance few people can judge tensions and tests have shown that errors of 10 mm in 30 m can be caused by exerting excessive pressure on the tape.

When the tape is allowed to sag there is even more error. The natural tendency is to apply excessive tension to correct the sag, and in many cases the error due to stretching of the tape is greater than that due to sag.

A constant-tension handle has now been developed which applies a compromise tension. The tension applied is such that the effect of sag is compensated by the effect of stretching the tape.

4. *Slope* It should be clear from error source 3 above that sagging of the tape should be avoided if possible, which suggests that measurements should be made along the ground. This, however, introduces slope errors, and so measurements of slope must be made and corrections applied.

A typical example of the application of tape corrections is shown in Chapter 8 Sec. 3(b) (page 144).

5. *Taping procedure* For highly accurate measurements the following procedure should be carried out:

(a) Use a good-quality tape which has never been broken.
(b) Lay the tape on wooden blocks if measuring on concrete, etc., to allow air to circulate around the tape.
(c) Whenever possible, do not allow the tape to sag.
(d) Measure the temperature.
(e) Use a spring balance or constant-tension handle.
(f) Measure the ground slope.
(g) Compute the various corrections and apply the correction to the distance set out.

Angular measurements
Angular measurements must always be made on both faces of the instrument because of the effect of instrumental maladjustments, which were fully explained in Chapter 7.

Procedure for setting out
Figure 12.8 shows the plan of a large factory building and office block. The columns of the office block are of reinforced concrete, the cladding being prefabricated panels. The factory columns are of steel supporting steel latticed roof trusses. The column centres must be placed at exactly the correct distance apart and must be perfectly in line.

The following procedure is necessary to ensure that the requirements are met:

1. Establish a line AB from the site traverse stations at a predetermined distance x metres from the centre line of the left-hand columns. Measure the distance accurately.
2. Set out peg D by face left and face right observations from A, positioning D at some predetermined distance from the centre line of the right-hand columns.
3. Set out C by double face observations from D, making CD = AB.
4. Finally, check angle C and distance CB to ensure that ABCD is a perfect rectangle.
5. Set out the column centres along each line on stout profiles or preferably on pegs embedded in concrete. The centres must be set out by steel tape with due allowance being made for temperature, etc.
6. The column centres are defined by the intersections of wires strung between the appropriate reference marks or by setting the theodolite at a peg on one side of the building, sighting the appropriate peg on the other side and lining in the columns directly from the instrument.

No hard and fast rules can be formulated for setting out since conditions vary greatly from site to site and some ingenuity is called for on the part of the engineer.

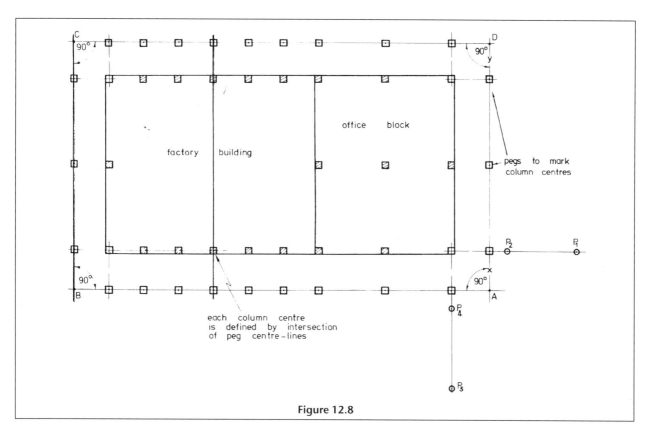

Figure 12.8

Exercise 12.3

1 Figure 12.9 shows the plan of a site where an access road and a building are to be set out.

(a) Show on the plan suitable setting-out points to locate the building and centre lines of the roadway straights on the ground.
(b) Using a scale rule, determine the coordinates of those setting-out points.
(c) Calculate the angles and distances required to locate the setting-out points on the ground.

(Scotvec – Higher National Certificate in Civil Engineering)

(b) Method 2 – by coordinates

Setting out points using total station instruments, either laser or infrared, has become standard practice on construction sites, since those instruments overcome all of the problems associated with taped measurements detailed in method 1.

Setting out by EDM methods requires that the coordinates of every proposed point be determined, usually by calculation. The coordinates of the point are compared with those of the setting-out control station and the horizontal distance and bearing between the two are computed. The proposed point is then set out from the control station.

Figure 12.10(a) shows the plan of an industrial estate

where the coordinates of survey stations A, B, C, D and E have been previously determined. Their coordinates are shown in Table 12.7. A large rectangular building X is to be set out on the estate.

Since the building is quite far removed from the survey points, a control station closer to the building would usually be established. The coordinates of the station, often called a 'free' station, are determined either by intersection or by resection.

Intersection of free station

In the intersection method of determining the co-ordinates of such an unknown point, the observed field data depend upon the kind of instrument being used. In Fig. 12.10(a), the coordinates of free station R are required.

(i) When using a conventional theodolite (non-EDM), it is necessary to measure two angles from known stations to the free station R. There is a choice, so perhaps the angles DCR and RDC would be chosen and observed from stations C and D respectively. The intersection of the lines CR and DR would then occur at point R. This is the true meaning of the term intersection.
(ii) When using a total station instrument, the field data required to compute the position of point R would all be collected at station R itself by measuring the lengths of lines RC and RD and the angle CRD.

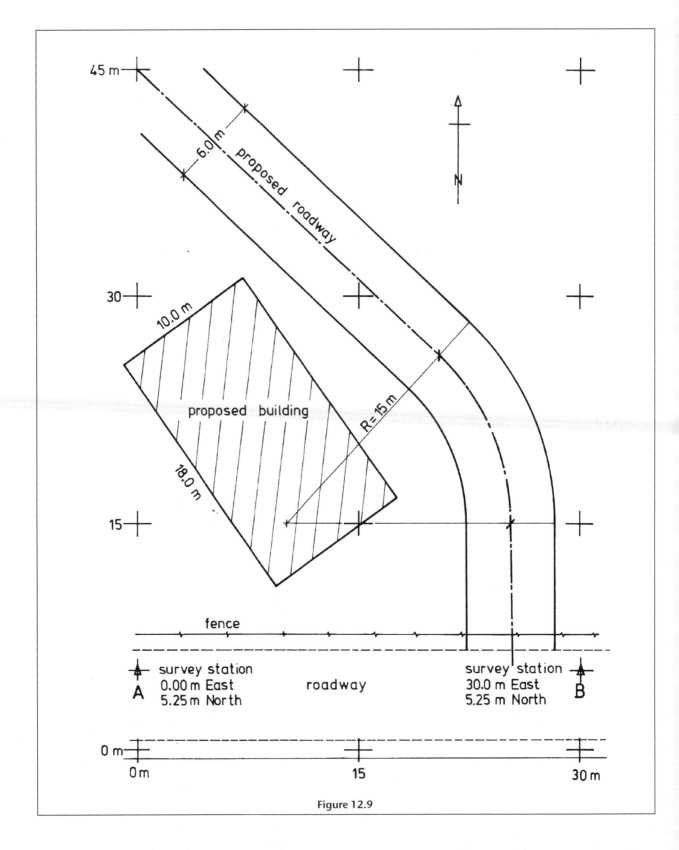

Figure 12.9

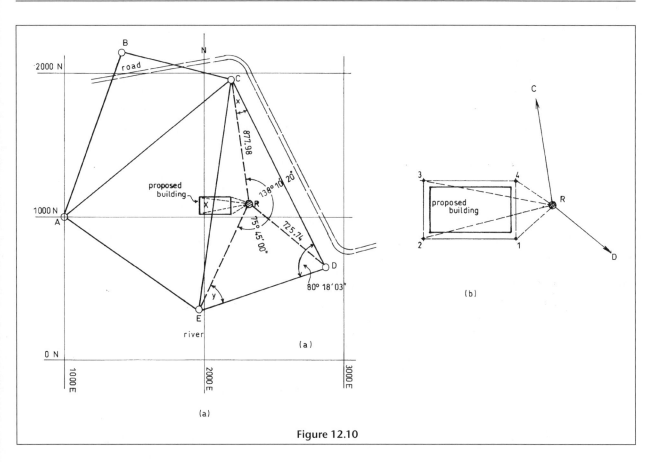

Figure 12.10

EXAMPLE

5 The following field data were observed in Fig. 12.10 from station R. [The remaining data should be ignored for this example.]

Angle CRD = 138° 10′ 20″
RC = 877.98 m
RD = 725.74 m

Calculate the coordinates of station R from the given data and from the data in Table 12.7.

Answer

The bearing from one known station (station C) to point R is required in order to calculate the coordinates of R.

1 In ΔDCR,

RD	= 725.74 m (measured)
CD	= 1499.07 m (known)

$$\text{Sin DCR} = \frac{RD \times \sin R}{CD}$$

$$= \frac{725.74 \times \sin 138° \ 10' \ 20''}{1499.07}$$

$$= 0.3228611$$

DCR   = 18° 50′ 10″

2 Bearing CD = 153° 03′ 25″ (known)
            + 18° 50′ 10″
∴ Bearing CR 171° 53′ 35″
   Length CR = 877.98 m

3 The coordinates of R are calculated from C using the length and bearing of line CR.

Table 12.7

Line	WCB	Length	East	North	Stn.
			1000.00	1000.00	A
AE	123° 49′ 14″	1158.270	1962.27	355.31	E
ED	72° 45′ 22″	962.683	2881.69	640.69	D
DC	333° 03′ 25″	1499.071	2202.45	1977.05	C
CB	281° 55′ 27″	798.726	1420.96	2142.08	B
BA	200° 14′ 00″	1217.190	1000.00	1000.00	A

$$E_R = E_c + RC \sin WCB\ RC$$
$$= 2202.45 + 877.98 \sin 171°\ 53'\ 35''$$
$$= 2202.45 + 123.84$$
$$= \underline{2326.29\ m}$$
$$N_R = N_c + RC \cos WCB\ RC$$
$$= 1977.05 + 877.98 \cos 171°\ 53'\ 35''$$
$$= 1977.05 - 869.20$$
$$= \underline{1107.85\ m}$$

Resection of free station

In the resection method of finding the position of the free station, only two angles need be observed to three known stations. This method is used where the known stations are inaccessible, e.g. church spires, flagpoles etc.

In Fig. 12.10(a), the angles CRD and DRE are observed from station R, using a total station or theodolite. The coordinates of point R are then calculated as in Example 6. (*Note*: The mathematics are somewhat complex, but the method works for all resection situations, whether the point R lies inside or outside the triangle formed by the three observed stations C, D and E.)

EXAMPLE

6 The following field data (two angles only) were observed in Fig. 12.10 from station R.

Angle CRD = 138° 10′ 20″
Angle DRE = 75° 45′ 00″

Calculate the coordinates of station R from the given data and from the data in Table 12.7.

Answer

The coordinates of point R are calculated using the bearing and length of line DR, the middle line of the resection. The calculation must be directed towards finding those two quantities.

1 In $\triangle DRC$

$$DR = \frac{CD \sin x}{\sin 138°\ 10'\ 20''} \text{ and in } \triangle DER$$

$$DR = \frac{DE \sin y}{\sin 75°\ 45'\ 00''}$$

$$\therefore \frac{CD \sin x}{\sin 138°\ 10'\ 20''} = \frac{DE \sin y}{\sin 75°\ 45'\ 00''}$$

$$\frac{\sin x}{\sin y} = \frac{DE \sin 138°\ 10'\ 20''}{CD \sin 75°\ 45'\ 00''}$$

$$= \frac{962.68 \sin 138°\ 10'\ 20''}{1499.07 \sin 75°\ 45'\ 00''}$$

$$= 0.441\ 864\ 9$$

$$\sin x = 0.441\ 864\ 9 \sin y$$

2 From previous fieldwork or from the coordinates of E, D, and C, the value of angle EDC is 80° 18′ 03″
In triangles EDR and RDC:

Angles $(x + y + EDC) = 360° - (138°\ 10'\ 20''$
$$+ 75°\ 45'\ 00'')$$
$$x + y + 80°\ 18'\ 03'' = 360° - 213°\ 55'\ 20''$$
$$= 146°\ 04'\ 40''$$
$$\therefore x + y = 65°\ 46'\ 37''$$
$$\therefore x = (65°\ 46'\ 37'' - y)$$

3 $\sin x = 0.441\ 864\ 9 \sin y$ (from 1 above).
Substitute $(65°\ 46'\ 37'' - y)$ for x giving $\sin (65°\ 46'\ 37'' - y) = 0.441\ 864\ 9 \sin y$
$$\sin 65°\ 46'\ 37'' \cos y - \cos 65°\ 46'\ 37'' \sin y$$
$$= 0.441\ 864\ 9 \sin y$$
$$0.911\ 955\ 1 \cos y - 0.410\ 290\ 0 \sin y$$
$$= 0.441\ 864\ 9 \sin y$$
$$0.911\ 955\ 1 \cos y = 0.441\ 864\ 9 \sin y$$
$$+ 0.410\ 290\ 0 \sin y$$
$$= 0.852\ 154\ 9 \sin y$$
$$\therefore \cos y = 0.934\ 426\ 4 \sin y$$

Divide by $\sin y$ giving
$$\cot y = 0.9344264$$
$$\therefore y = 46°\ 56'\ 29''$$
and since $x + y = 65°\ 46'\ 37''$
$$x = 18°\ 50'\ 08''$$

4 From 1 above
$$DR = \frac{CD \sin x}{\sin 138°\ 10'\ 20''}$$
$$= \frac{1499.07 \sin 18°\ 50'\ 08''}{\sin 138°\ 10'\ 20''}$$
$$= 725.722\ m$$
and $$DR = \frac{DE \sin y}{\sin 75°\ 45'\ 00''}$$
$$= \frac{962.68 \sin 46°\ 56'\ 29''}{\sin 75°\ 45'\ 00''}$$
$$= 725.717\ m \text{ (check)}$$

5 In triangle EDR:
Angle EDR $= 180° - (75°\ 45'\ 00'' + 46°\ 56'\ 29'')$
$$= 180° - 122°\ 41'\ 29''$$
$$= 57°\ 18'\ 31''$$
Bearing DE $= 252°\ 45'\ 22''$
$$+ 57°\ 18'\ 31''$$
$\therefore$ Bearing DR $= \underline{310°\ 03'\ 53''}$

6 Co-ordinates of R:
$$ER = E_D + DR \sin WCB\ DR$$
$$= 2881.69 + 725.72 \sin 310°\ 03'\ 53''$$
$$= 2881.69 - 555.41$$
$$= \underline{2326.28\ m}$$
$$N_R = N_D + DR \cos WCB\ DR$$
$$= 640.69 + 725.72 \cos 310°\ 03'\ 53''$$
$$= 640.69 + 467.11$$
$$= \underline{1107.80\ m}$$

Many total station instruments, e.g. the Sokkia Series 130R, calculate the coordinates of the free station using their built-in 3-D coordinate measurement programs. These programs can handle the data of more than the minimum number of stations (two for intersection, three for resection), e.g. the 130R can use the data from up to 10 stations. It displays the results on the control panel and if the result from any observed point is unacceptable, the point is re-observed or replaced with a new point.

Exercise 12.4

1 Figure 12.11 shows intersection observations from an uncoordinated point F to control stations C and D.

Calculate the coordinates of point F.

2 Figure 12.11 shows resection observations from an uncoordinated point R to control stations C, E and A.

Calculate the coordinates of point R.

Field procedure
It must be pointed out at this juncture that the mechanics of setting an EDM instrument along a certain line on a specific bearing varies with the type of instrument. Consequently, only the general principle can be described here.

1. The coordinates of four peg positions situated at known offsets from the corners of the building in Fig. 12.10(b) are obtained from the site plan.
2. The instrument is set up, centred and levelled at survey station R. The coordinates of R are entered via the keyboard. The bearing to the chosen back station, say C, is entered via the keyboard or the coordinates of the back station may be entered, whereupon the instrument will calculate the bearing.
3. The back station C is sighted and the instrument is clamped. The bearing and distance or coordinates of stakeout point no. 1 (Fig. 12.10(b)) are next entered and the instrument is set to tracking mode.
4. The assistant proceeds towards the approximate

position of the stakeout point and faces the instrument. Most instruments are fitted with a guide light unit composed of two lights, one coloured red, the other green, which emanate from an aperture below the telescope. The assistant sees green when he or she is left of the stakeout point and red when he or she is on the right. He or she sees both flashing simultaneously when he or she is on the set-out bearing.

5. The assistant turns the prism towards the instrument and the *x*, *y* and *z* coordinates are measured. The differences between these measured values and the design values are displayed on the instrument panel.
6. The assistant is directed, by radio, to the exact stakeout point and inserts a peg, re-checks the complete operation and, when satisfied that the peg is correctly set, proceeds to stakeout point no. 2.
7. The distance and bearing or *x*, *y*, *z* coordinates of point no. 2 are entered and the procedure repeated. The remaining pegs 3 and 4 are set out in the same manner.
8. Since small angular and distance errors inevitably arise when using EDM equipment, the column centres along the four sides of the building should *not* be similarly set out since their alignment is critical. It is common practice to remove the instrument to each of the four stakeout points and align the column centres exactly along each of the building sides, using the telescope. Where distances are short (less than 50 m) the column centres should always be set out by steel tape. EDM should only be used where unfavourable ground conditions prevail.
9. Alternatively, a laser instrument may be placed at the corner stakeout pegs and the columns aligned exactly, as and when required. (See method 4, following.)

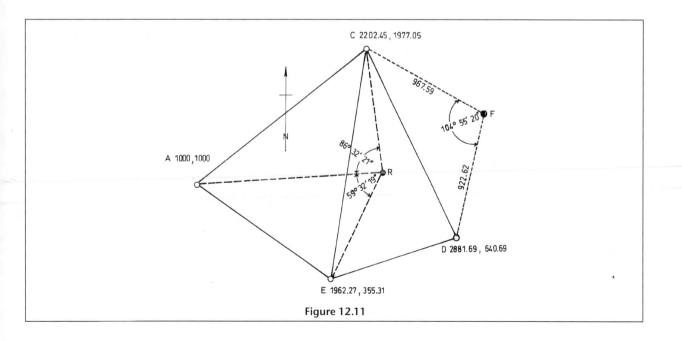

Figure 12.11

(c) Method 3 – using GPS

The Global Positioning System (GPS) was described in detail in Chapter 9 as primarily a system for establishing the positions of points anywhere on the surface of the Earth. However, the system can also be used to set out the positions of all manner of construction works.

Briefly, the GPS and GLONASS satellites orbit the Earth at a height of about 21 000 km. Points on Earth, determined by GPS, are defined in a coordinate system called WGS 84. However, points of a survey are defined on different land-based systems of coordinates and, in order to be compatible, a transformation of GPS to land-based coordinates must be made. This is done by placing a GPS receiver over a known land-based coordinated point (the base station) and comparing its fixed coordinates with those determined by GPS. The difference between the two is the GPS correction. The correction can be transmitted by radio to a second GPS instrument (the rover) which uses the correction to transform the coordinates of its position into land-based values. The principle is called real-time kinematic surveying (RTK).

In setting out construction works, e.g. the building shown in Fig. 12.10, the survey coordinates of the building are known. A GPS receiver (the base station) is set up over one of the survey control points (e.g. station C) and its GPS coordinates are determined and transformed via the system software.

The rover instrument is taken to a point close to the proposed building position and the stakeout coordinates are entered into the instrument via the keyboard. The GPS rover calculates the difference between its own position and the stakeout position and shows on the screen the direction in which the operator must move to reach the stakeout position. He or she then simply follows the direction arrow on the screen until the instrument indicates that the stakeout position has been reached.

The next set of coordinates are entered and the procedure is repeated until all points have been staked out.

(d) Method 4 – using laser instruments

Construction of instruments
About the beginning of the 1980s, the development of laser surveying resulted in a group of reliable instruments which are now in common use throughout the construction industry.

Wariness has always been the watchword with lasers. Surveying lasers have a very low power output and are harmless to the body and clothing but are hazardous to the eyes. Looking directly into a beam is dangerous.

Lasers are classed as: *Class 1* – intrinsically safe, no precautions required; *Class 2* – virtually harmless but looking down the beam is not recommended; *Class 3A* – sighting the reflection of a beam through any optical instrument requires special approval; *Class 3B* – Viewing the beam optically is forbidden and special safety glasses and clothing must be worn.

Two types of laser beam are used in surveying instruments, namely:

(i) visible (laser diode). The light produced is totally different from any other light in that it can be focused into a pure, monochromatic, highly coherent and very narrow beam. The diameter of the beam naturally increases with distance from the instrument. This is called beam divergence. The beam is characteristically red in colour, though Topcon have developed a green beam which is claimed to be four times brighter than conventional beams.

(ii) invisible (gallium arsenide semiconductor diode). Since the beam cannot be seen by the naked eye, it is detected by a laser receiver or detector. The receiver uses photocells to find the centre of the laser beam. It can also be used with visible beams. The detector has its own battery power source and can be hand held or mounted on a levelling staff using a clamping arrangement.

Quite how the beam is produced is of little relevance to the construction surveyor, he or she simply has to know that a laser surveying instrument is one in which the optical line of sight has been replaced by the laser beam.

Laser instruments are self-levelling, can be plumbed and levelled over a control point and can be adapted or specifically manufactured to set out horizontal, vertical or inclined lines of sight.

Two main types of laser instrument are used in setting out work.

(a) Fixed beam lasers which project a highly visible line in a direction controlled by the surveyor, e.g. laser theodolites and pipe laying lasers, for example the Topcon TP-L4 Series (Fig. 12.12), in which the line of sight is a laser beam directed by the instrument controls.

(b) Rotating beam lasers in which the laser beam, visible or invisible, is rotated continuously in a full circle at around 600 rpm. Since the beam diverges with distance, its centre has to be detected on a laser receiver. Figure 12.13(a) shows the Sokkia LP30A levelling laser, tripod mounted, together with its beam receiver, the LR100. Figure 12.13(b) shows the Topcon RL30 floor mounted on its side.

Pipe laying lasers and rotating beam lasers are primarily used in levelling operations, for example sewers, drains, floors, gradients etc. and their use is described in Sec. 9, following. However, most of these instruments can be laid on their side to produce horizontal and vertical beams which are used to set out structures in three dimensions.

Some manufacturers produce laser instruments for this sole purpose, for example the Hilti PM 24 multi direction laser, shown in Fig. 12.14.

The instrument is robustly built to withstand the rigours of construction site life. It can be tripod mounted over a corner point of a building and self levelled within three seconds, using a magnetically damped pendulum. A class 2 laser diode produces four highly visible mutually perpendicular beams in the x,y and z directions (both up and down).

Figure 12.12 (Courtesy Topcon)

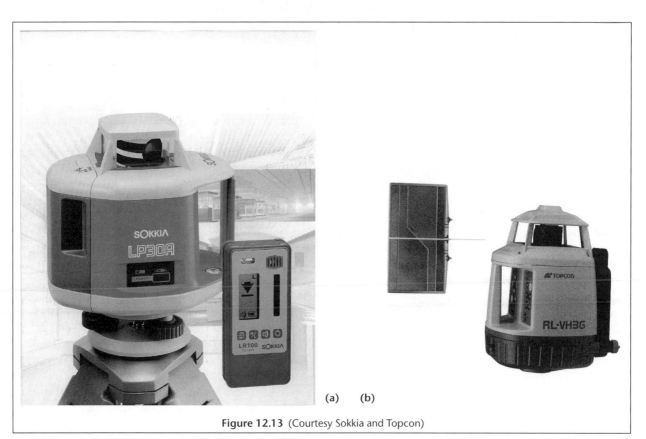

(a)   (b)

Figure 12.13 (Courtesy Sokkia and Topcon)

Figure 12.14 (Courtesy Hilti)

Field procedure

The Topcon RL 30 rotating laser, typical of this class, is laid on its side and using a built-in floor mount is positioned over a corner peg previously established by method 2 (total station). The plumbing is accomplished using the automatic downward-sighting laser plumb beam. The Hilti PM 24 is tripod mounted and is centred above a corner peg in a similar manner.

Each instrument, when switched on, self-levels within a few seconds and the main laser beam is aligned with a second corner peg. The 90° beam is emitted from the instrument thus establishing a clearly visible line along which columns etc. can be lined and plumbed with accuracy.

7. Checking verticality

In Fig. 12.15 an office block rises to a height of five storeys and the columns must be checked at every storey for verticality.

Although some older methods are still in use, for example heavy plumb-bobs, verticality is ensured by using either (a) a theodolite, which is slow, difficult and labour intensive or (b) a laser plummet.

(a) Theodolite method

1. When setting out the framework for horizontal control, further marks are established at the points P_1 to P_4, i.e. on two lines at right angles to each other. Four pegs are required for each corner. In Figs 12.8 and 12.15 the pegs are shown for the bottom right-hand corner only.

   The theodolite is set up on face left over the outer peg P_1 and sighted on to the inner peg P_2. The telescope is then raised to any level on the building and

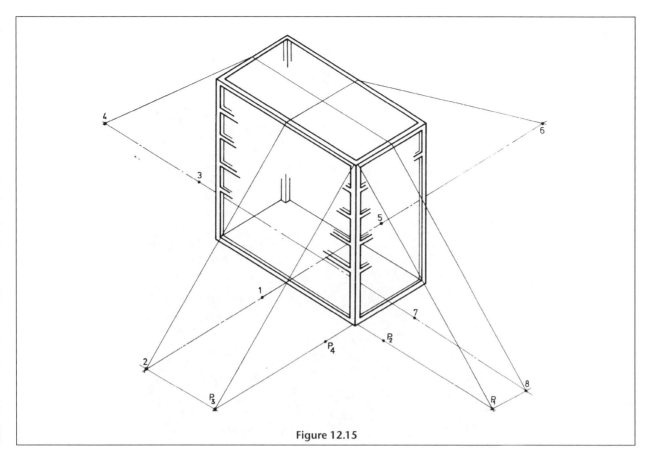

Figure 12.15

a mark made on the outside of the column. The procedure is repeated on face right. If the theodolite is in good adjustment the two marks will coincide. If not, the mean position is correct.

The instrument is removed to peg P_3 and the whole of the above procedure repeated to establish a second mark on the column. Thus, it is possible to determine the amount of deviation of the column from the vertical. Each corner is checked in this way.

Many of the observations in this method will be very steep and a diagonal eyepiece might have to be used. The eyepiece is interchangeable with the conventional eyepiece and is easily fitted to enable the surveyor to sight very high angles of elevation.

2. The centre lines of the building are established and extended on all four sides of the building. Two permanent marks are placed on each line. The marks are shown by the numbers 1 to 8 in Fig. 12.15.

The theodolite is placed on each of the outer marks, the telescope is raised to the required floor level and a mark established on all four sides by double face observations. The intersection of lines strung between the appropriate marks locates the centre of the building. From the lines measurements may be made to corners, etc.

(b) Laser plummet method

Several instrument companies produce instruments

that set out vertical lines of sight automatically, using laser beams. Fig. 12.14 shows the Hilti PM 24 instrument, which is widely used on smaller projects but has a limited range of around 30 m. Figure 12.16 shows the Sokkia LV1 which sets out a vertical beam with an accuracy of 5 seconds of arc (1 mm at 41 m) on upward sights and 1 minute of arc (1 mm at 3.5 m) on downward sights. The instrument can therefore locate points above and below any reference point with ease and accuracy. It has a range of 100 m upwards and 5 m downwards.

A compensator is used to make levelling fully automatic. An air and magnetic damping system ensures that the beam remains stable in any conditions.

In use, the instrument is mounted on a tripod and approximately centred and levelled over some reference point, using the footscrews. When switched on, the compensator automatically levels the instrument and the laser beams point vertically upwards and downwards. The shifting tripod head allows the instrument to be precisely centred over the reference mark. If, for any reason, the instrument is moved out of the auto-levelling range, a warning signal is given and the instrument shuts down after 30 minutes, if no action has been taken.

Arrangements must be made for leaving a hole in each floor for upward sighting. Sights are taken to a specially designed target which captures the narrow beam and measurements are taken from the target to the construction works.

Figure 12.16 (Courtesy Sokkia)

8. Setting out – vertical control

Whenever any proposed level is to be set out, sight rails (profiles) must be erected either at the proposed level in the case of a floor level or at some convenient height above the proposed level in cases of foundation levels, formation levels and invert levels. Suitable forms of sight rails or profiles are shown in Fig. 12.2. The rails should be set at right angles to the centre lines of drains, sewers, etc.

A traveller or boning rod is really a mobile profile which is used in conjunction with sight rails. The length of the traveller is equal to the difference in height between the rail level and the proposed excavation level. Figure 12.2 shows the traveller in use in a trench excavation.

9. Setting out a peg at a predetermined level

The basic principle of setting out a profile board at a predetermined level is shown in Fig. 12.17. Point A is a temporary bench mark (RL 8.55 m AD). Profile boards B and C are to be erected such that the level of board B is 9.000 m and that of board C is 8.500 m. These levels may represent floor levels of buildings or may represent a level of, say, 1.00 or 2.00 m above a drain invert level or a roadway formation level.

Setting up profile boards at different levels is the same operation and, once mastered, the methods may be used for any number of profiles on a site.

(a) Procedure – method 1

1. The observer sets up the levelling instrument at a height convenient for observing a site bench mark (RL 8.55 m AD) and takes a backsight staff reading (1.25 m). The height of collimation (HPC) is therefore

    RL bench mark + BS reading

    i.e. HPC = 8.55 m + 1.25 m = 9.80 m.

2. The assistant firmly hammers home a small peg, 300 mm long, beside the profile peg and a foresight reading (1.11 m) is taken to a staff held vertically upon it. The level of the top of the small peg is therefore

    RL = HPC – FS reading
    = 9.80 – 1.11 = 8.69 m

3. The difference in level between the top of the small peg (8.69 m) and the required profile level (9.00 m) is calculated:

    Difference = 9.00 m – 8.69 m = 0.31 m

Using a tape, the assistant measures this height against the profile peg and marks it in pencil.

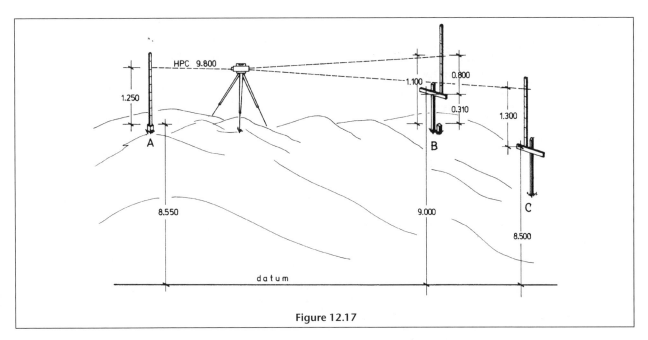

Figure 12.17

4. A profile board is nailed securely to the profile peg, such that the upper edge of the board is against the pencil mark, and is thus at a level of 9.00 m.

This method is widely used on construction sites because of its simplicity. However, it has the disadvantage that the observer has to rely upon the assistant (often untrained) to correctly mark the final height on the profile peg. The disadvantage is overcome by using the following method.

(b) Procedure – method 2

1. The observer sets up the levelling instrument, takes a backsight to the bench mark and computes the height of collimation (HPC) as before:

HPC = RL bench mark + BS reading
= 8.55 m + 1.25 m = 9.80 m

2. The staffman holds the staff against the profile peg and moves it slowly up or down until the base of the staff is at height 9.00 m, exactly. This will occur when the observer reads 9.80 m – 9.00 m = 0.80 m on the staff, since

HPC = 9.80 m
Required profile level = 9.00 m
Therefore staff reading = 0.80 m

3. The base of the staff is marked in pencil against the profile peg and the profile board is nailed securely to the peg, such that the upper edge of the board is against the pencil mark.

4. Profile board C is established in exactly the same manner, but since the board is to be erected at a different level, a new calculation is required:

HPC = 9.80 m
Required profile level = 8.50 m
Therefore staff reading = (9.80 – 8.50) m = 1.30 m

10. Setting out floor levels

Floor levels are set out on profile boards in exactly the manner described above in Sec. 9. Profile pegs are set around the perimeter of the house as required. A levelling instrument is set up and the height of collimation (HPC) of the level is determined. The proposed floor level is substracted from the HPC and the resultant staff reading used to set out the profile boards.

EXAMPLE

7 In Fig. 12.18 the floor level (17.30 m AD) of a building is to be set out from a nearby bench mark (16.830 m AD). Calculate the staff reading (x) required to set out a profile board at floor level.

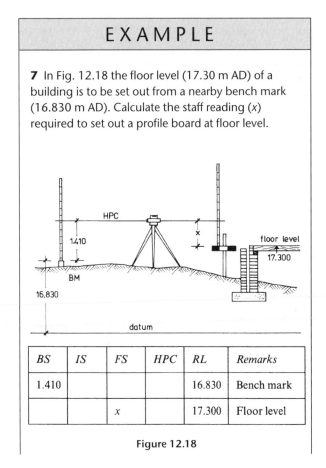

BS	IS	FS	HPC	RL	Remarks
1.410				16.830	Bench mark
		x		17.300	Floor level

Figure 12.18

Answer

$$HPC = 16.830 + 1.410$$
$$= 18.240$$
$$\text{Required reading } x = 18.240 - 17.300$$
$$= 0.940 \text{ m}$$

Exercise 12.5

1 The floor levels of a split-level house are upper level 25.500 m AD and lower level 24.300 m AD. They are to be set out in relation to a nearby bench mark (23.870 m AD). Table 12.8 shows the relevant readings. Calculate the staff readings x and y required to set out profiles at both floor levels.

Table 12.8

BS	IS	FS	HPC	Required level	Remarks
2.360				23.870	Bench mark
	x			25.500	Upper floor level
		y		24.300	Lower floor level

11 Setting out invert levels

(a) Setting out invert levels – office work

In setting out drains and sewers, it is not possible to set out the proposed levels of the drain, i.e. the invert levels, since they are always below the ground. Hence, profiles must be set at some convenient height above the invert levels. This height is chosen by the surveyor but should always be a multiple of 250 mm. The height is known as the traveller length and is determined as follows:

1. In Fig. 12.19, a drain, 30 metres long, is to be excavated, at a gradient of 1 in 40. The invert level at the start of the drain, chainage 0.00 m, is 44.320 m above datum.
2. Ground levels are taken at regular intervals along the centre line of the proposed drain. The levels, taken at 10 metre intervals, are shown in Table 12.9, column 2.

Table 12.9

1 Chainage (m)	2 Surface level	3 Invert level	4 Depth (m)
0	45.600	44.320	1.280
10	45.200	44.070	1.130
20	45.110	43.820	1.290
30	44.850	43.570	1.280

3. The invert levels of the drain are next calculated. In this case, the drain beginning at invert level 44.320 m falls at a gradient of 1 in 40. The fall over 10 metres is one-fortieth of 10 m = 0.250 m, and the invert level at chainage 10 m is therefore 44.320 – 0.250 = 44.070 m. Since the chainage intervals are regular, the fall must also be regular, resulting in the various invert levels shown in column 3 of Table 12.9.
4. The depth from surface to invert level is found by subtracting the invert level from the surface level. Thus, at 0 m chainage, the depth is 45.600 m (surface) – 44.320 (invert) = 1.280 m. The depths at the various chainage points are shown in column 4 of Table 12.9.
5. From the table, the maximum depth is 1.280 m and since profiles should be about 1 metre above ground level the length of traveller should be

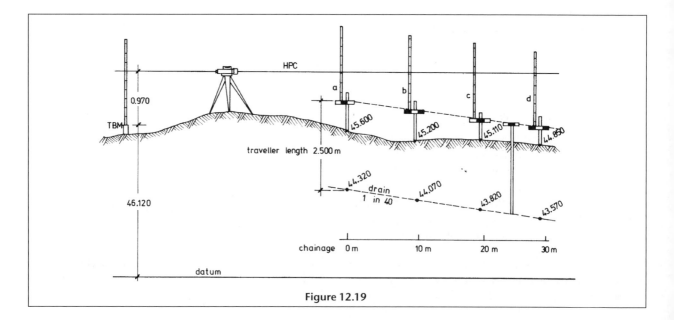

Figure 12.19

1.280 m + 1.00 m = 2.280 m. The traveller would probably be made 2.5 m long.

Having determined the length of the traveller, the levels of the profile boards to be erected along the line of the drain are next calculated. The profile level is 2.50 m greater than the invert level at any chainage. The profile levels are therefore as shown in Table 12.10, which becomes in effect the setting-out table.

Table 12.10

Chainage	Invert level	Profile level
0	44.320	46.820
10	44.070	46.570
20	43.820	46.320
30	43.570	46.070

(b) Setting out invert levels – fieldwork

In Fig. 12.19, the levelling instrument has been set up to sight to the bench mark and to the various profile board positions. Table 12.11 shows the relevant data. The staff readings a, b, c and d are required to set the profile boards at their correct levels.

The relevant readings are calculated as follows:

1. Calculate HPC = 46.120 + 0.970 = 47.090
2. Calculate:
   (a) = 47.090 – 46.820 = 0.270
   (b) = 47.090 – 46.570 = 0.520
   (c) = 47.090 – 46.320 = 0.770
   (d) = 47.090 – 46.070 = 1.020

Table 12.11

BS	IS	FS	HPC	Required level	Remarks
0.970				46.120	Bench mark
	(a)			46.820	Profile 0 m
	(b)			46.570	Profile 10 m
	(c)			46.320	Profile 20 m
		(d)		46.070	Profile 30 m

The profile boards are then set out by one of the methods described in Sec. 9.

8 Figure 12.20 shows a sewer, 55 metres long, which is to be set out on a gradient of 1 in 50 falling from chainage 0 m to chainage 55 m. The following data have been obtained:

Invert level at 0 m chainage = 24.210 m
Ground level at 0 m chainage = 25.690 m
BS reading 0.665 m to bench mark
(RL 25.685 m)

Calculate the staff readings required to set out sight rails at 0, 30, and 55 m chainages.

Answer

(a) Invert level at 0 m chainage = 24.210 m
    Fall = 1/50th of distance
    Therefore fall
        (0 to 30 m) = 1/50 × 30 = $\underline{-0.600 \text{ m}}$
            Invert level at 30 m = 23.610 m
    Fall 30 to 55 m = 1/50 × 25 = $\underline{-0.500 \text{ m}}$
            Invert level at 55 m = $\underline{23.110 \text{ m}}$

(b) Ground level at 0 m chainage
                        = 25.690 m
    Sight rail level
    (1 m above ground) = 26.690 m
        Therefore traveller = rail level – invert level
                        = 26.690 – 24.210
                        = 2.480 m
    This length of traveller is unsuitable and would be rounded up to 2.500 m.

(c) Sight rail level = invert level + traveller
    Therefore at 0 m chainage
        Rail level = 24.210 + 2.500 = 26.710 m
    At 30 m chainage
        Rail level = 23.610 + 2.500 = 26.110 m

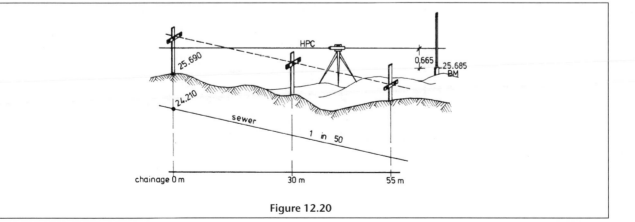

Figure 12.20

At 55 m chainage
Rail level = 23.110 + 2.500 = 25.610 m

(d) HPC = 25.685 + 0.665 = 26.350

(e) Staff readings required to set out rails =
HPC – rail level
At   0 m = 26.350 – 26.710 = –0.360
At 30 m = 26.350 – 26.110 = 0.240
At 55 m = 26.350 – 25.610 = 0.740

It should be noted that at chainage 0 m, the line of the sight through the telescope (HPC) is actually lower than the proposed sight rail level, and the resultant staff reading is therefore a negative value (–0.360 m). In order to establish the rail, an inverted staff reading of 0.360 m is required. In practice it usually proves impossible to turn the staff upside down since it would be resting on the ground, in which case the line of sight is marked against the profile peg in pencil and 0.360 m is measured upwards using a tape to establish the profile level.

Exercise 12.6

1 Figure 12.1 shows the proposed development at Lochview Road. Table 12.12 shows the data relating to proposed foul manholes F1 to F4. The foul sewer is to fall from manhole F4 to manhole F1 at a gradient of 1 in 40.

Table 12.12

Manhole number	Chainage (m)	Ground level (pegged)	Proposed invert level
F1	0.0000	6.660	2.950
F2	24.000	6.410	—
F3	48.000	5.870	—
F4	64.000	6.020	—

Calculate:

(a) the proposed invert level at manholes F2, F3 and F4,
(b) the profile board levels at all four manholes, given that the length of the traveller is 4.00 m,

(c) the heights to be measured up from the ground level pegs to establish the profile boards at their correct levels.

12. Setting out roadway levels

Setting out roadway profiles is essentially the same operation as setting out drain and sewer profiles. The profile board must be set at some predetermined height above the roadway formation levels in the same way as sewer profiles are set some height above the invert level.

The problem therefore lies simply in calculating the various formation levels along roadway slopes and vertical curves and adding a suitable height, say 1.0 metre, to all of them to produce profile levels. The profiles are then set out in the manner previously described.

Exercise 12.7

1 Table 12.13 and Fig. 12.21 show the formation and profile levels of a roadway vertical curve which was the subject of Chapter 11 Sect 7(e) (page 228). The profiles are 1m above Formation level.

Calculation the staff readings required to set out the profiles from a nearby BM (RL 262.065 m), given that the BS reading is 1.785 m.

Table 12.13

Chainage (m)	Formation level	Profile level (formation level + 1.000 m)
0	260.800	261.800
25	261.144	262.144
50	261.675	262.675
75	262.394	263.394
100	263.300	264.300

13. Setting out slope stakes

Roadway earthworks must inevitably form either embankments or cuttings, both of which forms of earthwork have side slopes. In order to construct these earthworks with correct side slope gradients, sloping

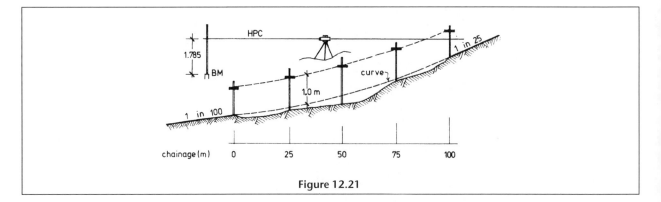

Figure 12.21

profiles, called slope stakes or batter boards, are erected close to, but clear of, the top edge of proposed cuttings and close to the toe of proposed embankments as in Fig. 12.22(a).

Chapter 6 describes how and where levels should be taken in order to produce suitable data for the drawing of longitudinal and cross sections. These levels are also used in the planning of slope staking and are shown in Fig. 12.22(a).

The first task in slope staking is to obtain a gradient of the existing ground surface across the line of the earthwork at the position where the slope stakes are to be erected.

Figures 12.22(b) and 12.22(c) show respectively an embankment and a cutting where ground levels have been surveyed at points A, B, C, D and E, from which the two existing gradients AB and DE are calculated. In both cases for ease of calculation a continuous ground slope gradient (1 in N), between points A and E, has been calculated from the levels as 1 in 10 (i.e. $N = 10$.)

(a) Embankment slope stakes

Consider firstly the embankment of Fig. 12.22(b) where the proposed formation level is 10.70 m AD and the side slopes have gradients of 1 in S, in reality 1 in 2 (i.e. $S = 2$).

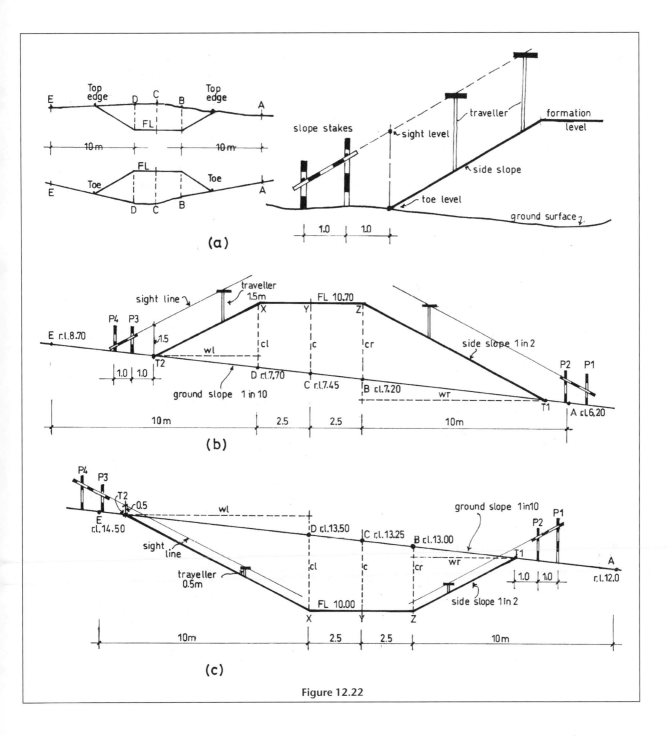

Figure 12.22

1. *Toe positions* As already stated, slope stakes should be placed clear of the toe of the embankment to avoid damage and would probably be erected at 1 metre and 2 metres from the toe. The toe positions T1 and T2 must therefore be established as a first step.

The principle of converging gradients is used to find those toe positions. [This principle is fully explained in Chapter 14 (Volumes) where it is most relevant and need not be repeated here.]

In Figure 12.22(a), height XD (called *cl*) = (10.70 – 7.70) = 3.00 m. By converging grades principle,

$$wl = cl \times \frac{NS}{(N+S)}$$
$$= 3.00 \times \frac{20}{12}$$
$$= \underline{5.00 \text{ m}}$$

Since the side slope gradient is 1 in 2, the proposed level of point T2 equals (10.70 – 5.00/2) = 8.20 m. As a check, level T2 = (7.70 + 5.00/10) = 8.20 m.

Height ZB (called *cr*) = (10.70 – 7.20) = 3.50 m.

$$wr = cr \times \frac{NS}{(N-S)}$$
$$= 3.50 \times \frac{20}{8}$$
$$= \underline{8.75 \text{ m}}$$

Since the side slope gradient is 1 in 2, the proposed level of point T1 equals (10.70 – 8.75/2) = 6.325 m. As a check, level T1 = (7.20 – 8.75/10) = 6.325 m.

2. *Sight line levels* The slope stake profiles could be placed at the embankment toes, and indeed frequently are, causing subsequent calculations to be greatly simplified. However, in this case, they are to be placed at distances of 1 m and 2 m from the toe.

Slope stakes are used in conjunction with travellers, which in this case are to be 1.5 m long. The sight line levels at T2 and T1 are therefore (8.20 + 1.50) m = 9.70 m and (6.325 + 1.50) m = 7.825 m respectively.

Since the sight line has a gradient of 1 in 2, the profile level P3 = (9.70 – 1.00/2) = 9.20 m and the profile level P4 = (9.70 – 2.00/2) = 8.70 m.

The sight rail levels of pegs P2 and P1 are similarly calculated and are as follows: profile level P2 = (7.825 – 1.00/2) = 7.325 m and profile level P1 = (7.825 – 2.00/2) = 6.825 m.

Nails are hammered into the pegs at these levels and a wooden sloping board is added to complete the slope staking operation.

(b) Cutting slope stakes

As already explained the cutting shown in Fig. 12.22(c) is to be constructed in a location where the existing ground slope gradients is 1 in *N*, in reality 1 in 10 (i.e. *N* = 10)

The formation level of the cutting is 10.00 m AD and the side slopes have gradients of 1 in *S*, in reality 1 in 2 (i.e. *S* = 2).

1. *Top edge positions* Slope stakes should be placed clear of the edge of the cutting and in this case are to be erected at 1 metre and 2 metres from the edge. The top edge positions T1 and T2 must therefore be established as a first step.

Height XD (called *cl*) = (13.50 – 10.00) = 3.50 m. Again by converging gradients principle,

$$wl = cl \times \frac{NS}{N+S}$$
$$= 3.50 \times \frac{20}{8}$$
$$= \underline{8.75 \text{ m}}$$

Since the side slope gradient is 1 in 2 the proposed level of point T2 equals (10.00 + 8.75/2) = 14.375 m. As a check, level T2 = (13.5 + 8.75/10) = 14.375 m.

Using a traveller of 0.5 m, the sight line level at T2 is (14.375 + 0.50) = 14.875 m and since the side slope gradient is 1 in 2, the profile level of peg P3 is (14.875 + 1.00/2) = 15.375 m. Similarly the profile level of peg P4 is (14.875 + 2.00/2) = 15.875 m.

As before, the profile levels are marked on the profile pegs and the top edge of a sloping board is nailed to the profiles at these points.

Exercise 12.8

1 In Fig. 12.22(c) calculate the position and level of top edge point T1 and the profile levels of pegs P1 and P2.

14. Vertical control using laser instruments

In setting out the proposed levels of construction works, a levelling instrument is set up in order to project a plane of collimation over the site. Measurements are then made downwards from the plane of collimation.

Since the early 1980s, laser instruments have gradually taken over the task of setting out and are now common place on construction sites. The instruments can either sweep out a plane of laser light over the site (rotating beam laser) or set out a narrow beam of laser light on a predetermined gradient and direction (pipe laser).

(a) Rotating beam laser

Figure 12.13(a) shows the tripod mounted Sokkia LP30A levelling laser and Fig. 12.13(b) shows the Topcon RL30 in its floor mounted position. All of the leading instrument companies manufacture some kind of rotating beam laser which all work on very similar principles.

Rechargeable Ni–Cd or alkaline batteries provide the

power to generate a visible (red or green) or invisible laser beam which rotates at speeds from 30 to 600 rpm. Some instruments may be operated externally using AC/12vDC battery power.

All of the instruments are self-levelling. Some require approximate levelling using footscrews; others are totally automatic and do not require levelling vials or footscrews.

Automatic self-levelling is achieved by means of either an internal electronic levelling mechanism or a precision pendulum compensator. The horizontal accuracy of the beam is in the order of 5 to 10 seconds of arc.

Field procedure
1. In use, the instrument is either set up on a tripod or is free standing and is levelled using two or three levelling screws and an integral levelling vial. Others, for example the Pentax PLP50 series, are simply set approximately upright by eye. Either action brings the instruments within their automatic self-levelling range. The instruments shut off if they are accidentally knocked out of position and have to be relevelled.
2. The unit is switched on and the beam is swept out across the site at a selected speed. The range varies with the type of instruments but is of the order of 200 to 300 metres. The beam is in reality acting as a plane of collimation, the height of which is as yet unknown.
3. In order to find the HPC of the beam, a backsight is taken to a levelling staff held on a site bench mark. Clamped to the staff is a rod adapter (Fig. 12.23) which clamps the beam detector to the staff. The

detector is moved slowly up or down the staff until the reference centre of the beam is acquired, whereupon the staff reading is noted and the HPC calculated in the normal fashion.
4. A peg is placed firmly in its stakeout position and the staff reading required to position a profile board at its correct design level, or at a chosen height above the design level, is calculated. The beam detector is moved to this position and re-clamped to the staff.
5. The assistant holds the staff against the profile peg and slowly moves it up and down the peg until the detector again acquires the beam. An audible signal is emitted when the reference centre of the beam is acquired. A mark is made against the peg and a profile board is nailed to the peg at this position. Example 8 numerically illustrates the procedure

(b) Pipe laser

Figure 12.12 shows the Topcon TP-L4 series of pipe laser. It is typical of this class of laser manufactured by most of the leading survey companies. It has a cast aluminium housing which is totally waterproof to a depth of 5 metres and has rubber bumpers to protect it from falls or other mishaps.

Rechargeable alkaline batteries or a 12V DC connection provide the power to generate a highly visible green laser beam up to a distance of 150 metres.

The display panel (LCD) is sloped at 45° so it can be read from the surface when the instrument is in position in a trench. It can be operated by hand using the finger touch panel, or by remote control from a distance of 200 metres in front of the instrument or from 45 metres behind it.

Field procedure
1. In use the instrument is placed in or lowered into the drainage trench, approximately level, and the design gradient of the trench is entered at the control panel. The gradient can vary between −15 per cent and +40 per cent and can be returned to 0 per cent automatically for quick set-up changes. The beam can also be moved to right or left and returned automatically to the centre of the range. The gradient and line settings are then locked to avoid disturbance.
2. The instrument can also be set along any arbitrary site gradient by simply setting up a target at the far end of any drain run and using the automatic alignment button on the LCD or remote control unit. The beam automatically centres on the target for perfect alignment. This arrangement can also be utilized above ground.
3. The drainage pipes are then simply placed in the trench so that the laser beam passes through them. For perfect alignment and gradient the target can be

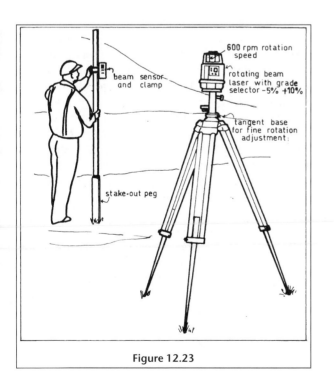

Figure 12.23

set on the inside of the pipe and the pipe moved to set the beam on the centre of the target.

15. Large-scale excavations

Consider Fig. 12.24 in which a large area of high ground is to be reduced to form a flat sports ground at formation level 50 m. Obviously heavy machinery has to be used to move large amounts of earth over a wide area and setting out by the normal processes requires either (a) the setting up of a great number of profiles, which is long, laborious, dangerous and very labour intensive, or (b) the automation of the setting-out process.

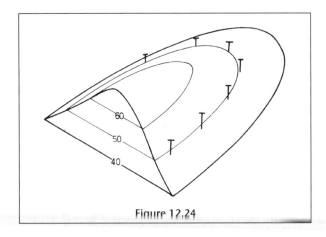

Figure 12.24

(a) Setting out profiles

The usual procedure in setting out profiles for large-scale work is as follows:

1. The sight rail level is calculated. This is normally 1.00 m above formation level = 51.00 m.
2. A levelling is made from a nearby TBM and the 50 m contour is traced on the ground. Sight rail up-rights are driven in around the contour at intervals. If large earth-moving plant is being used it is wise to move the upright outside the area of excavation.
3. The cross-pieces are nailed to the uprights at 51.00 m level using the collimation system of levelling.
4. The length of traveller in this case is 51.00 − 50.00 = 1.00 m. The length of the traveller is written on the back of the sight rail in paint or water-proof chalk.

Total station instruments are also frequently used for this work. The HPC of the instrument is obtained by an observation to a site bench mark. The prism height is set to same height as the instrument and moved around the site until the ground level of 50.00 m is obtained. (The survey team would have a fair idea of the location of such a point.) The data is stored in the memory of the particular program being used by the total station and when a further point is observed, the instrument is able to calculate the difference in height between it and the original point. The assistant is then easily guided to a point where the level is the required 50 m.

(b) Jobsite automation

On a construction site the process of moving around large quantities of earth has many different phases and in order to avoid repetition of staking and control, automation of the levelling process is desirable. The most obvious form of automation is to have the control system mounted on the machinery, so a laser receiver is mounted on the blade or bucket of an earth-moving machine.

The HPC of the rotating beam laser is computed in the usual manner and the signal is picked up by the detector mounted on the bulldozer as soon as it comes within range of the laser beam. The sensor's display arrows show the driver of the bulldozer the direction in which to move the blade of the machine. The simplest laser detectors can receive a signal within a 200° arc horizontally and 200 mm vertically. More advanced versions have a 360° horizontal capability.

The sensor is in reality equivalent to the profile board in a traditional slope-staking situation. The bottom of the blade of the bulldozer has to be at design level, however, so the sensor has to be set at a known height above the blade. The height is the difference between the two. The sensor is mounted on a telescopic mast and the height can be incremented to mm accuracy. By keeping the laser beam in the centre of the sensor, the driver steers the machine and the laser signal is used directly to control the hydraulics to steer the blade

Three-dimensional machine control has very recently been accomplished by Topcon. They have developed a 'Windows CE' three-dimensional control box. The complete jobsite design data are loaded via a portable PC or by a flashcard and are stored in the internal memory of the control box.

On site, a GPS receiver is set up at a known control station where its GPS coordinates are received and transformed into site coordinates. The GPS correction is therefore determined and this correction is sent by radio to a second GPS receiver 'rover' unit mounted on the machine. The rover now knows its position. However, GPS cannot provide the height with sufficient accuracy, so a rotating beam laser is required. Such an instrument can transmit only a very narrow beam (horizontal or tilted) which is difficult to detect in a moving bulldozer. Topcon have developed a 'Lazer-Zone' transmitter which transmits a 10 metre high zone of laser light. The machine sensor easily picks up this signal and computes its vertical position very accurately (Fig. 12.25).

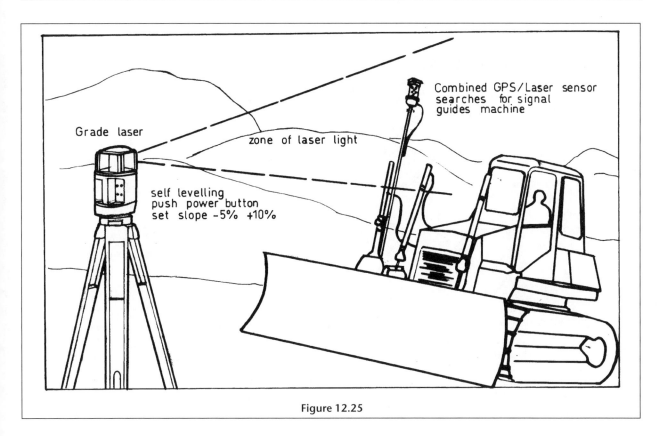

Figure 12.25

By comparing its real-time three-dimensional site position with the design data 3D position (already stored in the control box in the machine cabin), the operator can guide the machine to perform grading operations with millimetre accuracy. The system, not surprisingly, is called 'millimetre GPS' which, it is claimed, can accomplish grade finishing of roadways etc. with the benefits of GPS and the accuracy of a total station.

The system can also be used in real-time kinematic GPS surveys and stake-out work. The survey is carried out as for a normal RTK survey with base and rover stations operating as normal. The positioning zone laser, called the PZL-1, is set up at some convenient point and its collimation height calculated by reference to a site bench mark. A positioning zone sensor (the PZS-1) is mounted on the rover range pole where it receives the laser signal to give instant millimetre-accurate ground levels.

16. Answers

Exercise 12.1

1. Bearing R4–F = 123° 42′ 25″; Length = 11.223 m

Exercise 12.2

1. Long chord
$$\begin{aligned} &= 2(R \sin \theta/2) \\ &= 2(10 \sin 44.579\ 167°) \\ &= \underline{14.038\text{ m}} \end{aligned}$$

$$\begin{aligned} K &= (R^2 - 7.019^2)^{1/2} \\ &= 7.123 \text{ m} \end{aligned}$$
$$\begin{aligned} \text{Major offset} &= R - k \\ &= (10 - 7.123) \\ &= \underline{2.877 \text{ m}} \end{aligned}$$

$$\begin{aligned} \text{Quarter pt offset} &= (R^2 - x_n^2)^{1/2} - k \\ &= (10^2 - 3.509^2)^{1/2} - 7.123 \\ &= \underline{2.241 \text{ m}} \end{aligned}$$

Exercise 12.3

1. (a) See Fig. 12.26.

(b)

Point	East (m)	North (m)
A	0.00	5.25
B	30.00	5.25
C	25.40	3.50
D	25.40	21.70
E	0.00	45.00
F	9.40	10.80
G	17.50	16.70

(c)

Line	Δ East	Δ North	Length (m)	Bearing
AB	30.00	00.00	30.00	90° 00′ 00″
AC	25.40	−1.75	25.46	93° 56′ 29″
CD	00.00	18.20	18.20	360° 00′ 00″
DE	−25.40	23.30	34.47	312° 31′ 51″
AF	9.40	5.55	10.92	59° 26′ 29″
FG	8.10	5.90	10.02	53° 55′ 50″

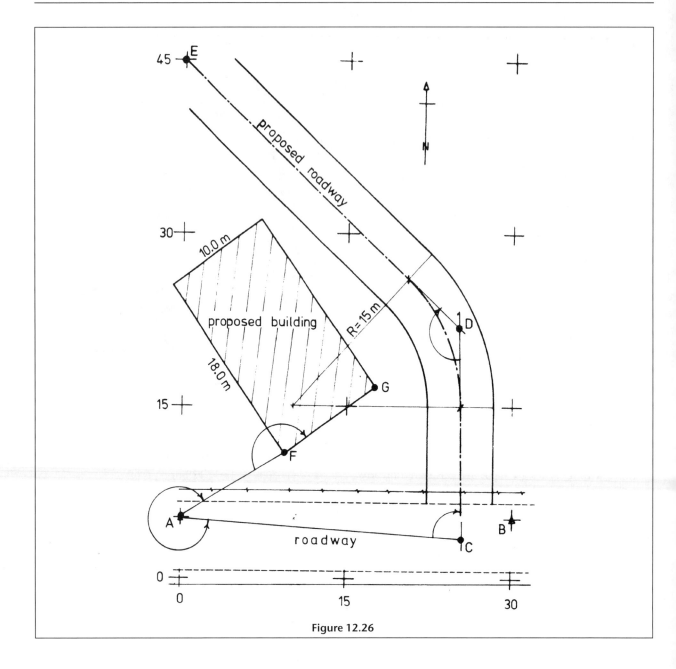

Figure 12.26

Angle	Value
BAC	3° 56′ 29″
ACD	86° 03′ 31″
CDE	132° 31′ 51″
CAF	325° 30′ 00″
AFG	174° 29′ 21″

Exercise 12.4

1. E_F = 3067.89 m
N_F = 1544.33 m

2. E_R = 2326.28 m
N_R = 1107.78 m

Exercise 12.5

1 Table 12.14

Table 12.14

BS	IS	FS	HPC	Reduced level	Remarks
2.360			26.230	23.870	Bench mark
	0.730		26.230	25.500	Upper floor level
		1.930	26.230	24.300	Lower floor level

Exercise 12.6

1 Table 12.15

Table 12.15

Manhole number	Chainage (m)	Ground level (pegged)	Proposed invert level	Profile level	Height from peg to profile
F1	0.00	6.660	2.950	6.950	0.290
F2	24.00	6.410	3.550	7.550	1.140
F3	48.00	5.870	4.150	8.150	2.280
F4	64.00	6.020	4.550	8.550	2.530

Exercise 12.7

1 Table 12.16

Table 12.16

BS	IS	FS	HPC	Reduced level	Profile level	Remarks
1.785			263.850	262.065		Bench mark
	2.050		263.850		261.800	Profile 0 m
	1.706		263.850		262.144	Profile 25 m
	1.175		263.850		262.675	Profile 50 m
	0.456		263.850		263.394	Profile 75 m
		−0.450	263.850		264.300	Profile 100 m

Note. The required staff reading to the profile peg (100 m) shows that an inverted staff reading 0.450 m is required.

Exercise 12.8

1. Top edge position is 7.5 m from centre of cutting
   Profile level P1 = 14.00 m
   Profile level P2 = 13.50 m

Chapter summary

In this chapter, the following are the most important points:

- Setting out in the construction sense means setting on the ground a series of stake-out pegs or profile boards to demarcate the exact horizontal and/or vertical positions of proposed construction works. These stake-out trappings take many forms and are shown in Figs 12.2 and 12.3 together with a description of how to physically insert a stake-out peg in a predetermined position.

- The complete range of surveying instruments (tapes, levels, theodolites, EDM and GPS) is used together with purpose-built rotating and beam lasers plus laser plummets, to set out proposed works.

- Roads and sewers are set out first to provide access and main drainage to the site. The centre line coordinates are obtained from the site plan and bearings and distances between any two points, say X and Y, are computed, if required, using the formulae:

  Distance $XY = (\Delta E_{XY}^2 + \Delta N_{XY}^2)^{\frac{1}{2}}$

  Bearing $XY = Sin^{-1} (\Delta E_{XY}/\Delta N_{XY})$

- The pegs denoting the centre lines and changes of direction of roads and sewers may be located by

  1. setting the theodolite over each peg as it is set out in the form of a traverse, or
  2. setting out coordinates from a remote control station by EDM tacheometric methods. The latter method is more open to error than the traverse method and stake-out points must be checked independently from a second control point.

- Buildings vary in size from small dwelling houses to huge commercial or industrial sheds and each requires its own stake-out method. Houses are usually set out by tape from the building line. The building is treated as an overall rectangle and right-angled corners are set out by tape. In any right-angled triangle ABC where angle C is 90°, the sides are in the ratio of:

  $a = (2n + 1)$, $b = 2n(n + 1)$ and $c = 2n(n + 1) + 1$, where n is any number.

  EDM is the preferred method of establishing corners of large buildings but is not sufficiently accurate for setting out steel columns, hence steel tapes are used. The measurements must be corrected for errors of standardization, tension (unless a constant-tension handle is used), temperature and slope (though most measurements will be horizontal).

- High-rise buildings obviously have to be built vertically. Verticality is achieved by using a theodolite, fitted with a diagonal eyepiece for steep sights, set up at strategic positions around the building. Lines of sight are then transferred upwards to each floor. Greater accuracy and speed are achieved by using a laser plummet. The instrument automatically sets out a vertical laser beam from which measurements can be made at any floor.

- Proposed levels of roads and buildings are called formation and foundation levels respectively. They are established on profiles (Fig. 12.2), set at a known height above the proposed levels, by level and staff.

  Since precision in vertical control is paramount, a new generation of laser instruments is used to achieve this with relative ease.

  A rotating beam laser is set up at a convenient location on site and its collimation height found by sighting a bench mark. Thereafter, proposed levels are established by simply measuring down (or up) from the laser plane by use of a beam sensor or locator.

- The bottom of the inside of a sewer pipe is its invert level. Since sewers are below ground, profile boards are set up at some known height above the invert level of the sewer using a level and staff. The invert level is then established using a traveller (Fig. 12.2). Alternatively, a beam laser may be set up in a sewer trench at a known invert level and set to the sewer gradient. The laser beam then defines the gradient along its complete length.

- The side slopes of cuttings and embankments are set out using slope stakes and batter boards, set to the gradient of the earthwork sides. The calculations involve the use of the principle of converging gradients and are best explained by reference to the text and diagram on pages 303–304.

- Large-scale excavations (e.g. industrial site clearance) may be controlled by profiles as before but the use of rotating lasers is much simple and more accurate. Three-dimensional machine control can now be achieved using a combination of GPS for horizontal control and laser for the vertical element. The complete three-dimensional real-time position of the earth-moving machine is compared with the previously loaded design position, held in a computer in the cabin. The machine then makes any necessary adjustments to achieve grade and position.

CHAPTER 13 **Mensuration – areas**

In this chapter you will learn about

- the formulae used to calculate the areas of regularly and irregularly shaped figures, including Simpson's rule, trapezoidal rule and prismoidal rule
- the methods of calculating areas from survey field notes and from coordinates
- the graphical and mathematical methods of calculating areas from plans and mechanical methods using mechanical and digital planimeters

---

On even the smallest site, calculations have to be made of a wide variety of areas and volumes, e.g. the area of the site itself, the volume of earthworks, cuttings, embankments, etc. Many of the figures encountered can be calculated by the direct application of the accepted mensuration formulae, but very often the figures are irregular in shape.

Regardless of the shape of the area, relevant data have to be obtained in some way in order to make the calculations:

1. The data are gathered in the field by some form of survey and properly recorded as survey notes.
2. The area is then calculated directly from the field notes.
3. The data are converted into coordinates or a plotted plan from which the area is computed.
4. The data already exist in the form of a map or plan, e.g. Ordnance Survey plans.

1. Regular areas

Figure 13.1 shows the common regular figures and the formulae required to calculate their areas. In modern construction practice, many developments include shapes comprised of several of the common geometrical figures.

EXAMPLES

1 The central piazza of a town centre development is shown in Fig. 13.2 (not to scale). The piazza has been split into six figures on the plan in order to calculate its area. Using the relevant surveyed dimensions, calculate the total area of the piazza in m².

Answer

(1) Square
    11.5×11.5 $\qquad$ = 132.25
(2) Rectangle
    12.1×20.8 $\qquad$ = 251.68
(3) Triangle
    $0.5 \times 12.1 \times 8.0$ $\qquad$ = 48.40
(4) Triangle
    $0.5 \times 5.1 \times 11.2$ $\qquad$ = 28.56
(5) Trapezoid
    $0.5 \times (5.1 + 14.7) \times 17.6$ = 174.24
(6) Trapezoid
    $0.5 \times (14.7 + 10.0) \times 11.5$ = 142.03

$$\text{Total} = \overline{777.16 \ m^2}$$

2 Figure 13.3 shows the dimensions of a small grassed plot of ground in a town centre garden. Calculate the area of the plot in square metres.

Answer

Area
(1) Rectangle
    15.00×10.00 $\qquad$ = 150.00 m²
(2) Trapezoid
    $1/2 \ (15.50 + 10.00) \times 9.50 = 121.13$ m²
(3) Quarter circle
    $\pi/4 \times 15.00^2$ $\qquad$ = 176.71 m²
(4) Triangle

$$\text{Area} = \sqrt{s(s-a)(s-b)(s-c)}$$

$a = 13.44$	$s - a = 6.27$
$b = 10.98$	$s - b = 8.73$
$c = 15.00$	$s - c = 4.71$
$2s = 39.42$	Check = 19.71
$s = 19.71$	

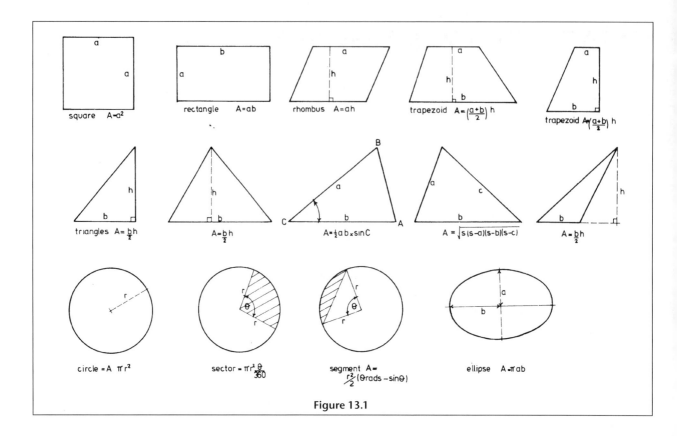

Figure 13.1

$$\text{Area} = \sqrt{s(s-a)(s-b)(s-c)}$$

$$= \sqrt{5081.47} \qquad = \underline{71.28 \text{ m}^2}$$

$$\text{Total area of plot} = 519.12 \text{ m}^2$$

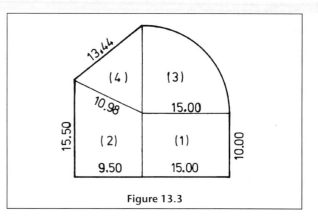

Figure 13.3

Figure 13.2

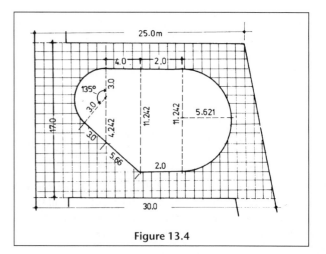

Figure 13.4

✍ Exercise 13.1

1 Figure 13.4 shows the surveyed dimensions of the concourse of a shopping centre which is to be floored with terrazzo tiles. The central area is to be a water feature with a fountain. Calculate the total area to be tiled.

2. Irregular areas

A plot of ground having at least one curved side is considered to be irregular in shape, unless, of course, the curved side forms part of a circle. The curved side or sides preclude the use of the regular geometric formulae and use has to be made of two rules, namely (a) the trapezoidal rule and (b) Simpson's rule, to calculate the area of the irregularly shaped plot.

(a) Trapezoidal rule

Figure 13.5 shows the field notes of the linear survey of a plot of ground, lying between a straight kerb PQ and a curved fence. The length of the kerb is 60 metres and the offsets are taken at regular intervals of 10 metres.

Calling each offset y, the area between *any* two offsets is calculated thus:

Area between chainage 20 m
and chainage 30 m $= \frac{1}{2}(y_{20} + y_{30}) \times 10$

Therefore total area

$$
\begin{aligned}
&= \tfrac{1}{2}(y_0 + y_{10}) \times 10 + \tfrac{1}{2}(y_{10} + y_{20}) \times 10 \\
&\quad + \tfrac{1}{2}(y_{20} + y_{30}) \times 10 + \ldots + \tfrac{1}{2}(y_{50} + y_{60}) \times 10 \\
&= \tfrac{1}{2} \times 10(y_0 + y_{10} + y_{10} + y_{20} + y_{20} + y_{30}) \\
&\quad + \ldots + y_{50} + y_{60}) \\
&= \tfrac{1}{2} \times 10(y_0 + y_{60} + 2y_{10} + 2y_{20} + 2y_{30} + 2y_{40} + 2y_{50}) \\
&= 10\left(\frac{y_0 + y_{60}}{2} + y_{10} + y_{20} + y_{30} + y_{40} + y_{50}\right)
\end{aligned}
$$

This is the *trapezoidal rule* and is usually expressed thus:

Area = strip width × (average of first and last offsets + sum of others)

In Fig. 13.5 the area is as follows:

$$
\begin{aligned}
\text{Area} &= 10\left(\frac{4 \times 4}{2} + 4.5 + 5.1 + 6.5 + 6.3 + 5.1\right) \\
&= 315.0 \text{ m}^2
\end{aligned}
$$

(b) Simpson's rule

The area can be found slightly more accurately by Simpson's rule. A knowledge of the integral calculus is required to prove the rule but it can be shown to be

Area $= \frac{1}{3}$ strip width (first + last offsets + twice sum of odd offsets + four times sum of even offsets).

Note. (a) There must be an *odd* number of offsets.
(b) The offsets must be at regular intervals.

Using Simpson's rule the area between the line PQ and the road is as follows:

$$
\begin{aligned}
\text{Area} &= 10/30[y_0 + y_{60} + 2(y_{20} + y_{40}) + 4(y_{10} + y_{30} + y_{50})] \\
&= 10/3[4 + 4 + 2(5.1 + 6.3) + 4(4.5 + 6.5 + 5.1)] \\
&= 10/3[8 + 2(11.4) + 4(16.1)] \\
&= 317.3 \text{ m}^2
\end{aligned}
$$

Example 3 illustrates the method of dealing with irregularly shaped areas. In general terms, ordinates, i.e. the offsets, are measured at right angles to a base line and Simpson's or trapezoidal rules applied.

EXAMPLE

3 Figure 13.6 shows an irregularly shaped pond, drawn to scale 1:500.

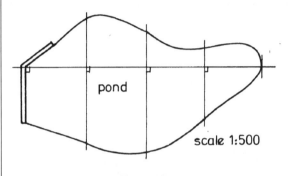

Figure 13.6

(a) Divide the figure into a suitable number of strips of the same width, using a 1:500 scale rule.
(b) Draw lines at right angles (offsets) at each strip with interval and measure the width of the pond along each offset.
(c) Using the trapezoidal rule and Simpson's rule, calculate the area of the pond.

Answer

(a) Since Simpson's rule is to be used, the pond must be divided into an even number of strips which produces an *odd* number of offsets, namely five. The strip width is therefore 8.0 metres.

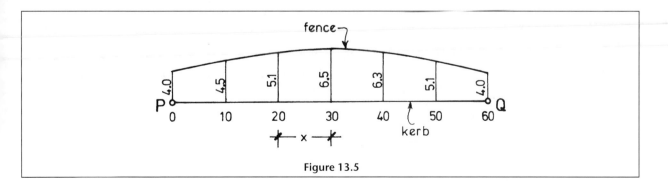

Figure 13.5

(b) The lengths of the offsets, measured by scale rule, are 7.6, 15.9, 16.3, 9.5 and 0.0 m.

(c) Using the trapezoidal rule,

$$\text{Area} = 8.0\left(\left(\frac{7.6 + 0.0}{2}\right) + 15.9 + 16.3 + 9.5\right)$$
$$= 8.0 + 45.5$$
$$= 364 \text{ m}^2$$

Using Simpson's rule,

$$\text{Area} = 8.0/3[7.6 + 0.0 + 2(16.3)$$
$$+ 4(15.9 + 9.5)]$$
$$= 8.0/3(141.80)$$
$$= 378.1 \text{ m}^2$$

3. Areas from field notes

(a) Linear surveys

In a linear survey the area is divided into triangles, and the lengths of the three sides of each triangle are measured. The area contained within any one triangle ABC is found from the formula

$$\text{Area} = \sqrt{s(s-a)(s-b)(s-c)}$$

where s is the semi-perimeter.

The boundaries of the linear survey are established by measuring offsets from the main lines. In Fig. 13.7 the area between the survey line and stream is composed of a series of trapezoids and triangles. The area of each figure must be computed separately.

The area between survey line PQ and the road is again composed of a series of separate figures. It must be noticed, however, that the offsets are at regular intervals of 15 metres in this instance and there is an *odd* number of offsets. Simpson's rule is therefore used to calculate the area.

Finally, along the line RP, there is an even number of offsets between R and P at regular 10 metre spacings. The area is calculated by the trapezoidal rule.

Exercise 13.2

1 In Fig. 13.7, PQR is the area of a proposed factory development taken from an OS map. PR is a new boundary fence denoting the northern limit of the site. A linear survey was made to determine the area of the site. The lengths of the sides are

PQ = 60.0 m
QR = 104.6 m
RP = 70.0 m

All offset dimensions are shown on the figure. Calculate the area of the site.

(b) Levelling

EXAMPLE

4 In Fig. 13.8 the survey of a proposed cutting shows that the depths at 20 m intervals are 0.0, 0.9, 1.5, 3.2 and 3.3 m. Given that the roadway is to be 5 m wide and that the cutting has 45° side slopes, calculate:

(a) the plan surface area of the excavation, ABCD,
(b) the actual area of the side slopes, ABE and CDF.

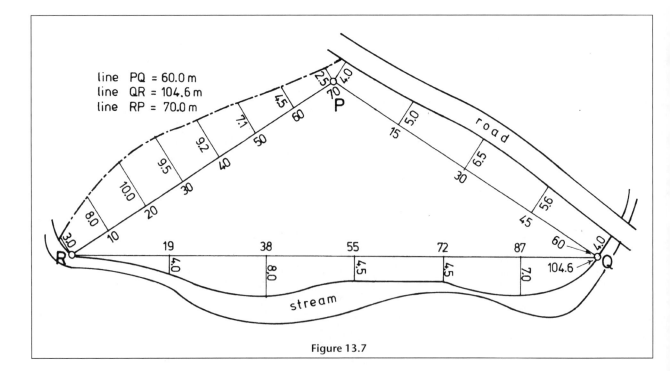

line PQ = 60.0 m
line QR = 104.6 m
line RP = 70.0 m

Figure 13.7

Answer

(a) Area ABCD

Chainage	Depth (d)	Top width [= (5.0 m + 2d)]
0	0.0	5.0
20	0.9	6.8
40	1.5	8.0
60	3.2	11.4
80	3.3	11.6

Using Simpson's rule:

$$\text{Area} = \frac{20}{3}[5.0 + 11.6 + 2(8.0)$$
$$+ 4(6.8 + 11.4)]\text{m}^2$$
$$= \frac{20}{3} \times 105.4$$
$$= 702.7 \text{ m}^2$$

(b) Plan areas ABE and CDF
  Area ABE + area CDF

$$= \text{area ABCD} - \text{area ADEF}$$
$$= 702.7 - (5 \times 80)$$
$$= 302.7 \text{ m}^2$$

Therefore area ABE = area CDF
$$= 151.35 \text{ m}^2$$
Actual side area = plan area ÷ cos 45°
$$= 151.35 \div 0.7071$$
$$= 214.04 \text{ m}^2$$

Alternatively, the side slope areas may be computed independently. At each chainage point the side width = depth since the slopes have 45° gradients.
Using Simpson's rule:

$$\text{Side area} = \frac{20}{3}[0.0 + 3.3 + 2(1.5) + 4(0.9 + 3.2)]$$
$$= \frac{20}{3} \times 22.7$$
$$= 151.33 \text{ m}^2 \text{ (plan)}$$

Using the trapezoidal rule:

$$\text{Side area} = 20\left(\frac{0.0 + 3.3}{2} + (0.9 + 1.5 + 3.2)\right)$$
$$= 145.0 \text{ m}^2 \text{ plan}$$
Actual area = 151.33 ÷ cos 45° = 214.02 m² or
145.0 ÷ cos 45° = 205.1 m²

Exercise 13.3

1 In Fig. 13.9, the survey of a proposed roadway earthwork shows the heights, at 30 metre intervals, to be 5.0 m (fill), 3.5 m (fill), 0.0 m, 2.4 m (cut) and 2.9 m (cut). The roadway is to be 5.0 m wide and the embankment sides are to slope at 45°. Calculate:

(a) the plan view surface area of the embankment,
(b) the actual area of the side slopes that have to be grassed.

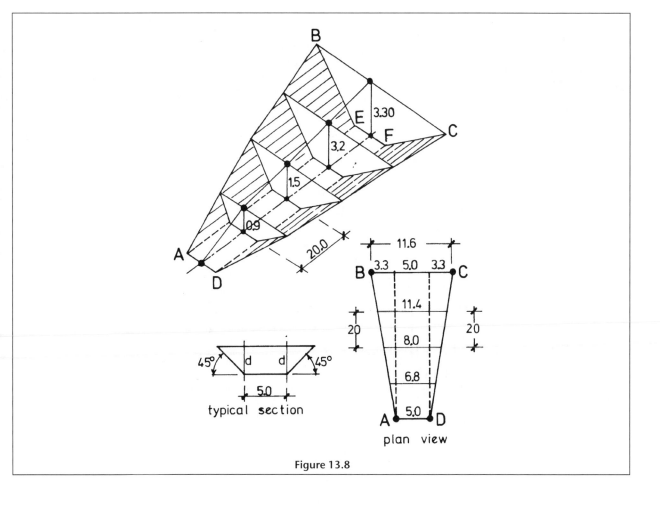

Figure 13.8

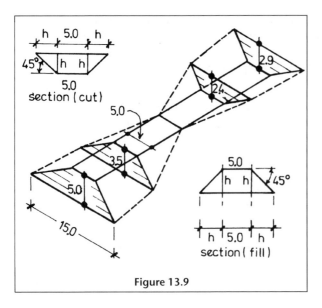

Figure 13.9

(c) Areas from coordinates

The calculation of areas is not made directly from the actual field notes but from the coordinate calculations made from the notes.

Method of double longitudes
The meanings of the terms, difference in easting (ΔE), difference in northings (ΔN), easting (E) and northing (N), have already been made clear in Chapter 8 and should be revised at this point.

One further definition is required, namely the longitude of any line is the easting of the mid-point of the line. In other words, the longitude of any line is the mean of the eastings of the stations at the ends of the line. In Fig. 13.10.

Longitude of line AB = $\frac{1}{2}$ (easting A + easting B)

Therefore,

Double longitude AB = (easting A + easting B)
$$= 30.0 + 90.0$$
$$= 120.0 \text{ m}$$

Figure 13.10 is an example of a coordinated figure where the coordinates of the stations are:

Station	Easting (m)	Northing (m)
A	30.0	60.0
B	90.0	100.0
C	120.0	20.0

Area of triangle ABC = area of trapezium 3BC1
$\qquad\qquad$ − area of trapezium 3BA2
$\qquad\qquad$ − area of trapezium 2AC1

i.e. triangle ABC $= \dfrac{E_3 - E_1}{2} \times (N_3 - N_1)$

$\qquad\qquad -\dfrac{E_3 + E_2}{2} \times (N_3 - N_2)$

$\qquad\qquad -\dfrac{E_2 + E_1}{2} \times (N_2 - N_1)$

Therefore 2 × triangle ABC
$$= (E_3 + E_1) \times (N_3 - N_1) - (E_3 + E_2) \times (N_3 - N_2)$$
$$- (E_2 + E_1) \times (N_2 - N_1)$$
$$= (210 \times 80) - (120 \times 40) - (150 \times 40)$$
$$\text{or} = (+210 \times +80) + (+120 \times -40)$$
$$+ (+150 \times -40)$$

The values of +210, +120 and +150 are the double longitudes of lines CB, BA and AC respectively, while the values +80, −40 and −40 are the partial latitudes of the same lines when moving around the figure in an anticlockwise direction.

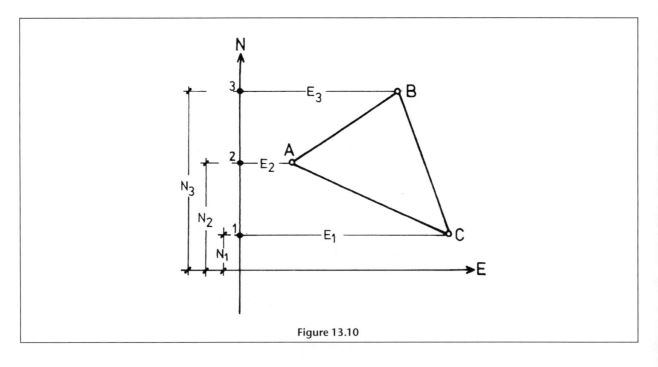

Figure 13.10

Table 13.1

Station	Total coordinates East (m)	Total coordinates North (m)	Line	Double longitude	Differences in northings	Double areas (m²) +	Double areas (m²) −
C	+120.0	+20.0					
B	+90.0	+100.0	CB	+210.0	+80.0	16 800	
A	+30.0	+60.0	BA	+120.0	−40.0		4 800
C	+120.0	+20.0	AC	+150.0	−40.0		6 000

$$16\ 800 \qquad 10\ 800$$
$$-10\ 800$$
$$=\ 6\ 000$$
$$\div 2 \ =\ 3\ 000\ \text{m}^2$$

Table 13.2

Station	Total coordinates East (m)	Total coordinates North (m)	Line	Double longitude	Differences in northings	Double areas (m²) +	Double areas (m²) −
A	+51.0	−150.2					
B	+300.1	−24.6	AB	+351.1	+125.6	44 098.16	
C	+220.1	+151.3	BC	+520.2	+175.9	91 503.18	
D	−50.0	+175.0	CD	+170.1	+23.7	4 031.37	
E	−125.2	−51.1	DE	−175.2	−226.1	39 612.72	
A	+51.0	−150.2	EA	−74.2	−99.1	7 353.22	

$$\text{Double area} = 186\ 598.65\ \text{m}^2$$
$$\text{Area} = 93\ 299.325\ \text{m}^2$$
$$= 9.3299\ \text{hectares.}$$

Therefore $2 \times$ triangle ABC = 16 800 − 4800 − 6000
$$= 6000\ \text{m}^2$$
$$\text{Triangle ABC} = 3000\ \text{m}^2$$

In order to find the area of any polygon, the following is the sequence of operations:

1. Find the double longitude and difference in northings (ΔN) of each line.
2. Multiply double longitude by ΔN.
3. Add these products algebraically.
4. Halve the sum.

The calculations are more neatly set out in tabular form as in Table 13.1, where the area of triangle ABC is calculated. It should be noted that a negative result is perfectly possible. For example, if the coordinates in Table 13.1 were written in a clockwise direction the area would have been −3000 m². In such cases the negative sign is ignored.

5 The coordinates shown below refer to a closed theodolite traverse ABCDEA:

Station	Easting (m)	Northing (m)
A	+51.0	−150.2
B	+300.1	−24.6
C	+220.1	+151.3
D	−50.0	+175.0
E	−125.2	−51.1

Calculate the area in hectares enclosed by the stations.

Answer Table 13.2

Method of total coordinate products
Figure 13.11 shows the same triangular area as Fig. 13.10, where the coordinates of the stations are:

Station	Easting (m)	Northing (m)
A	30.0	60.0
B	90.0	100.0
C	120.0	20.0

Area of triangle ABC = area of trapezium 1AB2
+ area of trapezium 2BC3
− area of trapezium 1AC3

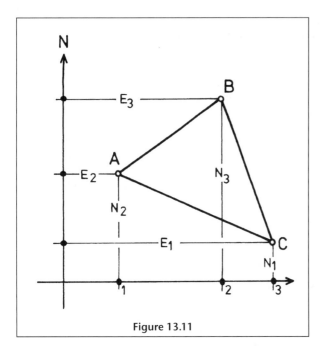

Figure 13.11

i.e.

$$\text{area of triangle ABC} = \left(\frac{N_2 + N_3}{2}\right) \times (E_3 - E_2)$$
$$+ \left(\frac{N_3 + N_1}{2}\right) \times (E_1 - E_3)$$
$$- \left(\frac{N_2 + N_1}{2}\right) \times (E_1 - E_2)$$

Therefore (2 × area) of triangle ABC

$$= (N_2 + N_3) \times (E_3 - E_2) + (N_3 + N_1) \times (E_1 - E_3)$$
$$- (N_2 + N_1) \times (E_1 - E_2)$$
$$= N_2E_3 - N_2E_2 + N_3E_3 - N_3E_2 + N_3E_1 - N_3E_3$$
$$+ N_1E_1 - N_1E_3 - N_2E_1 + N_2E_2 - N_1E_1 + N_1E_2$$

Six of these twelve terms cancel. Therefore, when the remaining six terms are rearranged, the double area of the triangle ABC is

$$2 \text{ ABC} = N_1E_2 + N_2E_3 + N_3E_1$$
$$- N_1E_3 - N_2E_1 - N_3E_2$$
$$= (E_1N_3 + E_2N_1 + E_3N_2)$$
$$- (E_1N_2 + E_2N_3 + E_3N_1)$$
$$= [\text{sum (easting of station}$$
$$\times \text{northing of preceding station)}$$
$$- \text{sum (easting of station}$$
$$\times \text{northing of following station)]}$$

The final result is easily remembered by writing the array in the following manner:

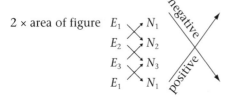

The following is the sequence of operations in the calculation of the area of any polygon:

1. Write the array of eastings and northings as shown above. Note that E_1 and N_1 are repeated at the bottom of the columns.
2. Multiply the departure of each station by the latitude of the preceding station and find the sum.
3. Multiply the departure of each station by the latitude of the following station and find the sum.
4. Find the algebraic difference between operations 2 and 3 above.
5. Halve this figure to give the area of the polygon. Again the results are always set out in tabular form as in Table 13.3.

Table 13.3

	Table coordinates		Areas	
	East (m)	North (m)	E_2N_1, etc.	E_1N_2, etc.
Station	(E)	(N)	+	−
A	+30.0	+60.0		
B	+90.0	+100.0	+5 400	+3 000
C	+120.0	+20.0	+12 000	+1 800
A	+30.0	+60.0	+600	+7 200
			+18 000	+12 000
			6 000	
		÷2 =	3 000 m²	

EXAMPLE

6 The coordinates listed below refer to a closed traverse PQRS:

Station	Easting (m)	Northing (m)
P	+35.2	+46.1
Q	+162.9	+151.0
R	+14.9	+218.6
S	−69.2	−25.2

Calculate the area in hectares enclosed by the stations.

Answer Table 13.4

Exercise 13.4

1 The coordinates listed below refer to a closed theodolite traverse.

Station	Easting (m)	Northing (m)
A	0.00	0.00
B	−34.39	34.39
C	−51.50	13.01
D	−12.07	−15.10
E	5.35	−28.78
F	16.34	−14.43

Calculate the areas enclosed by the stations, using
(a) the method of double longitudes,
(b) the method of products.

Table 13.4

Station	Easting (E)	Northing (N)	Area E_2N_1 (+ ve)	E_1N_2 (– ve)
P	+35.2	+46.1		
Q	+162.9	+151.0	7 509.69	5 315.20
R	+14.9	+218.6	2 249.90	35 609.94
S	–69.2	–25.2	–15 127.12	–375.48
P	+35.2	+46.1	–887.04	–3 190.12
			–6 254.57	37 359.54
			–37 359.54	
			–43 614.11	
		÷2 =	21 807.06 m²	

4. Measuring areas from plans

Several methods are available for calculating the area of a figure from a survey plot.

(a) Graphically

A piece of transparent graph paper is laid over the area, the squares are counted and the area is calculated by multiplying the area of a square by the number of squares.

(b) By Simpson's or trapezoidal rules

The area is divided into a series of equidistant strips.

The ordinates are measured and the rules applied as in previous examples.

(c) Mechanically

The area is measured, using a mechanical device known as a planimeter.

EXAMPLE

7 Figure 13.12 shows an irregular area drawn on a plan to a scale of 1:500. Calculate the area of the top of the embankment by the following methods:

(a) counting squares,
(b) Simpson's and trapezoidal rules.

Solution

(a) The graph paper shown superimposed on the area has squares of 10 mm side; therefore, for each square,

$$\text{Ground area} = (10 \times 500 \times 10 \times 500) \text{ mm}^2$$
$$= 100 \times 0.25 \text{ m}^2$$
$$= 25 \text{ m}^2$$
$$\text{Area} = (25 \times \text{number of squares}) \text{ m}^2$$
$$= (25 \times 21.75) \text{ m}^2$$
$$= 543.8 \text{ m}^2$$

(b) Consider the line marked xx as a base line and every line of the graph paper as an ordinate y, thereby producing seven in total (y_1 to y_7). The lengths of the respective ordinates are, by scaling, 16, 18.3, 20, 22.5, 23.8, 15.3 and 0 m, and the spacing of the ordinates is 5 m along the base line.

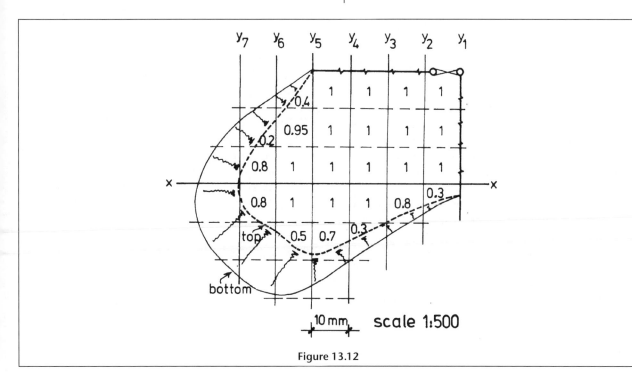

Figure 13.12

By Simpson's rule:

Area = $\frac{5}{3}[16 + 0 + 2(20 + 23.8)$
$+ 4(18.3 + 22.5 + 15.3)]$
$= 546.67$ m^2

By trapezoidal rule:

Area

$= 5\left(\frac{(16 + 0)}{2} + 18.3 + 20 + 22.5 + 23.8 + 15.3\right)$

$= 539.50$ m^2

(d) Planimeter

The area of any irregular figure may be found from a plan, by using a mechanical device for measuring areas known as a planimeter.

Three kinds of planimeter are available: (a) a fixed index model, (b) a sliding bar model and (c) a digital model.

Construction

The construction of the models is essentially the same (Fig. 13.13), consisting of:

1. An arm of fixed length, known as the polar arm. The polar arm rests within the pole block P, which, in turn, rests upon the plan in a stationary position.
2. A tracer arm carrying a tracer point, T, which can be moved in any direction across the plan.
3. Attached to both of these arms is the measuring unit, M, which is, in effect, a rolling wheel. As the tracer point moves, the rolling wheel rotates. The wheel is divided into ten units each of which is subdivided into ten parts. The drum therefore reads directly to hundredths of a revolution and a vernier reading against the drum allows thousandths of a revolution to be measured. A horizontal counting wheel is directly geared to the rolling wheel and records the number of complete revolutions. The counting wheel can be made to read zero by simply pressing the zero-setting button.

Use of planimeter

1. *Fixed index model* If only a small area is to be measured the following procedure is carried out:
   (a) Position the pole block outside the area.
   (b) Set the tracer point over a well-defined mark such as the intersection of two fences.
   (c) Press the zero-setting button. The instrument of course reads 0.000 revolutions.
   (d) Move the tracer point carefully around the boundary of the area being measured and return to the starting point.
   (e) Note the reading; let it be 3.250 revolutions.
   (f) Repeat all of the operations twice more and obtain a mean value of the number of revolutions of the wheel.

This particular model of planimeter reads the area in

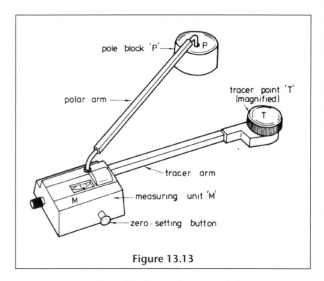

Figure 13.13

square centimetres. Each revolution of the measuring wheel is equivalent to 100 cm^2 of area.

In the example the area is therefore

(100×3.250)cm$^2 = 325.0$ cm^2

If the scale of the plan is full size, the actual area measured is 325 cm^2. If, as is likely, the scale is much smaller, for example 1:500, the actual area must be obtained by calculation.

On 1:500 scale, 1 cm = 500 cm
Therefore 1 cm$^2 = (500 \times 500)$ cm^2
$= \left(\frac{500 \times 500}{100 \times 100}\right)$ m^2
$= 25$ m^2
Therefore 325 cm$^2 = (25 \times 325)$ m^2
$= 8125$ m^2

If a large area is to be measured, account must be taken of the instrument's 'zero circle'. Every planimeter has a zero circle. When the polar arm and tracer arm form an angle of, say, 90° and the angle is maintained as the tracer point is moved round in a circle, the drum will not revolve and the area of the circle swept out on the plan by the polar arm will be zero. The manufacturer supplies the actual area of the zero circle in the form of a constant which is added to the number of revolutions counted on the drum.

If the enclosure in Fig. 13.14 is on a scale of 1:2500 and its area is required, the pole block is placed within the enclosure and the tracer point is moved around the boundary as before. The average number of revolutions after following the boundary three times is perhaps 5.290.

Number of revs = 5.290
Add zero circle constant = 22.300
Total revs = 27.590
Total cm$^2 = 27.59 \times 100$
$= 2759$ cm^2
On 1:2500 scale, 1 cm$^2 = \left(\frac{2500 \times 2500}{100 \times 100}\right)$ m^2
$= 625$ m^2

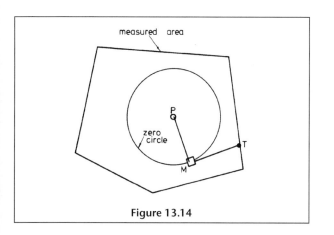

measured area

zero circle

P
Q

T

M

Figure 13.14

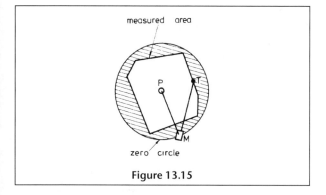

measured area

P
Q

T

M

zero circle

Figure 13.15

Therefore total area of

$$\text{enclosure} = 2759 \times 625$$
$$= 1\ 724\ 375\ \text{m}^2$$
$$= 1\ 72.438\ \text{hectares}$$

Great care must be taken when using the planimeter with the pole block inside the area. It is perfectly possible that the zero circle is larger in area than is the parcel of land being measured (Fig. 13.15). In such a case the rolling drum actually moves backwards and the second reading is subtracted from the first to obtain the area in square centimetres. If the first reading is called 10.000 instead of 0.000 the subtraction is simple; for example:

First reading = 10.000
Second reading = 7.535
Number of revs = 2.465

This area is, however, the area of the shaded portion of Fig. 13.15 and the true area is found by subtracting the number of revolutions from the zero circle constant:

Number of revs = 22.300 – 2.465
$$= 19.835$$
$$= 1983.5\ \text{cm}^2$$

If the scale of the plan is 1:100

True area = 1983.5 m^2

2. *Sliding bar model* On the sliding bar planimeter the tracer arm is able to slide through the measuring unit (Fig. 13.16) and can be clamped at any position.

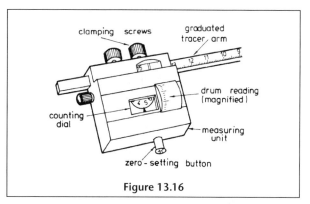

clamping screws

graduated tracer arm

drum reading (magnified)

counting dial

measuring unit

zero-setting button

Figure 13.16

The arm is graduated and any setting can be made against the index mark, which is attached to the measuring unit. A table on the base of the instrument gives the graduation reading for various scales in common use.

The number of revolutions is obtained by the methods previously described and the true area is calculated by multiplying by the appropriate conversion factor supplied by the manufacturer for the scale being used. For example, when the graduation setting for a scale 1:2500 is 9.92, one revolution of the wheel corresponds to a true area of 4 hectares.

If the pole block was positioned inside the enclosure, the mean readings for four measurements were:

First reading: 0.000   0.000   0.000   0.000
Second reading: 4.320   4.320   4.321   4.323

The true area is found as follows:

Mean of four measurements = 4.321
$$\text{Constant for zero circle} = \frac{25.730}{30.051}$$
$$\text{True area} = 30.051 \times 4\ \text{hectares}$$
$$= 120.204$$

3. *Digital model* Figure 13.17 illustrates the Sokkia N-Series Placom digital planimeter. The KP 80N model is the polar type version which closely resembles the traditional mechanical planimeter and the KP 90N is a roller type. Both models are powered by a rechargeable built-in NiCd battery or can be mains operated (100v to 240v) through an AC adapter.

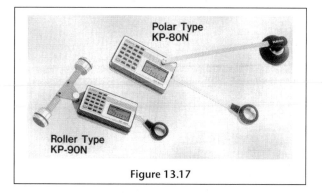

Polar Type
KP-80N

Roller Type
KP-90N

Figure 13.17

The planimeter is used in exactly the same manner as the mechanical model. It is placed outside the

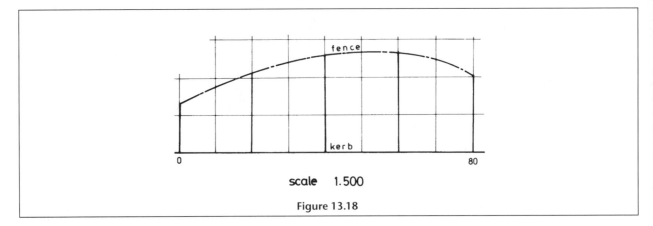

scale   1.500

Figure 13.18

boundary of a small area and the magnified tracer point is moved carefully around the site boundary to return to the starting point. The maximum perimeter of measurement is 300 mm.

On larger areas the instrument is placed inside the boundary area and the tracer arm is again moved around the boundary back to the initial point, in which case the maximum perimeter is increased to 800 mm.

The area of the enclosure is displayed digitally on a small screen in whatever units were chosen for the measurement, usually cm². Both models can store the measured data for future use.

It is not widely realized that digital instruments, like all planimeters, have a zero circle, which will still be present in measurements of a large area. However, the area of the circle is automatically included in the value shown on the screen and need not therefore be of any consequence to the user.

Together with an ability to measure areas, some digital planimeters can measure the length of the boundary traversed by the tracer point and may have the capability to compute the coordinates of points based on some chosen grid system. Other models can print out the results on a small integral printer.

Most survey companies manufacture some form of digital planimeter.

EXAMPLE

8 In Example 7, the area of the top of an embankment was calculated by (a) counting squares and (b) using trapezoidal and Simpson's rules. Check the area of the embankment in Fig. 13.12, which is drawn to scale 1:500, using a planimeter.

Answer

By planimeter (fixed index):

Number of revolutions = 0.2158
At 1:500, 1 rev = 2500 m²
Therefore area = 539.5 m²

Exercise 13.5

1 Figure 13.18 shows an irregular area of ground lying between a straight kerb and a curved fence drawn to scale 1.500. Calculate the area by:

(a) Simpson's and trapezoidal rules,
(b) counting squares,
(c) planimeter.

Exercise 13.6

1 Figure 13.19 is the sketch of a construction site, surveyed by linear means. The lengths of the sides are as follows:

AB 325 m, AD 195 m, DB 410 m,
DC 392 m, CB 260 m.

Calculate the area enclosed by the survey lines, in square metres.

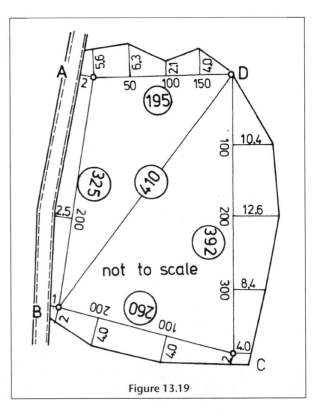

Figure 13.19

2 The site in Fig. 13.19 is bounded by fences along the sides AD, DC and CB and by a roadway along the side AB. The offsets to the fences and roadway are:

Line chainage (m)	Offset (m)	Line chainage (m)	Offset (m)
AB 0 (A)	2.0 right	AD 0 (A)	5.6 left
200	2.5 right	50	6.3 left
325 (B)	1.0 right	100	2.1 left
		150	4.0 left
		195 (D)	0.0
DC 0 (D)	0.0 left	CB 0 (C)	2.0 left
100	10.4 left	100	4.0 left
200	12.6 left	200	4.0 left
300	8.4 left	260 (B)	2.0 left
392 (C)	4.0 left		

Calculate the area of the site, in square metres.

3 Stations M, N, O, P and Q form a closed traverse. The following coordinates refer to the stations.

Station	Total latitude	Total departure
M	+2000	+2000
N	+3327	+1242
O	+4093	+2048
P	+3141	+3035
Q	+1192	+3572

Calculate the area in hectares enclosed by the stations.

4 Figure 13.20 shows an irregular parcel of ground bounded by a kerb on the south side and a fence on the north (scale 1.500). Calculate the area by:

(a) counting squares
(b) Simpson's rule
(c) trapezoidal rule
(d) planimeter.

5 Figure 13.21 shows a roadway cutting with cross sections drawn at 20 metre intervals. Calculate:

(a) the top width of each cross section
(b) the total plan area occupied by the cutting

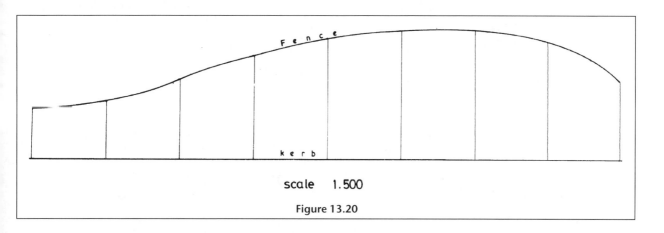

scale   1.500

Figure 13.20

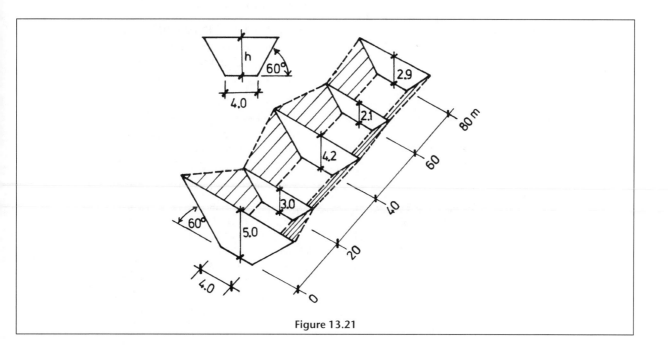

Figure 13.21

(c) the cross-sectional area of the trapezoidal sections at chainages 0.0 and 40.0 m.

5. Answers

Exercise 13.1

1 Overall area (trapezoid)

$$\text{Area} = 0.5(25.0 + 30.0) \times 17.0 = 27.5 \times 17$$
$$= 467.50^2$$

Water feature

1. Semi-circle
$$= \pi \times 5.621^2 \times 0.5 \qquad = 49.63$$
2. Rectangle
$$= 11.242 \times 2.0 \qquad = 22.48$$
3. Trapezoid
$$= 0.5 (11.242 + 7.242) \times 4.0 = 36.97$$
4. Triangle
$$= 0.5 (3 \times 3) \qquad = 4.50$$
5. Sector
$$= \pi \times 3^2 \times (135 \div 360) \qquad = \underline{10.60}$$

Total $= \underline{124.18 \text{ m}^2}$

Area to be tiled $= 467.50 - 124.18$
$$= 343.32 \text{ m}^2$$

Exercise 13.2

1 In triangle   PQ $= r = 60.0$ m $s - r = 57.3$
PQR   QR $= p = 104.6$ m $s - p = 12.7$
       RP $= q = 70.0$ m $s - q = 47.3$

Perimeter of PQR $= \underline{234.6 \text{ m}}$ Check $= \underline{117.3} = s$

Therefore semi-perimeter $s = 117.3$ m

Area of triangle PQR

$$= \sqrt{s(s - r)(s - p)(s - q)}$$
$$= \sqrt{117.3 \times 57.3 \times 12.7 \times 47.3}$$
$$= \sqrt{2009.3 \text{ m}^2}$$

Area between line RQ and stream is as follows:

Area of triangle (1) $= \frac{1}{2} \times 19 \times 4 = 38.0$
Trapezoid (2) $= \frac{1}{2}(4 + 8) \times (38 - 19) = 114.0$
Trapezoid (3) $= \frac{1}{2}(8 + 4.5) \times (55 - 38) = 106.25$
Rectangle (4) $= 4.5 \times (72 - 55) = 76.5$
Trapezoid (5) $= \frac{1}{2}(4.5 + 7) \times (87 - 72) = 86.25$
Triangle (6) $= \frac{1}{2}(104.6 - 87) \times 7 = 61.6$

Total $= \overline{482.6 \text{ m}^2}$

Area between line RP and fence is as follows:

$$\text{Area} = 10\left(\frac{3 + 2.5}{2} + 8 + 10 + 9.5 + 9.2 + 7.1 + 4.5\right)$$
$$= 510.5 + 5.0 \text{ (5.0 is the area of the small triangular parcel at P)}$$
$$= 515.5 \text{ m}^2$$

Area between PQ and road is as follows:
$$\text{Area} = \tfrac{15}{3}[4 + 4 + 2(6.5) + 4(5.0 + 5.6)]$$
$$= 317.0 \text{ m}^2$$

Total area of linear survey
$$= 2009.3 + 482.6 + 317.0 + 515.5$$
$$= 3324.4 \text{ m}^2$$

Exercise 13.3

1

Chainage (m)	Height h (m)	Bottom/top width (m) (5.0 + 2h)
0	5.0	15.0
30	3.5	12.0
60	0.0	5.0
90	2.4	9.8
120	2.9	10.8

Using Simpson's rule,

Plan area $= (30/3) [15.0 + 10.8 + 2(5.0)$
$+ 4(12.0 + 9.8)]$
$= 1230 \text{ m}^2$

Table 13.5

Station	Easting	Northing	Line	Double longitude	Difference in northings	Double area
A	0.00	0.00	–	–	–	–
B	−34.39	34.39	AB	−34.39	−34.39	1182.67
C	−51.50	13.01	BC	−85.89	21.38	−1836.33
D	−12.07	−15.10	CD	−63.57	28.11	−1786.95
E	5.35	−28.78	DE	−6.72	13.68	−91.93
F	16.34	−14.43	EF	21.69	−14.35	−311.25
A	0.00	0.00	FA	16.34	−14.43	−235.79
						−3079.58
					Area =	−1539.79 m²

Table 13.6

Station	Easting	Northing	Line	Areas	
				E_2N_1	E_1N_2
A	0.00	0.00	–	–	–
B	−34.39	34.39	AB	0.00	0.00
C	−51.50	13.01	BC	−1771.09	−447.41
D	−12.07	−15.10	CD	−157.03	777.65
E	5.35	−28.78	DE	−80.79	347.37
F	16.34	−14.43	EF	−470.27	−77.20
A	0.00	0.00	FA	0.00	0.00
				−2479.166	600.410
		−2479.166 − 600.410 =		−3079.576	
			Area =	1539.79 m²	

Road area = 120 × 5.0
= 600 m²

Therefore plan area of side slopes
= (1230 − 600) m²
= 630 m²

Actual area = 630/cos 45°
= 891 m²

Exercise 13.4

1 (a) Table 13.5
(b) Table 13.6

Exercise 13.5

1 (a) Using Simpson's rule:

Area = $\frac{10}{3}$ [6.5 + 10.0 + 2(12.8)
+ 4(10.5 + 13.2)] m²
= $\frac{10}{3}$ [136.9]
= 456.3 m²

Using the trapezoidal rule:

Area = $10\left(\dfrac{6.5 + 10.0}{2} + (10.5 + 12.8 + 13.2)\right)$
= 447.5 m²

(b) By counting squares (Fig. 13.17):

Ground area = 18 × 25 m² = 450 m²

(c) By planimeter:

Area = 25 × 18.1
= 452.5 m²

Exercise 13.6

1 80 300 m²
2 85 910.4 m²
3 319.5805 hectares
4 (a) 42.5 squares × 25 m² = 1062.5 m²
(b) Simpson's rule = 1062 m²
(c) Trapezoidal rule = 1047 m²
(d) 4.247 revs × 100 = 42.47 × 25 m²
= 1061.8 m²

5 (a)

Chainage (m)	Top width (m)
0	9.77
20	7.46
40	8.85
60	6.42
80	7.34

(b) Plan area occupied by cutting = 602 m²
(c)

Chainage (m)	Cross-sectional area (m²)
0	34.43
40	26.99

Chapter summary

In this chapter, the following are the most important points:

- Many calculations of area are required during the life cycle of a construction site. These include areas of tarmac and paving, grassed areas and gardens, roadway cuttings and embankments.

- The shapes are either regular or irregular and each shape requires the use of a formula or formulae to calculate its area. Regular area formulae include those of square, rectangle, triangle, trapezoid, circle and ellipse. Most of the common formulae are listed on page 280.

- The irregular area formulae are the trapezoidal and Simpson's rules. They can be used in a great many circumstances by simply drawing a baseline through the area and dividing it by a series of equally spaced right-angled ordinates into a number of trapezoids. The trapezoidal rule simply computes the areas of the trapezoids and adds them together to give the complete area of the irregular shape.

It is argued that a slightly more accurate value of the irregular area can be found by using Simpson's rule. Care is required to ensure that there is an *odd* number of ordinates in the shape of the irregular area.

The formulae and examples of their usage are shown on page 281.

- The areas of the shapes of surveyed sites can be calculated directly from field notes or from the coordinates of the points forming the survey. Usually, a combination of formulae is required. Examples are given on pages 282–287.

- Areas are generally calculated, however, from a scaled drawing. The methods available are:

  1. graphical, in which the site plan is overlain by a piece of transparent graph paper and the squares are counted. The answer is usually surprisingly accurate.
  2. by using Simpson's or trapezoidal rules in the manner previously described in this summary.
  3. by mechanical means, i.e. by using an instrument called a planimeter.

- The planimeter is a mechanical device which is laid on the site plan. A tracer point is used to follow the outline of the required area. A measuring unit in the shape of a rolling wheel counts the units of square centimetres covered by the tracer point. At any given scale 1 cm^2 is equivalent to a certain number of m^2. The required area is found by multiplying the square centimetre count by its value in square metres. The whole procedure is greatly enhanced by the use of a digital model planimeter, which computes the area directly and displays the answer on a small LCD screen.

CHAPTER 14 Mensuration – volumes

In this chapter you will learn about:

- the methods and formulae used to calculate the volumes of cuttings with vertical sides, typically trench excavations
- the methods and formulae used to calculate the cross-sectional areas of cuttings and embankments with various of configurations of sloping sides, typically roadway earthworks
- the methods of calculating the volumes of earthworks with sloping sides using Simpson's rule and the prismoidal formula
- the methods of calculating the volume of large-scale earthworks from spot level grids and from contour lines

---

On almost every construction site, some form of cutting or embankment is necessary to accommodate roads, buildings, etc. In general the earthworks fall into one of two categories:

(a) long narrow earthworks of varying depths – roadway cutting and embankments,

(b) wide flat earthworks – reservoirs, sports pitches, car parks, etc.

1. Cuttings (with vertical sides)

In Chapter 6, Fig. 6.4, reproduced below as Fig. 14.1, the longitudinal and cross sections of a proposed sewer are drawn to scale. The reader should revise the relevant section of Chapter 6, in order to appreciate the sources of the data used.

In Fig. 14.1 the sewer track has vertical sides. The depth varies along the length of the trench, the width is constant at 0.8 m and there is no ground slope across the section.

In this case, and *only* in this case, the volume may be calculated by *either* of the following methods.

(a) Method 1: side area

Computing the side area of the trench ABCD by Simpson's rule or the trapezoidal rule and multiplying the area by width 0.8 m.

(b) Method 2: cross sections

Computing the cross-sectional area of the trench at

each chainage point and entering the values into Simpson's rule to produce the volume directly. The rule is the same as the rule for area except that cross-sectional areas are substituted for the ordinates in the formula. Therefore, in an earthwork having five cross sections, A_1 to A_5, at a chainage interval of d metres, the formula is

$$\text{Volume} = \frac{d}{3}[A_1 + A_5 + 2 \times A_3 + 4 \times (A_2 + A_4)] \text{ m}^3$$

EXAMPLE

1 In Fig. 14.1, the accompanying longitudinal and cross sections show that the depth of the trench, at 8 metre intervals, is as follows;

Chainage (m)	Depth (m)
0	3.71
8	3.44
16	3.40
24	2.85
32	2.52
40	2.25
48	1.72
56	1.45
64	1.47

Given that the trench is to be 0.8 metre wide and has vertical sides, calculate the volume of material to be removed to form the excavation.

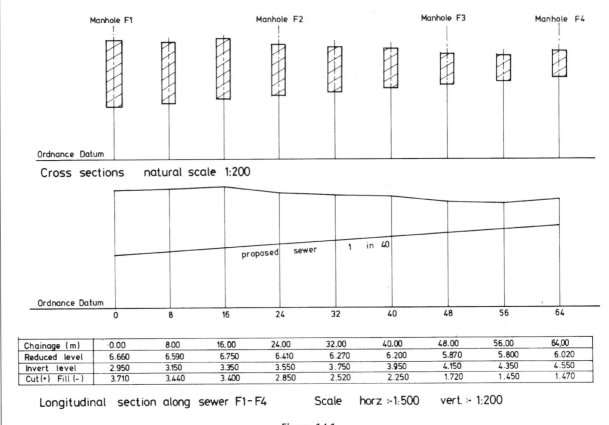

Longitudinal section along sewer F1–F4   Scale   horz :-1:500   vert. :- 1:200

Chainage (m)	0.00	8.00	16.00	24.00	32.00	40.00	48.00	56.00	64.00
Reduced level	6.660	6.590	6.750	6.410	6.270	6.200	5.870	5.800	6.020
Invert level	2.950	3.150	3.350	3.550	3.750	3.950	4.150	4.350	4.550
Cut (+) Fill (-)	3.710	3.440	3.400	2.850	2.520	2.250	1.720	1.450	1.470

Figure 14.1

Answer

Method I

(a) Using Simpson's rule and calling the chainage interval D, the area A of the side of the trench is found from the formula:

$A = (D/3)$[first + last + 2(odds) + 4(evens)] m²
$= (8/3)$[(3.71 + 1.47) + 2(3.40 + 2.52 + 1.72) + 4(3.44 + 2.85 + 2.25 + 1.45)]
$= (8/3)$[5.18 + 2(7.64) + 4(9.99)]
$= (8/3)$[60.42]
$= 161.12$ m²

   Volume = (161.12×0.8) m
   $= 128.9$ m³

(b) Using the trapezoidal rule.

$A = D$[average of first and last depths + sum of others]
$= 8$[(3.71 + 1.47)/2 + 3.44 + 3.40 + 2.85 + 2.52 + 2.25 + 1.72 + 1.45]
$= 161.76$ m²

   Volume = (161.76×0.8)
   $= 129.4$ m³

Method 2

The cross-sectional area at each chainage point is calculated and the values entered directly into Simpson's rule to produce the volume.

Chainage (m)	Depth (m)	Area $(D \times 0.8)$ m²
0	3.71	2.97
8	3.44	2.75
16	3.40	2.72
24	2.85	2.28
32	2.52	2.02
40	2.25	1.80
48	1.72	1.38
56	1.45	1.16
64	1.47	1.18

Volume = $(8/3)$[(2.97 + 1.18)
   + 2(2.72 + 2.02 + 1.38)
   + 4(2.75 + 2.28 + 1.80 + 1.16)] m³
   $= (8/3)$[4.15 + 2(6.12) + 4(7.99)]
   $= (8/3)$[48.35]
   $= 128.9$ m³

Exercise 14.1

1 Figure 6.12, reproduced here as Fig. 14.2, shows the longitudinal and cross sections of the proposed storm drain of a building site. Using the chainages and depths shown in the data table and given that the drain is 0.75 metres wide, calculate the volume of material to be removed in excavating the drain track.

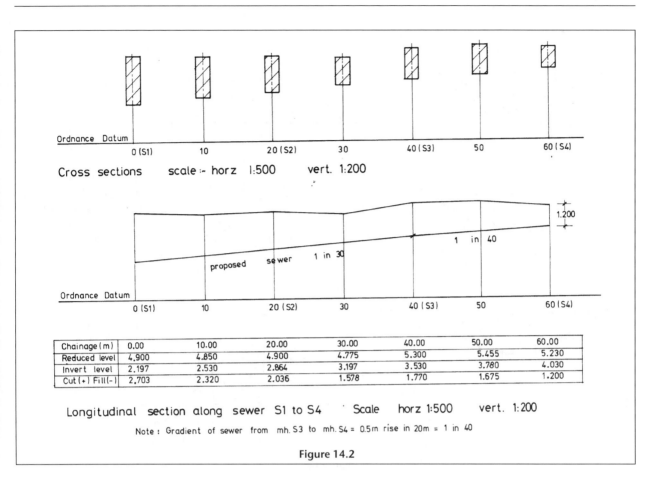

Cross sections   scale :– horz 1:500   vert. 1:200

Chainage (m)	0.00	10.00	20.00	30.00	40.00	50.00	60.00
Reduced level	4.900	4.850	4.900	4.775	5.300	5.455	5.230
Invert level	2.197	2.530	2.864	3.197	3.530	3.780	4.030
Cut (+) Fill(–)	2.703	2.320	2.036	1.578	1.770	1.675	1.200

Longitudinal section along sewer S1 to S4   Scale   horz 1:500   vert. 1:200

Note : Gradient of sewer from mh. S3 to mh. S4 = 0.5m rise in 20m = 1 in 40

Figure 14.2

2. General rule for calculating volume

Example 1 and Exercise 14.1 clearly demonstrate that two methods are available for calculating volumes of earthworks.

It must be emphasized, however, that method 1, i.e. calculating the area of the side of the trench and multiplying that area by the width, may be used *only* when the trench has vertical sides. In effect, the vertical side of the trench is the base of a prism, the shape of which remains constant over the width of the trench. This area may therefore be multiplied by any width to produce the volume of the trench.

In every other case, i.e. where trench *sides are not vertical*, method 2 must be used. Using this method, the cross-sectional area of the earthwork at regular intervals must be calculated first of all. These areas are then used directly in Simpson's rule to produce the volume.

The use of Simpson's rule presents a problem where there is an even number of cross sections, since the rule only works when there is an *odd* number of sections. In such cases, Simpson's rule is applied to the maximum odd number of sections, leaving an end portion that is easily computed using the prismoidal formula.

The trapezoidal rule, in any form, should not be used, since major errors occur in cases where cross-sectional areas differ substantially.

In general, therefore, the steps in calculating the volume of trenches, cuttings and embankments are:

1. Calculate the cross-sectional area of the earthwork at regular intervals.
2. In cases where there is an odd number of sections, use those sections in Simpson's rule.
3. In a case where there is an even number of sections, use the maximum odd number of cross-sectional areas in Simpson's rule and use the prismoidal rule to calculate the volume of the remaining prismoid. The prismoidal rule is the subject of Sec. 5(b) (page 306).

3. Cuttings and embankments (with sloping sides)

From the foregoing, it is clear that a major part of any volume calculation is the calculation of the cross-sectional areas of the earthworks. There are three distinct types of cross section:

1. *One-level section*   This type of section is developed from the longitudinal section of any earthworks. The one level which is known is the centre line ground level at any particular chainage. In Fig. 14.3(a), showing a cutting, and in Fig. 14.3(b), showing an embankment, the ground level (*g*) has been obtained by levelling. On the cross section, the ground surface is treated as being horizontal through the centre line level. The outline of the earthwork is added from a knowledge of the formation level (*f*), *formation width (w)* and side slope

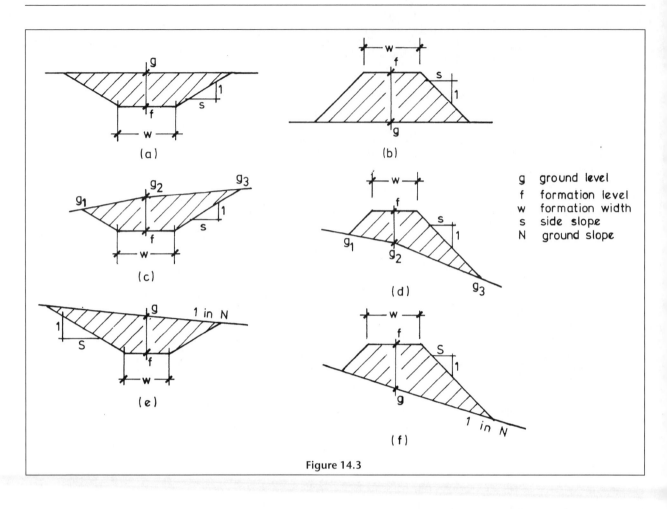

Figure 14.3

gradients (s). The drawing of cross sections is treated fully in Secs (3a) and (3b) of Chapter 6.

2. Three-level section This type of section is usually developed from contour plans or from fieldwork levelling, where levels are taken on either side of the centre line. The three known levels (g) are shown in Fig. 14.3(c) and (d), where they are denoted as g_1, g_2 and g_3. Again the outline of the cutting or embankment is added from a knowledge of the formation level (f), the formation width (w) and side slope gradients (s). The drawing of this type of section is treated fully in Exercises 6.4 and 6.5.

3. Cross fall section This type of section is developed from the longitudinal section of the earthworks. The centre line level at any particular chainage is observed and the gradient N of the slope across the line of the proposed earthwork is measured using a clinometer. Figure 14.3(e) shows a cutting, and Fig. 14.3(f) an embankment, drawn from a knowledge of the gradient 1 in N. This method of obtaining cross sections is not normal practice and is seldom used.

4. Calculation of cross-sectional areas

(a) One-level section

In Fig. 14.4, the longitudinal and cross sections of

Glasgow Metropolitan College Outdoor roadway R1–R10 at the Centre are shown at chainage intervals of 10 metres.

At each chainage point, only the centre line level was known. Each cross section was produced by assuming that the ground across the section was horizontal. By adding the roadway formation level and side slopes, a trapezoidal cross section was produced.

In Fig. 14.5 the cross section is trapezium shaped. Dimensions w and c are known, as explained above, and D is the only unknown. Since the sides slope at 1 to s:

$$D = cs + w + cs$$
$$= 2cs + w$$
$$\text{Area of trapezium} = \left(\frac{D+w}{2}\right) \times c$$
$$= \left(\frac{2cs + w + w}{2}\right) \times c$$
$$= (cs + w) \times c$$

Alternatively, the side slope gradient may be expressed as an angle of θ degrees. In Fig. 14.5, dimension $cs = c/\tan \theta$ and the area of the trapezium is as follows:

$$\text{Trapezium } (cs + w)c = (c/\tan \theta + w)c$$
$$= (c \cot \theta + w)c$$

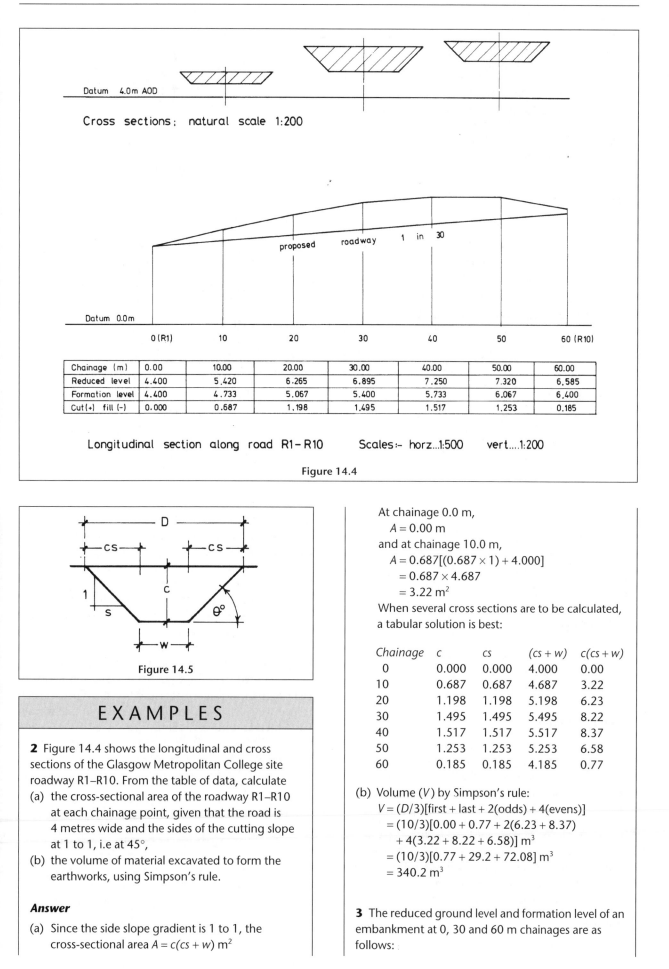

Datum 4.0m AOD

Cross sections: natural scale 1:200

proposed   roadway   1 in 30

Datum 0.0m

0 (R1)   10   20   30   40   50   60 (R10)

Chainage (m)	0.00	10.00	20.00	30.00	40.00	50.00	60.00
Reduced level	4.400	5.420	6.265	6.895	7.250	7.320	6.585
Formation level	4.400	4.733	5.067	5.400	5.733	6.067	6.400
Cut (+) fill (−)	0.000	0.687	1.198	1.495	1.517	1.253	0.185

Longitudinal section along road R1– R10   Scales:– horz...1:500   vert....1:200

Figure 14.4

Figure 14.5

EXAMPLES

2 Figure 14.4 shows the longitudinal and cross sections of the Glasgow Metropolitan College site roadway R1–R10. From the table of data, calculate
(a) the cross-sectional area of the roadway R1–R10 at each chainage point, given that the road is 4 metres wide and the sides of the cutting slope at 1 to 1, i.e at 45°,
(b) the volume of material excavated to form the earthworks, using Simpson's rule.

Answer

(a) Since the side slope gradient is 1 to 1, the cross-sectional area $A = c(cs + w)$ m^2

At chainage 0.0 m,
  $A = 0.00$ m
and at chainage 10.0 m,
  $A = 0.687[(0.687 \times 1) + 4.000]$
    $= 0.687 \times 4.687$
    $= 3.22$ m^2
When several cross sections are to be calculated, a tabular solution is best:

Chainage	c	cs	(cs + w)	c(cs + w)
0	0.000	0.000	4.000	0.00
10	0.687	0.687	4.687	3.22
20	1.198	1.198	5.198	6.23
30	1.495	1.495	5.495	8.22
40	1.517	1.517	5.517	8.37
50	1.253	1.253	5.253	6.58
60	0.185	0.185	4.185	0.77

(b) Volume (V) by Simpson's rule:
  $V = (D/3)[\text{first} + \text{last} + 2(\text{odds}) + 4(\text{evens})]$
    $= (10/3)[0.00 + 0.77 + 2(6.23 + 8.37)$
      $+ 4(3.22 + 8.22 + 6.58)]$ m^3
    $= (10/3)[0.77 + 29.2 + 72.08]$ m^3
    $= 340.2$ m^3

3 The reduced ground level and formation level of an embankment at 0, 30 and 60 m chainages are as follows:

Chainage (m)	0	30	60
RL (m)	35.10	36.20	35.80
FL (m)	38.20	38.40	38.60

Given that the formation width of the top of the embankment is 6.00 m, that the transverse ground slope is horizontal and that the embankment sides slope at 1 unit vertically to 2 units horizontally, calculate the cross-sectional areas at the various chainages.

Answer (Fig. 14.6)

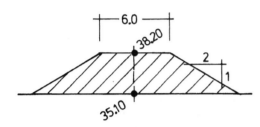

Figure 14.6

Chainage 0 m

$$c = 38.20 - 35.10$$
$$= 3.10 \text{ m}$$
$$w = 6.00 \text{ m (given)}$$
$$D = 2cs + w$$
$$= (2 \times 3.10 \times 2) + 6.00$$
$$= 12.40 + 6.00$$
$$= 18.40 \text{ m}$$
$$\text{Area} = \left(\frac{D + w}{2}\right) \times c$$
$$= \left(\frac{18.40 + 6.00}{2}\right) \times 3.10$$
$$= 12.2 \times 3.10$$
$$= 37.82 \text{ m}^2$$

Alternatively,
$$\text{Area} = (cs + w) \times c$$
$$= (3.10 \times 2 + 6.00) \times 3.10$$
$$= 12.20 \times 3.10$$
$$= 37.82 \text{ m}^2$$

Tabular solution

Chainage	RL	FL	c	cs	(cs + w)	(cs + w)c
0	35.10	38.20	3.10	6.20	12.20	37.82 m²
30	36.20	38.40	2.20	4.40	10.40	22.88 m²
60	35.80	38.60	2.80	5.60	11.60	32.48 m²

4 In Fig. 14.7, the depths of excavation required to form a cutting at 20 m intervals are 0.00, 0.90, 1.50, 3.20 and 3.30 m. Given that the roadway is 5.0 m wide and the side slopes at gradient 30°, calculate the cross-sectional areas of the cutting and hence the volume of excavated material.

Answer

Cross-sectional areas

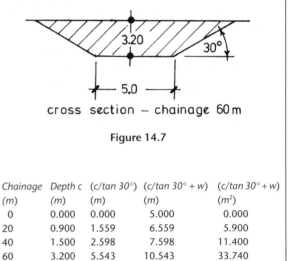

cross section – chainage 60 m

Figure 14.7

Chainage (m)	Depth c (m)	(c/tan 30°) (m)	(c/tan 30° + w) (m)	(c/tan 30° + w) (m²)
0	0.000	0.000	5.000	0.000
20	0.900	1.559	6.559	5.900
40	1.500	2.598	7.598	11.400
60	3.200	5.543	10.543	33.740
80	3.300	5.716	10.716	35.360

Volume (V) by Simpson's rule:

$$V = (20/3)[0.00 + 35.36 + 2(11.40)$$
$$+ 4(5.90 + 33.74)] \text{ m}^3$$
$$= (20/3)[35.36 + 22.80 + 158.56) \text{ m}^3$$
$$= 1444.8 \text{ m}^3$$

Computer spreadsheet solution

In practice there are usually a large number of cross sections in any earthworks. The volume is most easily calculated using a computer spreadsheet. Excel and Microsoft Works are probably the best known of the spreadsheets. Table 14.1 is the solution of Example 3, using Works in a very basic manner as an example of the possibilities of spreadsheet programming. It is not intended, however, that this textbook become or replace a good computing textbook. However, a little explanation is obviously desirable for readers who are not familiar with spreadsheets.

Table 14.2 is Table 14.1 with the addition of a grid of letters and numbers. In Table 14.2, the side slope gradient is entered in cell C4, and the formation width in cell C5. All other *known* data and text are typed in their respective cells. *The calculations* require simple formulae, which are entered as shown in Table 14.3.

Table 14.3

Cell	Formula entry
D10	= C10–B10
D11 and D12	Copy formula of D10
E10	= D10 * C4
E11 and E12	Copy formula of E10
F10	= E10 + C5
F11 and F12	Copy formula of F10
G10	= F10 * D10
G11 and G12	Copy formula of G10

Table 14.1 Volume calculation using Microsoft Works Spreadsheet
Side slope grade = 1 in 2
Formation width (m) = 6

Chainage	Reduced level	Formation level	Calculations			
			c	cs	(cs + w)	(cs + w)c
0	35.1	38.2	3.1	6.2	12.2	37.82
30	36.2	38.4	2.2	4.4	10.4	22.88
60	35.8	38.6	2.8	5.6	11.6	32.48

Table 14.2

	A	B	C	D	E	F	G
1							
2	**Volume calculation using Microsoft Works Spreadsheet.**						
3							
4	Side slope grade = 1 in		2				
5	Formation width (m) =		6				
6							
7	Chainage	Reduced	Formation	Calculations			
8		level	level	c	cs	(cs + w)	(cs + w)c
9							
10	0	35.1	38.2	3.1	6.2	12.2	37.82
11	30	36.2	38.4	2.2	4.4	10.4	22.88
12	60	35.8	38.6	2.8	5.6	11.6	32.48
13							
14							
15							

Exercise 14.2

1 Figure 14.8 shows part of a roadway, 5.0 m wide, which is partly in cutting and partly on embankment. The relevant earthworks data are given below:

Chainage (m)	Depth of cutting	Height of filling	Side slope gradient
0	1.1	—	1 to 2
10	2.5	—	1 to 2
20	0.0	0.0	—
30	—	1.0	1 to 1
40	—	2.3	1 to 1

Calculate:
(a) the cross-sectional area at each chainage point,
(b) the volume of material excavated from the cutting,
(c) The volume of material infilled to form the embankment.

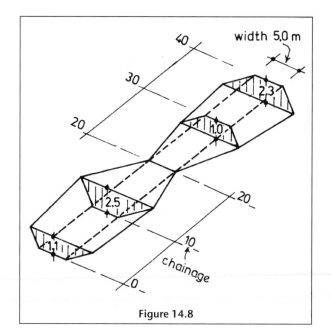

Figure 14.8

(b) Three-level section

Chapter 5 dealt with the subject of contouring and intersection of surfaces. Figure 5.23 shows a contoured plan of a sloping area of ground across which an embankment is to be built. The reader should revise, in Chapter 5, Exercise 5.3, Question 1 (page 93), the method of drawing the outline of the embankment and the cross sections at points A, B and C.

Figure 14.9 is a reproduction of these cross sections and shows clearly (a) the formation level of the

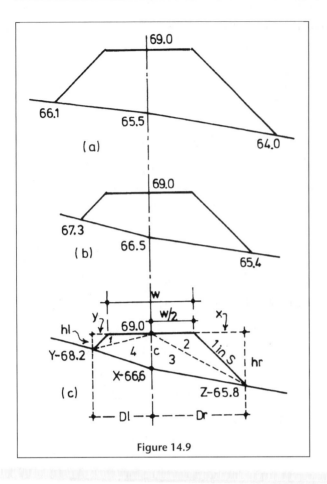

Figure 14.9

embankment, (b) the ground level on the centre line of the embankment and (c) the two ground levels at the base of the side slopes of the embankment. (The figure is reproduced from Fig. 5.27.)

Since the figure is not a regular geometrical figure, it has to be split into triangles where the base and altitude can be readily calculated. The differences between the formation level and reduced levels X, Y and Z produce the dimensions c, hl and hr. Since the sides slope at 1 to s,

$$x = (hr \times s) \qquad \text{and} \qquad y = (hl \times s)$$

and

$$Dr = x + w/2 = (hr \times s) + w/2$$
$$Dl = y + w/2 = (hl \times s) + w/2$$

In all four triangles, area $= \frac{1}{2}$ base $\times$ altitude

Area triangle 1 $= \frac{1}{2}(w/2 \times hl) = (w/4 \times hl)$
Area triangle 2 $= \frac{1}{2}(w/2 \times hr) = (w/4 \times hr)$
Area triangle 3 $= \frac{1}{2}(c \times Dr) = c/2 \times Dr$
Area triangle 4 $= \frac{1}{2}(c \times Dl) = c/2 \times Dl$

Therefore

$$\text{Total area} = (w/4 \times hl) + (w/4 \times hr) + (c/2 \times Dr)$$
$$+ (c/2 \times Dl)$$
$$= (w/4)(hl + hr) + (c/2) (Dr + (Dl)$$

EXAMPLE

5 In Fig. 14.9(c), the width of the roadway is 5.0 m and the sides slope at gradient 1 to 1. Calculate the cross-sectional area of the embankment.

Solution

Cross section 15.9(c)

At Y, $hl = 69.0 - 68.2 = 0.80$ m
At X, $c = 69.0 - 66.6 = 2.40$ m
At Z, $hr = 69.0 - 65.8 = 3.20$ m

Since the sides slope at 1 to 1 and $w/2 = 2.50$ m,

$$Dl = (0.80 + 2.50) = 3.30 \text{ m}$$
$$Dr = (3.20 + 2.50) = 5.70 \text{ m}$$

Area of triangle 1
$= (w/4 \times hl) = 1.25 \times 0.80 = 1.000$
Area of triangle 2
$= (w/4 \times hr) = 1.25 \times 3.20 = 4.000$
Area of triangle 3
$= \frac{1}{2}(c \times Dr) = 1.20 \times 3.30 = 3.960$
Area of triangle 4
$= \frac{1}{2}(c \times Dl) = 1.20 \times 5.70 = 6.840$
Total area $= \overline{15.800 \text{ m}^2}$

Alternatively,

$$\text{Total area} = (w/4)(hl + hr) + (c/2)(Dr + Dl)$$
$$= 1.25(0.80 + 3.20)$$
$$+ 1.20(3.30 + 5.70) \text{ m}^2$$
$$= 15.80 \text{ m}^2$$

6 In Fig. 14.9, calculate the areas of the (a) and (b) cross sections, given that the roadway is 5.0 m wide and the sides slope at gradient 1 to 1.

Answer

Cross section (a) (top)

$hl = 2.9$	$Dl = 5.4$	Area $= 32.45 \text{ m}^2$
$c = 3.5$	$Dr = 7.5$	
$hr = 5.0$		

Cross section (b) (middle)

$hl = 1.7$	$Dl = 4.2$	Area $= 19.50 \text{ m}^2$
$c = 2.5$	$Dr = 6.1$	
$hr = 3.6$		

Exercise 14.3

1 [Figure 5.28 is the solution to Exercise 5.3 question 2 (page 93). In this exercise a roadway cutting is to be constructed across an area of undulating ground and cross sections are to be drawn at points X, Y and Z.] Figure 14.10 is a reproduction of the solution showing the three

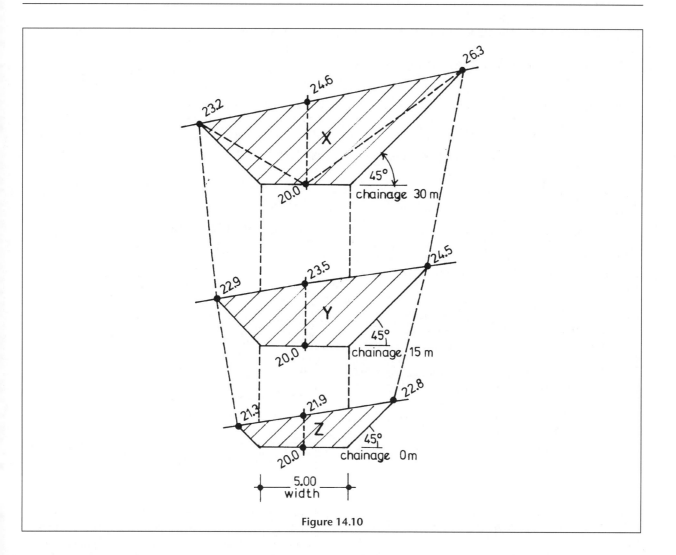

Figure 14.10

cross sections with their respective ground levels and formation levels.

Calculate:

(a) the areas of the cross sections at X, Y and Z,
(b) the volume of material excavated in constructing the cutting.

(c) Cross fall section

In Fig. 14.11 a roadway cutting ABCD is to be formed in ground that has a regular gradient across the longitudinal line of the roadway. From the field work, the reduced ground level F on the centre line is known, together with the formation level E, side slope values I in S, and transverse ground slope 1 in N.

Since the figure is not a geometrically regular figure, it must be split into figures whose areas are easily calculated. Perpendicular lines are drawn through B and C to split the figure into two triangles AHB and GDC and one trapezium BCGH. Horizontal lines are also drawn through A and D to meet the perpendiculars at J and K respectively.

The central height c is the difference between the reduced level F and formation level E. Thus w and c are

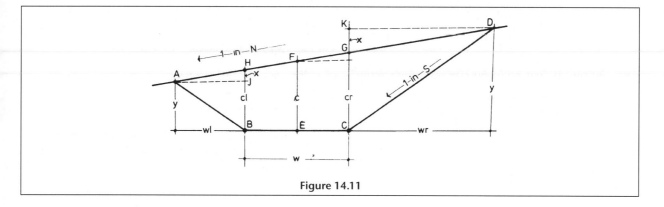

Figure 14.11

known and the area of the trapezium BCGH = $(c \times w)$ m².

The areas of the triangles AHB and GDC are of course $\frac{1}{2}$(base × altitude).

Therefore area $\triangle AHB = \frac{1}{2}(cl \times wl)$

and area $\triangle GDC = \frac{1}{2}(cr \times wr)$

In Fig. 14.11 the following information is known:

Formation width = 5 m
Central height c =4 m
Side slopes are 1 to 2
Transverse ground slope is 1 to 5

Since the ground slopes at 1 to 5, cr = 1/5 of $w/2$ greater than c, while cl is the same value less than c. Therefore,

$cr = c + (1/5 \times 2.5)$ and $cl = c - (1/5 \times 2.5)$
$\quad = 4.0 + 0.5$ $\qquad\qquad = 4.0 - 0.5$
$\quad = 4.5$ m $\qquad\qquad\quad = 3.5$ m

In $\triangle ABJ$, $y/wl = \frac{1}{2}$ and therefore $y = wl/2$ and in $\triangle AHJ$, $x/wl = \frac{1}{5}$ and therefore $x = wl/5$. Now

$$cl = (y + x) = \frac{wl}{2} + \frac{wl}{5}$$
$$= \frac{5wl + 2wl}{5 \times 2}$$
$$= \frac{wl(5 + 2)}{5 \times 2}$$

Substitution of the general S for 2 and N for 5 shows that

$$cl = wl\,\frac{N + S}{NS}$$
$$\text{and } wl = cl\,\frac{NS}{N + S}$$

Therefore $wl = 3.5 \times \dfrac{(5 \times 2)}{(5 + 2)} = 5.00$ m

Similarly, in $\triangle KDC$, $y/wr = \frac{1}{2}$ and therefore $y = wr/2$ and in $\triangle GDK$, $x/wr = \frac{1}{5}$ and therefore $x = wr/5$.

Now $cr = (y - x) = \dfrac{wr}{2} - \dfrac{wr}{5}$
$$= \frac{5wr - 2wr}{5 \times 2}$$
$$= \frac{wr(5 - 2)}{5 \times 2}$$

Again substitution of S for 2 and N for 5 shows that

$$cr = \frac{wr \times (N - S)}{N \times S}$$
$$\text{and } wr = cr \times \frac{NS}{N \times S}$$

Therefore $wr = 4.5 \times \dfrac{(5 \times 2)}{5 - 2} = 15.0$ m

A general formula can be derived showing that

$$\text{Horizontal length} = \text{vertical length} \times \frac{NS}{N \pm S}$$

The positive sign applies when the ground slope and side slope run in opposite directions, i.e. one up and one down, while the negative sign applies when both gradients are in the same direction, i.e. either both up or both down.

The areas of the three component figures in Fig. 14.11 are:

Trapezium HGCB area = $(c \times w)$
$\qquad\qquad\qquad = 4 \times 5$
$\qquad\qquad\qquad = 20$ m²

Triangle GDC = $cr/2 \times wr$
$\qquad\qquad = \frac{1}{2}(4.5 \times 15.0)$
$\qquad\qquad = 33.75$ m²

Triangle ABH area = $cl/2 \times wl$
$\qquad\qquad\quad = \frac{1}{2}(3.5 \times 5)$
$\qquad\qquad\quad = 8.75$ m²

Total cross-sectional area = 62.5 m²

EXAMPLES

7 In Fig. 14.12, a roadway is to be built on ground having a transverse ground slope of 1 in 8. The road is 8.0 m wide, has a central height of 3.5 m and 1 to 4 side slopes. Calculate the cross-sectional area of the embankment.

Solution

$c = 3.5$, $w = 8.0$ m (given); therefore $w/2 = 4.0$ m, $cr = c + (\frac{1}{8} \times 4.0) = 4.0$ m and $cl = c - (\frac{1}{8} \times 4.0) = 3.0$ m.

$wr = \dfrac{4.0 \times NS}{N - S}$ $\qquad$ $wl = \dfrac{3.0 \times NS}{N + S}$
$\quad = \dfrac{4.0 \times 32}{4}$ $\qquad\qquad = \dfrac{3.0 \times 32}{12}$
$\quad = 32.0$ m $\qquad\qquad\quad = 8.0$ m

Area HGCB = $c \times w$ $\qquad$ Area ABH = $\dfrac{cr}{2} \times wr$
$\qquad\qquad = 3.5 \times 8.0$ $\qquad\qquad = 2 \times 32.0$
$\qquad\qquad = 28$ m² $\qquad\qquad\quad = 64.0$ m

Area CGD = $\dfrac{cl}{2} \times wl$
$\qquad\qquad = 1.5 \times 8.0$
$\qquad\qquad = 12.0$ m²
Total area = 104.0 m²

8 Figure 14.13 shows a side hill section where the ground slope gradient is known. Given the following information, calculate the area of cutting and area of filling.

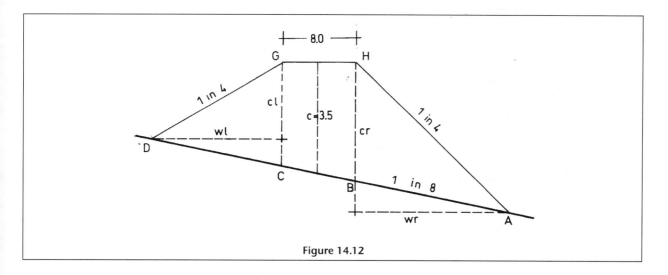

Figure 14.12

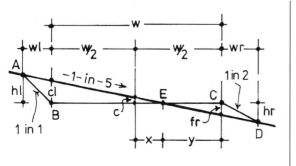

Figure 14.13

Formation width $w = 15$ m
Central height $c \quad = 0.5$ m
Side slopes
   Cutting $= 1$ in 1
   Embankment $= 1$ in 2
Ground slope 1 in $N = 1$ in 5

Answer

Cutting $cl = c + (\frac{1}{5}$ of $w/2)$ m as before
$$= 0.5 + (\tfrac{1}{5} \times 7.5) \text{ m}$$
$$= 2.0 \text{ m}$$

Point of no cutting or filling occurs at x metres from the centre line of the formation:

$$x = 5 \times c$$
$$= 2.5 \text{ m}$$
Therefore $y = 5.0$ m
Filling $fr = \frac{1}{5}$ of y
$$= 1.0 \text{ m}$$

By principle of converging gradients:

$$wl = \frac{cl \times (5 \times 1)}{5-1} \qquad wr = \frac{fr \times (5 \times 2)}{5-2}$$
$$= \frac{2.0 \times 5}{4} \quad \text{and} \qquad = \frac{1.0 \times 10}{3}$$
$$= 2.5 \text{ m} \qquad\qquad = 3.33 \text{ m}$$

Slope of cutting $\quad = 1$ in 1
Therefore $hl \qquad = 2.5$ m

Slope of embankment $= 1$ in 2
   Therefore $hr \qquad = 1.67$ m

Area of cutting $\qquad =$ triangle ABE
$$= \tfrac{1}{2}(w/2 + x) \times hl$$
$$= \tfrac{1}{2} \times 10.0 \times 2.5$$
$$= 12.5 \text{ m}^2$$

Area of filling $=$ triangle ECD
$$= \tfrac{1}{2} y \times hr$$
$$= \tfrac{1}{2} \times 5.0 \times 1.67$$
$$= 4.17 \text{ m}^2$$

Exercise 14.4

1 In Fig. 14.14, a roadway is to be built on ground that has a transverse ground slope of 1 in 10. The road is to be 5.00 metres wide, with a central height of 2.00 metres and 1 to 2 side slopes. Calculate the cross-sectional area of the embankment.

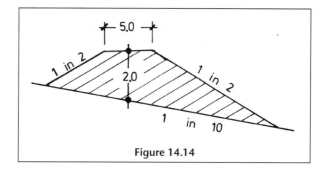

Figure 14.14

5. Calculation of volume

(a) Calculation of volume by Simpson's rule

Once the various cross sections have been calculated the volume of material contained in the embankment is calculated by Simpson's volume rule. The rule has already been used in Examples 1, 2 and 3, where there were nine, seven and five cross sections respectively.

   The rule applies only when there is an *odd* number

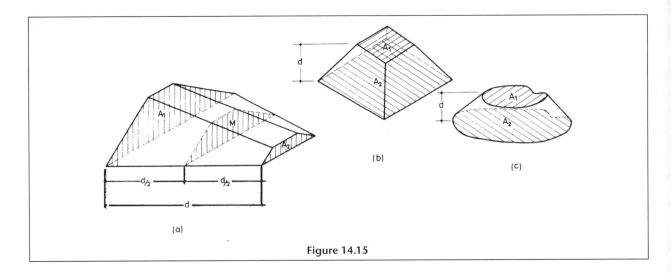

Figure 14.15

of cross sections. Should there be an *even* number. Simpson's rule is used to calculate the maximum odd number of sections and the prismoidal rule applied to the remainder of the earthworks.

(b) Calculation of volume by the prismoidal rule

A prismoid is defined as any solid having two plane parallel faces, regular or irregular in shape, which can be joined by surfaces either plane or curved on which a straight line may be drawn from one of the parallel ends to the other. Examples of prismoids are shown in Fig. 14.15.

Consider Fig. 14.15(a). In order to determine the volume by Simpson's rule, it is necessary to split the figure such that there is an odd number of equidistant cross sections. Three is the minimum number which fulfills this condition.

Calling the mid-section M, the volume by Simpson's rule is

$$\text{Volume} = (\tfrac{1}{3} \times d/2) [A_1 + A_2 + 2(\text{zero}) + 4M]$$
$$= (d/6) [A_1 + A_2 + 4M]$$

This is Simpson's prismoidal rule which can be used to find the volume of any prismoid, provided the area M of the central section is determined.

Note. The area of M is *not* the mean of areas A_1 and A_2. The area must be computed from its own dimensions, which are the means of the heights and widths of the two end sections.

EXAMPLES

9 Figure 14.16 shows a proposed cutting where the following information is known:

$$\begin{aligned}
\text{Length of cutting} &= 30 \text{ m} \\
\text{Formation width} &= 8 \text{ m} \\
\text{Depth at commencement} &= 8 \text{ m} \\
\text{Depth at end} &= 5 \text{ m} \\
\text{Side slopes} &= 1 \text{ in } 1
\end{aligned}$$

Using the prismoidal formula, calculate the volume of material to be removed.

Solution

Section A_1

$$\begin{aligned}
\text{Formation width} &= 8 \text{ m} \\
\text{Top width} &= (8 + 2c) \text{ m} \\
\text{Central depth } c &= 8 \text{ m} \\
\text{Therefore top width} &= (8 + 16) \text{ m} \\
&= 24 \text{ m}
\end{aligned}$$

Section A_2

$$\begin{aligned}
\text{Formation width} &= 8 \text{ m} \\
\text{Top width} &= (8 + 2c) \text{ m} \\
\text{Central depth } c &= 5 \text{ m} \\
\text{Therefore top width} &= (8 + 10) \text{ m} \\
&= 18 \text{ m}
\end{aligned}$$

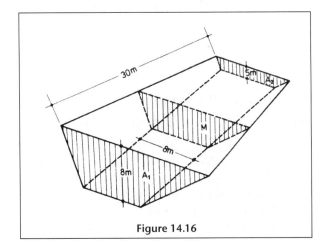

Figure 14.16

Section M

Formation width $= 8$ m

Top width $= (8 + 2c)$ m

Central depth $c =$ average of depths of A_1
and A_2
$= \frac{1}{2}(8 + 5)$ m
$= 6.5$ m

Therefore top width $= 8 + 13$ m
$= 21$ m
$=$ average of widths of A_1
and A_2

Cross-sectional areas (trapezia)

$A_1 = \frac{1}{2}(8 + 24) \times 8 \quad = 128$ m^2
$A_2 = \frac{1}{2}(8 + 18) \times 5 \quad = \ 65$ m^2
$M = \frac{1}{2}(8 + 21) \times 6.5 = \ 94.25$ m^2

Volume $= (30/6)[128 + 65 + (4 \times 94.25)]$ m^3
$= 2850$ m^3

10 Figure 14.15(b) shows a mound of excavated material stored on site in the form of a truncated pyramid (prismoid). Top area A_1 is 4.00 metres square, base area A_2 is 10.000 metres square, while height d is 4.50 metres. Calculate the volume of material contained in the pyramid using the prismoidal rule.

Answer

Top area $= 4.00 \times 4.00 = 16.00$ m^2
Base area $= 10.0 \times 10.0 = 100.00$ m

Mid dimensions M
$= \frac{1}{2}(4 + 10)$ m square $= 7.00$ m

Mid area $M = (7.00 \times 7.00)$ m^2
$= 49.00$ m^2

Volume
$= (d/6)[A_1 + A_2 + 4M]$ m^3
$= (4.5/6)[16.00 + 100.00 + (4 \times 49.00)]$ m^3
$= 234$ m^3

Note. It is fairly common practice on site to use the end areas formula to calculate the volume of a prismoid. Using this method, the mean of the two end areas is multiplied by the height to produce a volume.

In Example 10, the volume, as computed by this method, is $4.5[(100 + 16)/2] = 261$ m^3, which produces an error of 12 per cent. As the top area tends towards zero, the error becomes progressively larger and the use of the end areas formula is not recommended.

Exercise 14.5

1 A proposed service roadway, 5.5 m wide, is to be built along a centre line XY. The embankment is to have side slopes of 1 to 2. Given the following data, calculate the volume of material required to construct the embankment:

Chainage (m)	40	60	80	100	120	140
Ground level (m)	10.00	9.60	9.50	10.40	7.30	10.40
Formation level (m)	11.00	11.10	11.20	11.30	11.40	11.50

6. Volumes of large-scale earthworks

Whenever the volumes of large-scale earthworks have to be determined, e.g. the formation of sports fields, reservoirs, large factory buildings, the fieldwork consists of covering the area by a network of squares and obtaining the reduced levels. The volume is then determined either from the grid levels themselves or from the contours plotted therefrom.

(a) Volumes from spot levels

Figure 14.17 shows a small section of a grid. The total area is to be excavated to a formation level of 90.00 m to form a car park. The sides of the excavation are to be vertical.

The solid formed by each grid square is a vertical truncated prism, i.e. a prism where the end faces are not parallel.

Volume of each prism $=$ mean height $\times$ area of base

Mean height of each truncated prism above 90.00 m level is

Prism $1 = (1.0 + 3.0 + 2.0 + 2.0) \div 4 = 2.0$ m
$2 = (3.0 + 4.0 + 3.0 + 2.0) \div 4 = 3.0$ m
$3 = (2.0 + 3.0 + 2.0 + 1.0) \div 4 = 2.0$ m
$4 = (2.0 + 2.0 + 1.0 + 3.0) \div 4 = 2.0$ m

Area of base of each truncated prism $= 10 \times 10$
$= 100$ m^2

Therefore

Volume of $1 = 100 \times 2.0 = 200$ m^3
$2 = 100 \times 3.0 = 300$ m^3
$3 = 100 \times 2.0 = 200$ m^3
$4 = 100 \times 2.0 = 200$ m^3

Total volume of excavation $= \overline{900 \text{ m}^3}$

Alternatively, the volume can be found thus:

Volume $=$ mean height of excavation $\times$ total area

The mean height of the excavation is the mean of the mean heights of the truncated prisms. It is *not* the mean of the spot levels.

Mean height excavation $= (2.0 + 3.0 + 2.0 + 2.0) \div 4$
$= 2.25$ m

Total area of site $= 20 \times 20 = 400$ m^2
Therefore total volume $= 2.25 \times 400 = 900$ m^3

When examined closely, it is seen that spot level A is used only once in obtaining the mean height of the excavation, spot level B twice, while spot level E is used four times in all. This mean height, and hence the volume, can be readily found by tabular solution as in Table 14.4.

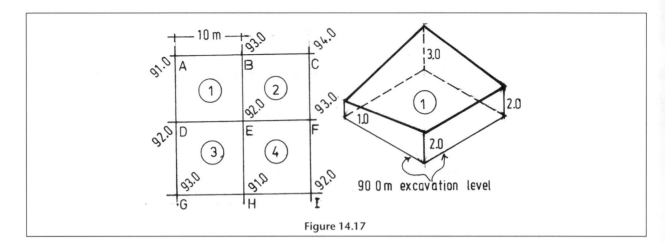

Figure 14.17

Table 14.4

Grid station	Height above formation level	Number of times used	Product
A	1.0	1	1.0
B	3.0	2	6.0
C	4.0	1	4.0
D	2.0	2	4.0
E	2.0	4	8.0
F	3.0	2	6.0
G	3.0	1	3.0
H	1.0	2	2.0
I	2.0	1	2.0
		Sum 16	Sum 36.0

Mean height of excavation = 36.0/16 m
= 2.25 m as before

The various spot heights are tabulated in column 2 and the number of times they are used in column 3. Column 4 is the product of columns 2 and 3. The mean height of the excavation is found by dividing the sum of column 4 by the sum of column 3.

EXAMPLE

11 Figure 14.18 shows spot levels at 20 metre intervals over a site which is to be excavated to 47.00 m to accommodate three tennis courts. Calculate the

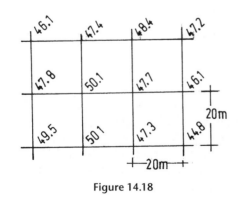

Figure 14.18

volume of material to be removed, assuming that the excavation has vertical sides.

Answer

Height above formation level	Number of times used	Product
−0.9	1	−0.9
0.4	2	0.8
1.4	2	2.8
0.2	1	0.2
0.8	2	1.6
3.1	4	12.4
0.7	4	2.8
−0.9	2	−1.8
2.5	1	2.5
3.1	2	6.2
0.3	2	0.6
−2.2	1	−2.2
	24	25.0

$$\text{Mean height of excavation} = \left(\frac{25}{24}\right) \text{m}$$
$$= 1.042 \text{ m}$$
$$\text{Total area of site} = 60 \times 40$$
$$= 2400 \text{ m}^2$$
$$\text{Therefore volume of excavation} = 1.042 \times 2400$$
$$= 2500 \text{ m}^3$$

Exercise 14.6

1 Figure 14.19 shows spot levels at 10 metre intervals over a small site, which is to be made level at 5.70 m AOD. Calculate the volume of material to be excavated, assuming the excavation has vertical sides.

(b) Volumes from contours

Figure 14.20 shows a mound that has been contoured. If the mound is to be removed the volume of material can be calculated by considering the solid to be split along the contours into a series of prismoids. The volume can then be calculated by successive appli-

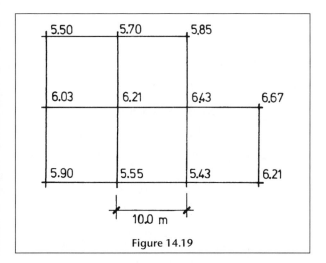

5.50	5.70	5.85	
6.03	6.21	6.43	6.67
5.90	5.55	5.43	6.21

10.0 m

Figure 14.19

cations of the prismoidal rule or, where circumstances are favourable, by direct application of Simpson's rule.

When using the prismoidal rule, three contours are taken at a time and the central one is used as the mid-area. The accuracy of the volume depends basically on the contour vertical interval. Generally, the closer the contour interval, the more accurate is the volume.

Taking the prismoid formed by contours 110 m and 130 m, the areas enclosed by the contours are determined from the plan by planimeter. The mid-area enclosed by the 120 m contour is likewise determined and the volume of the prismoid is therefore

$$\text{Volume} = \frac{2h}{6}[A_1 + 4A_2 + A_3]$$

Similarly, the volume between contours 130 and 150 m is

$$\text{Volume} = \frac{2h}{6}[A_3 + 4A_4 + A_5]$$

Adding these results gives the volume between the 110 and 150 m contours:

$$\text{Volume} = \frac{2h}{6}[A_1 + 4A_2 + A_3] + \frac{2h}{6}[A_3 + 4A_4 + A_5]$$

$$= \frac{h}{3}[A_1 + A_5 + 2A_3 + 4(A_2 + A_4)]$$

which is the volume by Simpson's rule.

The part of the solid lying above the 150 m contour is not included in the above calculations. It must be approximated to the nearest geometrical solid and calculated separately. In general, the nearest regular solid is a cone or pyramid where the volume = $\frac{1}{3}$ base area × height.

EXAMPLE

12 Figure 14.21 shows ground contours at 1 metre vertical intervals. ABCD is a proposed factory building where the floor level is to be 32.00 m. The volume of material to be excavated is required. The side slopes of any earthworks are 1 in 2.

(a) The earthwork contours are drawn at 1 metre vertical intervals, i.e. 2 metres horizontally apart.

(b) The surface intersections are found and the outline of the cutting drawn (broken line).

(c) The area enclosed by each contour is obtained by planimeter. The 32 m contour is bounded by AICD, the 33 m contour by all points numbered 2, the 34 m contour by points numbered 3, the 35 m contour by points numbered 4, while the 36 m contour (point 5) has no area.

(d) The respective areas are:

Contour	32	33	34	35	36
Area (m²)	315.0	294.5	125.0	30.0	0.0

(e) Volume by Simpson's rule:

$$V = \tfrac{1}{3}[315.0 + 0.0 + (2 \times 125.0) + 4(294.5 + 30.0)] \text{ m}^3$$
$$= 621.0 \text{ m}^3$$

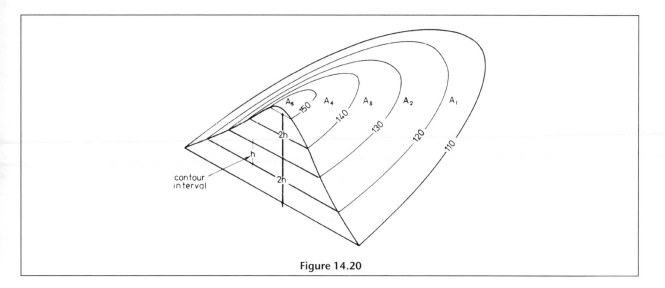

Figure 14.20

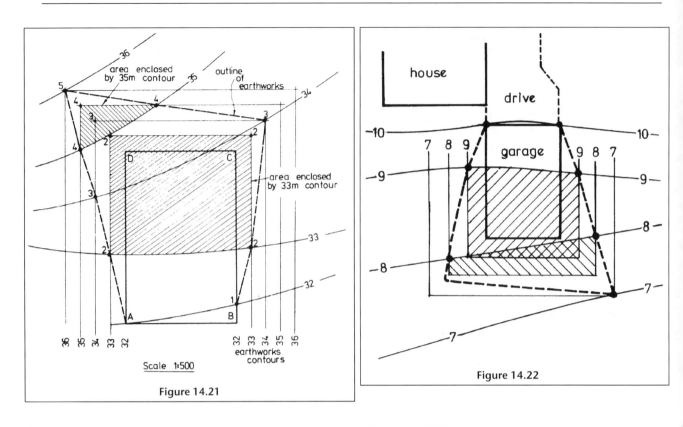

Figure 14.21

Figure 14.22

Exercise 14.7

1 Figure 14.22 shows ground contours at 1 metre vertical intervals. A lockup garage, measuring 6.0 m × 4.0 m, is to be built to the rear of an existing house, with a floor level of 10.00 m. The garage base requires to be built up on an embankment, the sides of which are to slope at 45°. Calculate:

(a) the areas of the contour planes 10, 9 and 8 m.
(b) the volume of material required to construct the embankment.

Exercise 14.8

1 A sewer 0.75 m wide is to be excavated along a line AB. The sides are to be vertical. Given the following data, calculate the volume of material to be excavated to form the sewer track.

Chainage (m)	0	20	40	60	80	100	120	140
Depth (m)	1.20	1.70	0.95	2.21	2.27	2.21	0.95	1.82

2 The reduced ground level and formation level of an embankment at 0, 30, 60 and 90 m chainages are shown on page 311.

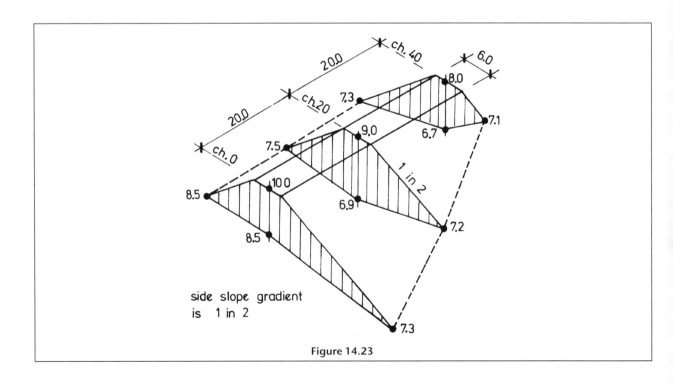

Figure 14.23

Chainage (m)	0	30	60	90
RL (m)	55.30	56.40	56.00	58.00
FL (m)	58.40	58.60	58.80	59.00

Given that the formation width of the top of the embankment is 6.00 m, that the transverse ground slope is horizontal and that the embankment sides slope at 1 unit vertically to 2 units horizontally, calculate:

(a) the cross-sectional areas at the various chainages,
(b) the volume of material contained in the embankment.

3 Calculate the volume of earth required to form the embankment shown in Fig. 14.23.

4 The central heights of an embankment at two points A and B 90 m apart are 4.0 and 6.5 m respectively. The embankment is built on ground where the maximum slope is 1 in 10 at right angles to the line of the embankment.

Given that the formation width of the embankment is 6 m and the side slope at 1 in 2, calculate the volume of material in cubic metres contained between A and B.

5 Figure 14.24 shows a rectangular grid with levels at 10 m intervals. The whole area is to be covered with waste

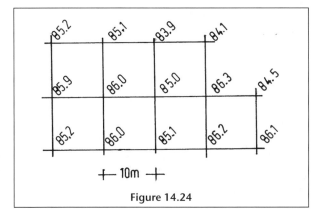

+— 10m —+

Figure 14.24

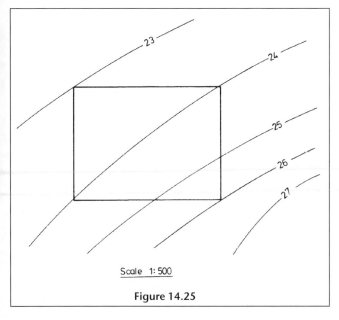

Scale 1: 500

Figure 14.25

material to form a car park formation level 86.5 m. Calculate the volume of material to be deposited.

6 Figure 14.25 shows contours over an area where it is proposed to erect a small building with a formation level of 23.00 m AOD. Draw on the plan the outline of any earthworks required and thereafter calculate the volume of material required to be cut and filled to accommodate the building. All earthworks have side slopes of 1 in 1.

7. Answers

Exercise 14.1

Chainage (m)	Depth d (m)	Area (d × 0.75) (m²)
0.00	2.703	2.027
10.00	2.320	1.740
20.00	2.036	1.527
30.00	1.578	1.184
40.00	1.770	1.328
50.00	1.675	1.256
60.00	1.200	0.900

Volume (Simpson's rule)

$$V = (10/3)[(2.027 + 0.900) + 2(1.527 + 1.328) \\ + 4(1.740 + 1.184 + 1.256)] \text{ m}^3 \\ = 84.5 \text{ m}^3$$

Exercise 14.2

1 (a) Chainage (m) | 0 | 10 | 20 | 30 | 40

Cross-sectional area (m²) 7.92 25.00 0.00 6.00 16.79

(b) Volume of cutting = 359.7 m³
(c) Volume of embankment = 136.0 m³

Exercise 14.3

1 Cross-sectional area X = 45.23 m³
Cross-sectional area Y = 30.95 m³
Cross-sectional area Z = 13.77 m³
Volume = 914 m³

Exercise 14.4

1 Cross-sectional area = 18.9 m²

Exercise 14.5

1

Chainage (m)	Depth c (m)	Width w (m)		Area (w + 2c)c (m²)
5 sections (40 m–120m)				
40	1.00	5.50	7.50	7.50
60	1.50	5.50	8.50	12.75
80	1.70	5.50	8.90	15.13
100	0.90	5.50	7.30	6.57
120	4.10	5.50	13.70	56.17
2 sections (120 m–140 m)				
120	4.10	5.50	13.70	56.17
130 (mid)	2.60	5.50	10.70	27.82
140	1.10	5.50	7.70	8.47

Volume

Sections (40–120 m) by Simpson's rule:

$$\text{Volume} = \frac{20}{3}[7.50 + 56.17 + 2(15.13)$$

$$+ 4(12.75 + 6.57)]$$

$$= \frac{20}{3}[63.67 + 30.26 + 77.28]$$

$$= 1141.40 \text{ m}^3$$

Sections (120–140 m) by prismoidal rule:

$$\text{Volume} = \frac{20}{6}[56.17 + 4(27.82) + 8.47]$$

$$= 586.4 \text{ m}^3$$

Total volume = 1727.8 m³

Exercise 14.6

1

Station	Height above or below FL	Weighting	Product
1	–0.2	1	–0.2
2	0.0	2	0.0
3	0.15	1	0.15
4	0.33	2	0.66
5	0.51	4	2.04
6	0.73	3	2.19
7	0.97	1	0.97
8	0.20	1	0.20
9	–0.15	2	–0.30
10	–0.27	2	–0.54
11	0.51	1	0.51
		20	5.68

Mean height of excavation = 5.68/20 = 0.284 m

Area = 5 × 100 = 500 m²

Volume = 500 × 0.284 m

= 142 m³

Exercise 14.7

Area of 10 m contour plane = 24.0 m²

Area of 9 m contour plane = 34.8 m²

Area of 8 m contour plane = 14.4 m²

Volume (10–8 m) = (2/6)[24.0 + 14.4 + 4(34.8)] m³

= 59.2 m³

Volume (8–7 m) = 1/3 × 1.0 × 14.4

= 4.8 m³

Total volume = 64 m³

Exercise 14.8

1 Area of side of trench (trapezoidal rule) = 236 m²

Volume = 236 × 0.75

= 177 m³

2

Chainage	RL	FL	c	cs	(cs + w)	(cs + w)c
0	55.30	58.40	3.10	6.20	12.20	37.82 m²
30	56.40	58.60	2.20	4.40	10.40	22.88 m²
60	56.00	58.80	2.80	5.60	11.60	32.48 m²
90	58.00	59.00	1.00	2.00	8.00	8.00 m²
75	57.00	58.90	1.90	3.80	9.80	18.62 m²

Volume 0–60 m (Simpson's rule) = 1618.20 m³

60–90 m (prismoidal rule) = 574.80 m³

Total volume = 2193 m³

3 Area 0 m = 17.10 m²

Area 20 m = 18.18 m²

Area 40 m = 8.38 m²

Volume = 654.7 m³

4 8236 m³

5 740.0 m³

6 430 m³ (see Fig. 14.26)

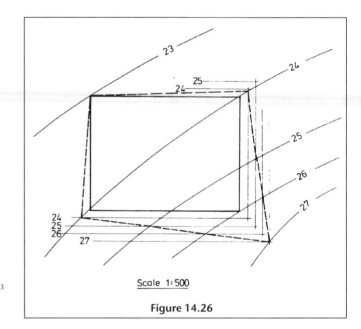

Scale 1:500

Figure 14.26

Chapter summary

In this chapter, the following are the most important points:

- Many calculations of volume are required during the life cycle of a construction site. These include the volumes of trenches, cuttings, embankments and site clearance earthworks.

- Cuttings have either vertical sides, e.g. drainage trenches, or sloping sides, e.g. roadways. The methods of computing their volumes differ.

  With vertically sided trenches, the volume (m³) is computed by

  1. calculating the area (m²) of the side of the trench by Simpson's rule, then multiplying that result by the trench width (m) to give the volume (m³).

2. calculating the cross-sectional areas (m²) of the trench at regular intervals (m) and entering the values into Simpson's rule to give the volume (m³) directly.

It is important that method 2 is used to calculate the volume of cuttings and embankments that are formed with sloping sides.

- In earthworks, there are three possible shapes of cross section, namely one-level sections, three-level sections and cross-fall sections. The methods of calculating their areas are shown in detail on pages 298–305.

- Simpson's rule is used in all volume calculations. However, it should be remembered from Chapter 13 that the rule can only be used when there is an *odd* number of cross sections. If there is an even number, Simpson's rule is used to calculate the maximum odd number of cross sections and the prismoidal rule is applied to the prismoid forming the remainder of the earthworks.

In a prismoid there are two parallel cross sections A_1 and A_2, at a distance d metres apart. A third mid section M is introduced to produce an odd number of sections, namely three. The volume of this part of the earthwork can therefore be computed using Simpson's rule once again. The volume (V) is therefore:

$$V = (1/3 \times d/2) \, [A_1 + A_2 + 2(\text{zero}) + 4M]$$
$$\quad = d/6 \, [A_1 + A_2 + 4M]$$

This formula is known as the prismoidal formula.

- The volumes of large-scale earthworks are calculated from

  1. Spot level grids in which the area is covered by a regular network of levels. The area formed by four adjacent levels is a square and the solid formed by the square and its formation level is a truncated prism. The mean height of this truncated prism is calculated and when multiplied by the area of the solid gives the volume.
  2. The contoured area of a site, which is considered to be a solid and can be split along the contours into a series of prismoids. The volume is then computed by successive applications of the prismoidal formula. Three contours are taken at a time and the central one is used as the mid area M in the formula derived previously in this summary.

CHAPTER 15 Surveys of existing buildings

In this chapter you will learn about:

- the surveying principles of measuring existing buildings
- the preparation, equipment and procedure required to measure buildings in order to produce plans, elevations and sectional drawings
- the methods used in measuring higher rise buildings using basic measuring equipment and the use of total station instruments to measure elevations of buildings and positions of inaccessible points on buildings
- the automated methods of measuring building elevations, using photographic and laser scanning means
- the method of plotting buildings using first angle orthogonal projection

The building technician and, in particular, the building surveyor will at some time in their careers be concerned with the extension, repair, alteration or demolition of existing buildings. In all of these cases, planning departments and building control departments of local authorities require accurate plans of the existing and proposed buildings, and it is the surveyor's task to take sufficient measurements to enable these to be made. The survey of even a small building will necessitate a large number of measurements being taken.

Besides being capable of conducting the survey, the surveyor must also be aware of (a) the Building Regulations and (b) current building construction practice.

(a) Building Regulations

Before any building works can proceed, the local authorities must be satisfied that the Building Regulations have been observed.

The surveyor does not require a detailed knowledge of these regulations but should certainly be aware of their implications.

(b) Building construction

The surveyor must have a sound knowledge of building construction and must understand thoroughly the construction of foundations, solid and cavity walls, roof, floors and windows in order to be able to draw a building convincingly.

1. Classification of drawings

In all construction schemes, several classes of drawings are required, the classification depending upon the particular information that is to be disseminated to users.

In *Recommendations for Building Drawing Practice* (British Standards Institution), the following classification of drawings is recommended.

(a) Design stage

Sketch drawings show the designer's general intentions.

(b) Production stage

1. *Location drawings*
   (a) Block plans to identify the site and locate outlines of buildings in relation to the town plan wherever possible. It is recommended that this plan should be made from the appropriate OS 1:1250, although most authorities accept a scale of 1:2500.
   (b) Site plans to locate the position of buildings in relation to setting-out points, means of access, general layout of site. The plans should also contain information on services, drainage, etc. The recommended scale of these drawings is 1:200, but again local authorities accept 1:250 or 1:500.
   (c) General location drawings to show the position occupied by the various spaces in the building, the general construction, the overall dimensions of new extensions, alterations, etc. The recommended scale of these plans is 1:50 or 1:100.

2. *Component drawings* This classification includes ranges of components, details of components and assembly drawings, which are not really the concern of this chapter.

2. Principles of measurement

In general, the principles involved in measuring a building are those used in the measurement of areas of land. In particular, the principles of linear surveying are applied most often, since buildings can usually be measured completely by taping.

Complex buildings may require the use of a theodolite or reflectorless total station instrument.

When floor levels are to be related to outside ground and drainage levels, a level and staff are required.

3. Conducting the survey

Figure 15.1 shows the location of a holiday cottage situated in pleasant rural surroundings on the bank of a loch (or lake). The cottage is old and requires renovation and extension. No drawings exist and a complete survey of the premises is to be made.

The following sections describe and illustrate some of the survey work required.

At the conclusion of the chapter the reader should attempt to plot the survey.

(a) Preparation

Before surveying any properties, it is good practice to study the OS sheet and any old drawings that may

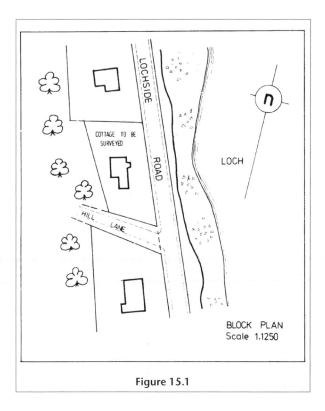

BLOCK PLAN
Scale 1.1250

Figure 15.1

exist. From these plans, a knowledge of the north direction can usually be gained and a list of adjoining addresses compiled. The proprietors or tenants of these adjoining properties may have to be contacted before permission to build any extension will be given by local authorities.

From the study, it may also be possible to gain some knowledge of difficulties that may arise during the subsequent survey.

(b) Reconnaissance

As in all surveys, time spent in reconnaissance is well spent. During the 'recce', attention is paid to the shape of the building, the number of floors, type of roof, position of doors and windows.

Squared paper is a necessity, and all sketches made during the reconnaissance should be roughly to scale. The scale depends upon the complexity of the building, but in general 1:100 scale proves adequate.

In measuring buildings, the relationship between rooms is all important, and a plan view of each floor is preferable to a room-by-room sketch.

Elevations will also be required, and here again the whole side of a building should be sketched in preference to a floor-by-floor elevation. The rule of working from the whole to the part is thereby adhered to.

(c) Equipment

In most cases the method of taping will be used, in which case the following equipment will be required:

(a) 20 m steel tape with locating hook,
(b) 5 m steel tape,
(c) 2 m folding rule,
(d) plumb-bob and string, chalk, light hammer and short nails,
(e) measurement book or loose-leaf pad containing a supply of squared paper,
(f) soft pencils, hard pencils and eraser.

The ideal number of surveyors is two, one of whom should be fairly experienced.

Modern equipment now includes:

(g) laser range meters,
(h) sonic range meters,
(i) hand-held laser line tools.

(d) Laser range meters

Figure 15.2 shows one of the Hilti laser range meters, the PD32 model, which is a hand-held instrument measuring $(120 \times 80 \times 30)$ mm. It is designed to measure up to 70 m to most construction surfaces (brick stone, concrete, plasterboard etc. and up to 200 m using a reflective target, with an accuracy of ±1.5 mm. (Many other survey companies produce similar

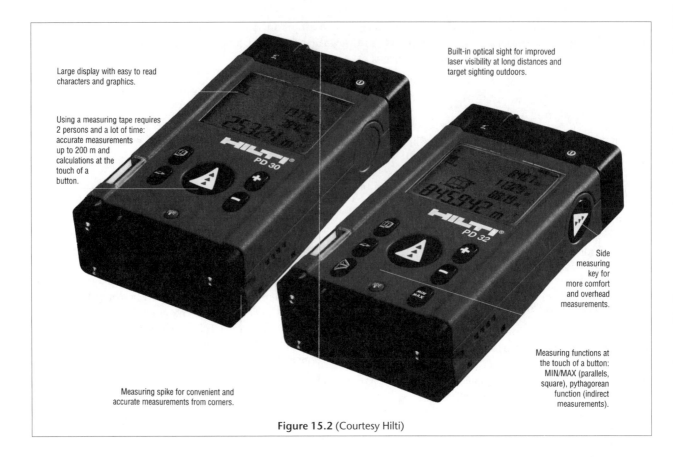

Large display with easy to read characters and graphics.

Using a measuring tape requires 2 persons and a lot of time: accurate measurements up to 200 m and calculations at the touch of a button.

Built-in optical sight for improved laser visibility at long distances and target sighting outdoors.

Side measuring key for more comfort and overhead measurements.

Measuring functions at the touch of a button: MIN/MAX (parallels, square), pythagorean function (indirect measurements).

Measuring spike for convenient and accurate measurements from corners.

Figure 15.2 (Courtesy Hilti)

instruments e.g. the Pacific Laser Systems PLSI, distributed by Topcon.)

The PD32 instrument utilizes a class 2 laser, powered by two AA alkaline or two NiMH rechargeable batteries.

In order to measure the dimensions of a room, the base of the instrument is held against one wall; the measure key (white arrow) is pressed and the distance is measured to a red laser spot on the opposite wall. This model has, additionally, a side measuring key which facilitates use of the in-built daylight sight for outdoor measurement. In use, the sighting aperture is held to the eye and the beam is activated, whereupon a dot is seen in the window. The dot is lined onto the point to which a measurement is required and by pressing the side measuring key the distance is obtained. The maximum range is a considerable 200 metres. The use of laser glasses improves the laser spot visibility significantly. The measured dimension appears on the display and five measurements can be stored via a data entry key but they cannot be downloaded. Other room dimensions are similarly entered. The key pad then allows the area and volume of the room to computed, if required.

When the measurement across a room is critical the tracking mode facility is engaged and the instrument is slowly swung in a horizontal plane. The minimum reading on the display is the shortest distance across the room. In addition, the PD32 is fitted with a direct entry function button marked 'min/max'. Pressing this button automatically puts the tool in tracking mode

and simultaneously displays the longest and shortest measured distances and the difference between them.

The instrument incorporates an ingenious spike device which is very useful in measuring from inaccessible corners etc. The distance is automatically corrected for the additional length of the spike. Diagonal dimensions are therefore easily measured from floor to ceiling across a room to check for perpendicularity.

The PD28 model incorporates a setting-out function and a memory function holding up to 1000 dimensions which can be tagged with a unique numeric location and downloaded to a computer in an Excel spreadsheet format for post processing and plotting.

(e) Procedure

1. *Site survey* The site survey uses the principles of linear surveying, namely trilateration and offsetting. On most small sites the details can be surveyed by trilateration alone.

In many cases it is possible to use the building sides as base lines and extend them to the boundaries, supplementing the dimensions with additional diagonal checks (Fig. 15.3).

Whenever possible, running dimensions should be taken, as this procedure is normally physically easier and leads to fewer errors.

The drainage arrangements may be shown on this plan if the system is simple; otherwise a separate sketch is drawn.

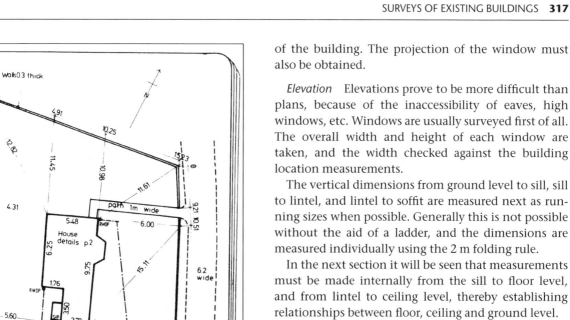

Walls03 thick

path 1m wide

House details p2

RWDP

RWDP

6.2 wide

LOCHSIDE ROAD

path 3.1 wide

SITE SURVEY P1.
08.02.80

Figure 15.3

of the building. The projection of the window must also be obtained.

Elevation Elevations prove to be more difficult than plans, because of the inaccessibility of eaves, high windows, etc. Windows are usually surveyed first of all. The overall width and height of each window are taken, and the width checked against the building location measurements.

The vertical dimensions from ground level to sill, sill to lintel, and lintel to soffit are measured next as running sizes when possible. Generally this is not possible without the aid of a ladder, and the dimensions are measured individually using the 2 m folding rule.

In the next section it will be seen that measurements must be made internally from the sill to floor level, and from lintel to ceiling level, thereby establishing relationships between floor, ceiling and ground level.

Heights are then measured to the eaves, either from ground level or from a convenient window lintel. A ladder is usually necessary to reach the eaves but, if it is dangerous, a 4 or 5 m levelling staff is usually long enough to enable the measurement to be made.

From the paragraphs above it will be obvious that vertical measurement is difficult, sometimes dangerous. However, a simple, convenient, accurate method of determining heights exists and yet is seldom used. A level is set up in a convenient position and, in a very few minutes, staff readings can be made to ground levels and window sills. The staff is then inverted and reading taken to lintels, soffits and eaves. If the level

2. Building location

Plan During the reconnaissance survey, a plan view detailed sketch of the outside of the building is drawn on squared paper (Fig. 15.4(a)). The measuring procedure is arranged in a systematic, orderly manner, so that every feature of the building is recorded. The measurements are taken in a clockwise direction and, wherever possible, running sizes are taken along a complete side of the building.

The tape must be fitted with a locating hook; otherwise one surveyor will have to be employed simply to hold the tape at the zero chainage point of each side. When hooked, the tape is run out along the side of the building; each feature is noted in turn and booked as in Fig. 15.4 (see page 318 for Fig.15.4(a)).

Bay windows present a problem, since the running chainage must be terminated against the window and individual dimensions taken round the window. In Fig. 15.4 these measurements are clearly shown. A check dimension is made by holding the tape in line with the front of the building and measuring the running dimensions to the window and then to the back

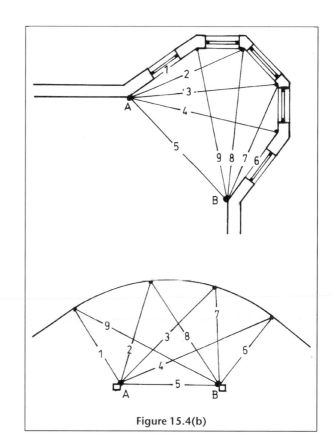

Figure 15.4(b)

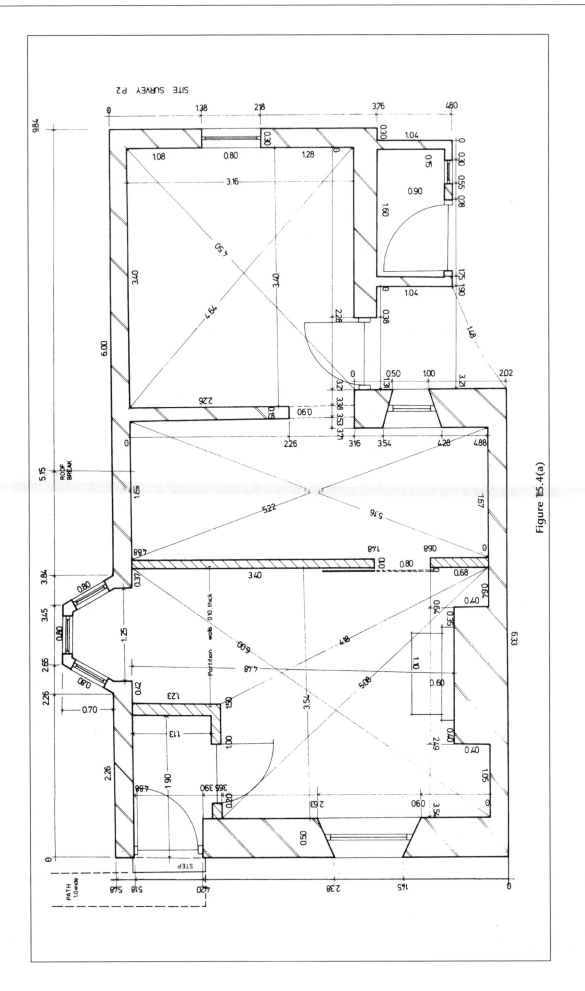

Figure 15.4(a)

Table 15.1

BS	IS	FS	Rise	Fall	Reduced level	Remarks	
1.51					10.00	A	Ground level
	0.62		0.89		10.89	B	Top of wall
	1.29			0.67	10.22	C	Path
	1.14		0.15		10.37	D	Step
	1.41			0.27	10.10	E	Ground level
	1.15		0.26		10.36	F	Ground level
	−1.85		3.00		13.36	G	Soffit
	−3.99		2.14		15.50	H	Ridge
	−4.70		0.71		16.21	I	Chimney stack
−2.65		−2.65		2.05	14.16	J	Roof change
	−1.11			1.54	12.62	K	Roof of extension
	−0.96			0.15	12.47	L	Roof of WC
	−1.01		0.05		12.52	M	Roof of WC
	1.15			2.16	10.36	N	Floor level extension
	0.20		0.95		11.31	O	Window WC
	1.21			1.01	10.30	P	Ground level
	1.20		0.01		10.30	Q	Manhole cover
	2.40			1.20	9.11	R	Invert level
	1.31		1.09		10.20	S	Ground level
	0.31		1.00		11.20	T	Top of wall
	0.99			0.68	10.52	U	Floor level living room
	−1.88		2.87		13.39	V	Ceiling level living room
		−2.10	0.22		13.61	W	Ceiling level extension
−1.14		−4.75	13.34	9.73	13.61		
−(−4.75)			−9.73		−10.00		
+3.61			+3.61		+3.61		

has been judiciously placed, a reading can be taken through an open window to floor level, and an inverted reading taken to ceiling level.

The disadvantage of this method is, of course, the fact that time must be spent in reducing the levels. However, the ease with which the levels are obtained (and their accuracy) is a great advantage and the method should be used whenever possible.

Table 15.1 and Fig. 15.5 show the levels obtained along the western gable and extension of the house.

3. Internal survey

Plan Once again a neat accurate sketch of the interior of the building should be prepared on squared paper. In the case of a simple building, the sketch is combined with the building location sketch (Fig. 15.4).

The measuring procedure is arranged to start at the entrance hall and proceed in a clockwise direction around the hall. Each room is then taken in a clockwise direction and similarly measured.

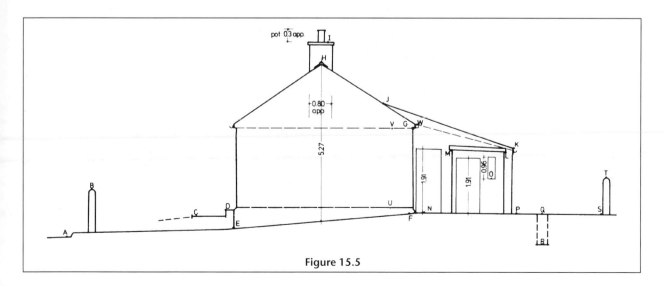

Figure 15.5

Whenever possible, running dimensions are taken, but individual measurements of window openings, cupboards etc. may have to be made in case paintings, pictures or curtains are damaged by stretching a long tape along a wall. Furthermore, it may prove impossible to hook a tape handle to a point that is convenient for running dimensions.

A sonic or laser meter (Fig. 15.2) is used to measure the length and breadth of each room. When using the meter, care must be taken to ensure that measurements are made between walls and not to projections, pictures or furniture etc.

Particular attention must be paid at door and window openings when obtaining the thickness of internal and external walls. A sound knowledge of building construction practice is very useful in this context.

The booking of the results is most important and time must be taken to ensure that cross-checks etc. are properly dimensioned. Figure 15.4(a) shows the booking required in this example.

Internal heights Heights must be taken for the purposes of drawing cross sections and for determinating ceiling heights in possible dormer extensions, etc. A floor-to-ceiling height is taken in each room, in the middle of the room and at each window. The dimensions should be measured from the ceiling downwards to internal lintels and sills of windows (Fig. 15.6).

A spirit level is a useful item of equipment when measuring heights to sills, since sills are not horizontal. The spirit level is laid across the bottom rail of the window and appropriate dimensions measured to the sill internally and externally.

All door heights are also measured.

When a building consists of more than one floor, the floor-to-floor heights are usually easily measured at the stair well. If difficulty is experienced, a levelling may have to be made up the stairs. Checks may be made externally by hanging a tape from a window on one floor down to a window on the next.

Measurements must also be taken from the floor of the uppermost room to the ceiling and into the roof space. These measurements are made through the ceiling hatch to the top of the ceiling joist, and from there to the apex of the roof.

Irregularly shaped rooms. In many Victorian buildings and in very modern buildings the surveyor encounters problems with awkwardly shaped rooms. Figure 15.4(b) (page 317) shows two such rooms where the room shapes are difficult to determine. In most cases, intersection solves the problem.

In both rooms sizes 1 to 5 are measured from a point A, which may be the corner of a concrete column, a room projection or even a tripod erected in the room. The sizes 6 to 9 are then measured from point B to intersect the previous measurements. The laser meter (Fig. 15.2) is the most useful tool for this occasion. All dimensions are stored on the memory of the instrument and are downloaded to a computer for post processing. The results are displayed on the screen and printed in graphical form on some form of plotting machine.

4. Sections, services, etc. Depending upon the purpose of the survey, sections may be required through the roof space, ground floor window heads and door thresholds.

The construction details are accurately sketched on squared paper and the relevant sizes obtained using a short 5 m tape. Figure 15.6 shows the relevant survey details for drawing a section through the roof space.

Services are traced individually and separate sketches

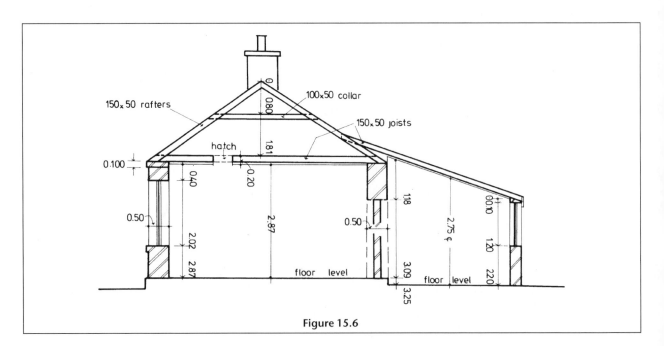

Figure 15.6

made to show the run of electrical cables and conduits, and all waste and soil pipes.

4. Surveying higher-rise buildings

(a) Tape and level survey

Figure 15.7 shows a three-storey building, typical of much local authority housing in any country.

The principles of surveying this type of building are much the same as have been described for a low-rise building, indeed the plan measurement of each floor is identical. The problem with higher-rise buildings lies in obtaining accurate height measurements for drawing elevations, since the use of ladders is dangerous and the cost of scaffolding is not cost effective.

Windows on each floor of high-rise buildings either lie vertically above each other or visibly do not and it is usually a case of measuring the ground floor only, in order to obtain their plan positions. The heights of these first floor windows are obtained using a 2 metre

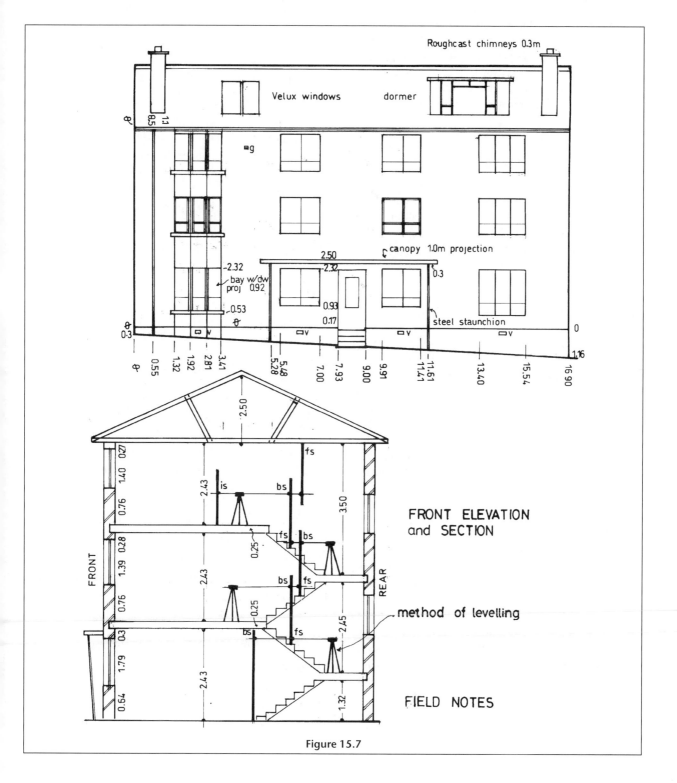

Figure 15.7

rigid rule. The heights of second floor windows may be accessible by using a level and inverted staff readings but third and higher floors cannot be directly measured.

One method of reaching higher floors is to physically measure up the stairwell using a tape. The easiest way to use the tape is actually to measure down the stairwell by hooking the tape at top floor level and letting the tape hang. The running dimensions to every floor can then simply be read by walking down the stairs. The sizes of windows etc. are then measured from each floor level.

Occasionally, the stairwell construction is such that it is not possible to use a tape, so a levelling must be made up the stairs. Figure 15.7 shows the process which is usually fairly awkward, since sights are short and unbalanced, leading to difficulties in focusing. Collimation error is irrelevant since the sights are so short.

(b) Theodolite intersections survey

1. Plan position

Some buildings are awkwardly shaped with protrusions (e.g. gas flues, dormers etc.) and curved sections, presenting problems of location. In such cases a theodolite or reflectorless total station is required.

In Fig. 15.7 the dormer lies in an irregular position. It would be possible to fix its position (with difficulty) from inside measurements. Figure 15.8 shows how its position is surveyed from outside the building.

1. *Fieldwork* In Fig. 15.8(a) a baseline AB is set up near the building. *It need not lie parallel to the building* but it does provide a reassuring and neater solution, when plotting, if it does. To achieve parallelism the x distances should be equal. They should also be some way back from the building if steep sights are to be avoided. The distance AB is accurately measured as 16.900 m. Assuming AB to be the North direction and point A to be local origin, the coordinates of the baseline are:

A	0.000E	0.000N
B	0.000E	16.900N

The theodolite is then set up at A and B in turn and used to measure angles 1, 2, 3 and 4. The two front corner positions of the dormer are calculated as an intersection survey as in Example 1.

In practice the measured data is captured on a data logger attached to the electronic theodolite and post processed on appropriate software (as in the example) to produce the coordinates of points C and D. These in turn are plotted by the associated drawing package.

2. *Calculation* The method of calculation is shown in Example 1.

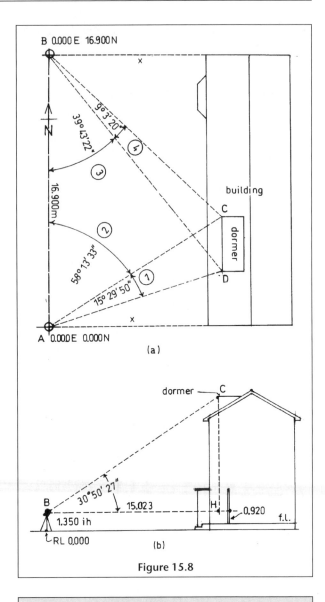

Figure 15.8

EXAMPLE

1 In Fig. 15.8, the following angles were measured to points C and D on the dormer window of Figure 15.7.

Angle	Measured value
1	15° 29′ 50″
2	58° 13′ 33″
3	39° 43′ 22″
4	09° 03′ 20″

The distance AB is 16.900 m and the local coordinates of A and B are:

A (0.000E, 0.000N) B (0.000E, 16.900N).

Calculate the coordinates of C and D.

Answer

Using the sine rule,

$$BC = \frac{\sin 58° \ 13′ \ 23″ \times 16.9}{\sin 72° \ 59′ \ 45″}$$

$$= 15.023 \text{ m}$$

$$BD = \frac{\sin 73° \ 43' \ 13'' \times 16.9}{\sin 66° \ 33' \ 24''}$$

$$= 17.682 \text{ m}$$

Bearing BC $= 180° - (9° \ 3' \ 20'' + 39° \ 43' \ 22'')$
$= 131° \ 13' \ 18''$

Bearing BD $= 180° - 39° \ 43' \ 22''$
$= 140° \ 16' \ 38''$

Easting C $= 0.000 + (15.023 \sin 131° \ 13' \ 18'')$
$= 11.300 \text{ m}$

Northing C $= 16.900 + (15.023 \cos 131° \ 13'18'')$
$= 7.000 \text{ m}$

Easting D $= 0.000 + (17.682 \sin 140° \ 16' \ 38'')$
$= 11.300 \text{ m}$

Northing D $= 16.900 + (17.682 \cos 140° \ 16' \ 38'')$
$= 3.300 \text{ m}$

Exercise 15.1

The coordinates of points C and D should be checked from station A. As an exercise the reader should compute the coordinates from the information given in Example 1.

2. Elevation heights
Reduced levels on selected points of the elevation may be obtained simultaneously with plan positions, if vertical angles are observed.

(i) *Fieldwork*
   (a) One of the baseline points, say B, is chosen as datum 0.000 m (Fig. 15.8(b)).
   (b) The theodolite is set over peg B and the instrument height is measured by tape. The floor level of the ground floor is obtained by setting the telescope of the theodolite to read zero degrees and taking a staff reading on the floor level.
   (c) Vertical angles (or zenith angles, depending on the instrument) are observed to the chosen elevation points C and D.

(ii) *Calculation of reduced levels* The observed readings are as follows:

Reduced level of point B	0.000 m
Instrument height at B	1.350 m
Staff reading on floor level	0.920 m
Vertical angle to C	30° 50′ 27″
Horz distance BC –	
(by calculation in Example 1)	15.023 m

In Δ BCH

$$\frac{CH}{BH} = \tan 30° \ 50' \ 27''$$

$$CH = 15.023 \ (\tan 30° \ 50' \ 27'')$$
$$= 8.970 \text{ m}$$

Red.Lev.floor $= 1.350 - 0.920$
$= 0.430 \text{ m}$

Red.Lev. C $= 0.430 + 0.920 + 8.970$
$= 10.320 \text{ m}$

Exercise 15.2

1 The reduced level of point D on the dormer window of Figure 15.8 is required. The following observations were made from station B.

Vertical angle	$= 26° \ 53' \ 55''$
Horz dist BD (by calculation in Ex. 1)	$= 17.682 \text{ m}$

Calculate the reduced level of point D.

In practice the measured data (vertical angles) is captured with the horizontal angles on a data logger attached to the electronic theodolite and post processed on appropriate software to produce the reduced levels of points C and D.

(c) Total station survey
The task of finding the three-dimensional coordinates on inaccessible points of buildings (such as the gas flue in Fig. 15.7) is made much easier with the use of a reflectorless total station, assuming of course that the surface being sighted is capable of returning the signal.

The three-dimensional coordinates of the instrument station are loaded into the instrument; the instrument height is measured and loaded and the observations (horizontal angle, vertical angle and slope length) are made, recorded and processed by the instrument to give an instant result. Full details of the process are given in Chapter 7.

5. Automated methods of building frontage survey

(a) Photographic

(i) *Precise photogrammetric methods* use pairs of overlapping terrestrial photographs (stereo pairs) from a calibrated camera set upon fixed baseline points opposite the building. Control point targets are fixed at intervals on the building. Measurement is carried out by the methods outlined in Chapter 10. A conventional plot or an ortho-rectified photograph can be produced.

(ii) *Simple photographic methods* using digital cameras are now relatively cheap and give adequate accuracy. Software can produce an elevation plot or a 3D model if all sides of the building are surveyed.

The Topcon P3000 Digital Image Measurement System uses overlapping images. The camera orientation elements are established by a bundle adjustment of at least six control points on the front of the building. The digital image is then ortho-rectified. Detail can then be plotted or the image used to form a 3D model. Accuracies of 4 mm at 10 m distance are possible.

Single photographic systems can provide a cheap

solution with reasonable accuracy. The image is rectified to remove distortion due to tilt of the camera axis, using four control points, before being imported into a CAD package where the detail can be traced. Accuracy will be reduced if there are major architectural protrusions in the frontage. Multiple exposures may minimize errors.

An example of this type of system is CURAMESS from LazerCAD. A 3D model can be produced as well as elevations, and the photo image used to render the model. A report can be generated on window and door openings.

(b) Reflectorless total station measurement

As described in Chapter 7, direct reflex measurement is possible to any surface at ranges up to about 200 m depending on the instrument. A building elevation can be fixed by radial methods from fixed points along the front of a building in the same way as a detail survey. Openings and architectural details are measured. The more sophisticated instruments can automatically scan a face at a set grid interval. This is more applicable to a quarry face, but it may indicate future integration of total stations with the laser scanners of the method following.

(c) Laser scanning

This is the terrestrial equivalent of the LIDAR system of producing digital terrain models mentioned in Chapter 10. It can be used for external or internal surveys.

Radial position fixing from a fixed base station is again used, but a fine mesh of points (point cloud) is automatically scanned with a resolution of 3–6 mm. The point cloud is georeferenced using the coordinates and orientation from the base station. Pulsed laser 'time of flight' distance measurement is used in normal instruments, giving centimetre accuracy. Phase measurement short range instruments giving mm accuracy are used for precise industrial surveys. Costs are extremely high, about ten times that of a total station, therefore only very high value projects or companies who specialize in this type of work make this method economical.

The instrument uses a rotating mirror or prism to direct the laser beam vertically and rotates horizontally. Angle sensors record the angle.

Fig. 15.9 shows a Riegl LMS Z2101.

Listed below are some common instruments:

Leica/Cyrax:
Pulsed

HDS 2500:	40° ×	40° scan.
HDS 3000	360° × 270° scan.	

Phase

HDS 4500	360° × 310° scan.	

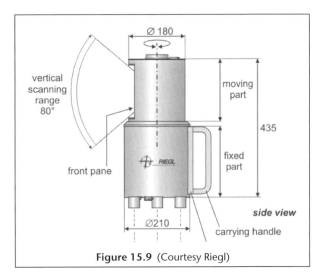

Figure 15.9 (Courtesy Riegl)

Figure 15.10 (Courtesy Riegl)

Trimble:
Pulsed

	GS 200	360° × 70° scan.

Riegl
Pulsed

	LMS Z series	360° × 90° scan.

Fig. 15.10 shows a 3D model produced from a scan of building frontages.

6. Plotting the survey

In drawing buildings, several views from different angles are required. These include a plan and elevations of all sides of the building. It is essential that these views be presented in a systematic manner. It is recommended that only orthogonal first-angle projection (Fig. 15.11) be used for building drawings.

Plotting of the survey information begins with the plan view, and from it all other views are constructed. In deciding where the plan view is to be placed on the sheet, the surveyor must decide which elevation is to be the principal view. Generally this is the front elevation. If it is possible to place all of the views on

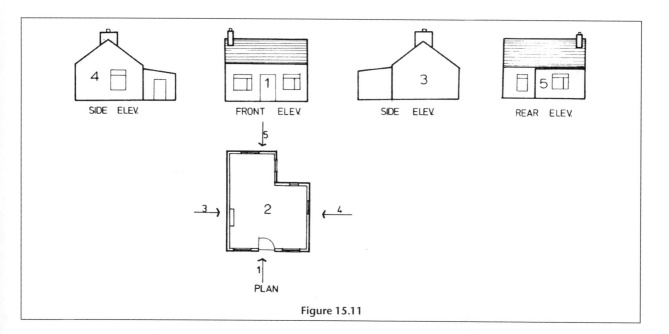

Figure 15.11

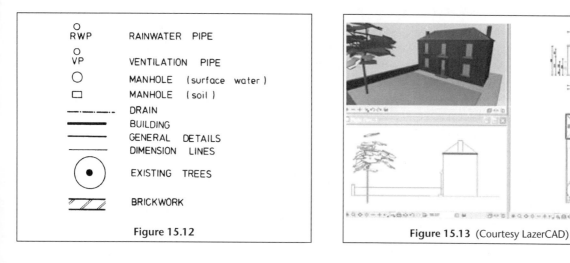

Figure 15.12

Figure 15.13 (Courtesy LazerCAD)

one sheet, the principal view is placed left of centre (position 1) in the top half of the sheet (Fig. 15.11).

The plan view is then positioned immediately below the front view (position 2). The side views from left and right are placed on the right (position 3) and left (position 4) respectively of the front view. The view from the rear is usually placed on the extreme right (position 5), although, if more convenient, it may be placed on the extreme left.

As far as possible, the principal symbols used on building drawings should be graphical. Figure 15.12 shows a selection of these symbols, which will be adequate to allow the reader to complete Exercise 15.3, following.

Nowadays, the final drawing will most likely be produced using CAD software.

AutoCAD and MicroStation are the most widely used, but cheaper packages such as CAD 3D MAX and LazerCAD Allplan will do an adequate job. The more sophisticated packages have a library of the standard

building drawing symbols such as in AutoCAD Architectural Desktop.

Figure 15.13 shows a survey from photographs measured with CURAMESS software and plotted in Allplan software.

Exercise 15.3

1 Using the survey data shown in Figures 15.1, 15.3, 15.4(a) and 15.5, draw:

(a) a site plan to a scale of 1:250
(b) a general location drawing of the building to a scale of 1:50, showing the internal arrangements
(c) an elevation of the western side of the building to a scale of 1:50
(d) a north–south section through the building to a scale of 1:50.

2 Figure 15.7. shows the front elevation of a three-storey local authority flatted dwelling. The front elevation is to be renovated.

The various floor levels were located by measuring up the stairwell. The windows were observed to be vertically aligned and chimney dimensions were estimated. Plot the front elevation to a scale of 1:50.

7. Answers

Exercise 15.1

1. Easting C 11.300 m   Northing C   7.000 m
   Easting D   11.300 m Northing D 3.300 m

Exercise 15.2

1. Reduced level D   10.320 m

Exercise 15.3

1. The site plan, elevation and sections should mirror the relevant Figures 15.1, 3, 4(a) and 5, since the latter were drawn to scale but sized to fit the page size of this textbook.
2. The elevation should mirror Figure 15.7, which was drawn to scale but sized to fit the page size of this textbook.

Chapter summary

In this chapter, the following are the most important points:

- In this, the age of rehabilitation and renovation, the aspiring surveyor has to be able to make a survey of a building and therefore requires a fair knowledge of building construction and possibly of Building Regulations. This textbook is not a guide to either of these subjects.

- In surveying a property for possible renovation, the surveyor should be aware of the Building Standards Regulations regarding the types of drawings that are required. These comprise sketch drawings, location drawings, site plans, construction drawings and component drawings. The surveyor is mainly concerned with the site plan and construction drawings.

- Site plan drawings usually involve the survey of a small area of land surrounding the building which is to be surveyed. A knowledge of linear surveying is required. Usually a tape survey is the only viable method of making a survey of an enclosed area where bushes and outhouses etc. present problems in EDM surveys.

- In measuring a building, it is essential to be methodical and meticulous. Many dimensions have to be taken and the survey begins with an exterior survey of the ground floor. Attention should be paid to higher floors during this survey. The positions of windows etc. will probably be vertically above those of the ground floor.

- This is followed by a survey of the interior of the building. Each room should be measured in an orderly fashion, generally by working clockwise around the room. Extra care must be exercised when linking the surveys of adjacent rooms. Overall widths, heights and diagonals are required and are reliably measured by some form of laser meter. Irregularly shaped rooms are best surveyed by intersection methods as illustrated in Fig. 15.4(b).

- Relative levels in and around the building are very important and use should be made of a level and staff. A knowledge of inverted staff readings is essential since the relative vertical positions of floors, ceilings, window sills and chimneys are usually critical in renovations. In higher-rise buildings a levelling may have to be made up a stairwell.

- Irregularly shaped or inaccessible exteriors at higher floors are surveyed by means of theodolite intersection methods or EDM tacheometry using a reflectorless total station instrument.

- Automated methods of surveying building facades are becoming more common. Photogrammetric techniques using pairs of overlapping photos taken from fixed base positions provide three-dimensional coverage. Laser scanning is also used to provide 3D images.

Glossary

Compensator Level sensor used in some electronic theodolites and total stations.

CORS Continuously recording reference station network for DGPS.

DGPS Differential GPS. GPS survey with corrections from a reference station.

Digital level An automatic levelling instrument which reads a barcode staff and displays readings and results on an LCD screen without manual input.

Draping Superimposing an aerial photograph on a digital terrain model.

DTM Digital terrain model.

DXF Drawing exchange format for transferring graphics data.

EDM Electronic distance measurement.

EGNOS European geographic overlay system. A reference system for DGPS.

Ellipsoid A best fit ellipse for the shape of the earth's crust.

Ephemeris Satellite orbit parameters.

ETRF European terrestrial reference frame defined by precise GPS reference stations.

ETRS European terrestrial reference system. Coordinate system based on ETRF.

Face left Observing position of the theodolite with the vertical circle on the left of the telescope.

Face right Observing position of the theodolite with the vertical circle on the right of the telescope.

Feature code Series of letters or numbers to identify the detail feature being surveyed.

Flying levelling A procedure in levelling whereby the unknown elevation of a remote point is computed from the elevation of a known point through a series of backsights and foresights only.

Galileo European GPS system.

GDOP Geometric dilution of precision of GPS fix resulting from available satellite positions.

Geoid The shape of the earth at sea level defined by a line at right angles to the direction of gravity.

GIS Geographic information system. Databases linked to digital maps.

GLONASS Russian GPS system.

Graticule Glass disc carrying theodolite cross-hairs.

Kinematic GPS GPS survey with a moving receiver.

LIDAR Airborne laser scanning to produce a terrain model.

Local scale factor (LSF) The mathematical factor by which any distance on the ground is multiplied to produce the projection length on the map projection.

Map projection System of converting positions on the globe to the flat paper map.

Mean Sea Level Correction (MSL) The mathematical correction applied to any distance on the ground to produce the equivalent sea level value.

Micrometer Optical device for reading the theodolite circle accurately.

Multipath Reflections of GPS satellite signal from surrounding buildings etc.

NAVSTAR US Dept of Defense GPS constellation.

NTF Ordnance Survey digital map data transfer format.

Optical plummet Device for centring a theodolite over a survey mark.

Orientation of surveys The procedure used to relate a survey to a known direction (some form of north).

Orthorectify Conversion of aerial photograph to remove tilt and relief distortion.

OS Ordnance Survey, the British national mapping organisation.

OSDN OS height datum based on Newlyn tide gauge.

OSGB36 OS coordinate system based on the National Grid and trig pillars.

OSGM02 OS geoid model 2002 used for GPS heights.

OSGM91 OS geoid model for heights based on fundamental benchmarks.

OSNET OS reference network of points with permanently recording GPS receivers.

OSTN02 Transformation program for converting GPS coords to OS coords.

Passive stations OS network of fixed GPS points.

PDL radio Type of radio used for communication between receivers in DGPS.

Photogrammetry Measurement from (usually aerial) photographs.

Radial survey Fixing of position by measuring an angle and distance to a point.

Raster Data in a grid cell format of pixels.

Reciprocal levelling A rare procedure used in levelling where it is physically impossible to make the length of a backsight equal to that of a foresight.

Rectify Remove tilt displacement from a photograph.

RINEX Receiver independent exchange format for GPS data.

SQL Structured query language for interrogating a database.

String Multiple points defining a linear feature in detail survey.

Superelevation (Cant) A technique used to counter-balance the effects of centrifugal force in which the outer rail of a railway or outer side of a roadway is raised to some predetermined extent.

TIN Method of defining a terrain model for contouring.

Track light A light beam emitted by a total station to aid alignment in setting out.

Transformation Conversion between coordinate systems.

Transition curve A parabolic curve used in the horizontal plane to effect a smooth change from a straight section of road or railway to a curved section.

TRF Terrestrial reference frame. The basis of a coordinate system.

Tribrach Base of surveying instrument.

TRS Terrestrial reference system. The practical realization of the TRF.

Vector A line with distance and direction.

Vertical curve A parabolic curve used in the vertical plane to effect a smooth change from one road or railway gradient to another.

WAAS Wide area augmentation system. DGPS correction system.

Well conditioned triangle A triangle in which all three angles lie between 30° and 120°.

WGS84 Global coordinate system for GPS.

Index

Digital Terrain Model (DTM), 206–8
  applications of terrain models, 207–8
  data acquisition, 206–7
  digital contours, 207
  draping display, 207
  grid based methods, 207
  ground survey for, 206
  layer shading, 207
  LIDAR, data acquisition by, 206–7
  model display, 207
  modelling methods, 207
  NEXTMap, 208
  OS Profile, 208
  photogrammetry for, 206
  relief shading, 207
  rendering, application of, 207
  scale of terrain models, 207–8
  shading, application of, 207
  Synthetic Aperture Radar (SAR), 207
  terrain analysis, 208
  terrain visualization, 208
  triangular irregular network (TIN), 207
  wire frame display, 207

E